Re-presenting GIS

Re-presenting GIS

Edited by

PETER FISHER

City University, London, UK

DAVID J. UNWIN

Birkbeck College, London, UK and University College, London, UK

John Wiley & Sons, Ltd

Telephone (+44) 1243 779777

Email (for orders and customer service enquiries): cs-books@wiley.co.uk
Visit our Home Page on www.wiley.com

Other Wiley Editorial Offices

John Wiley & Sons Inc., 111 River Street, Hoboken, NJ 07030, USA

Jossey-Bass, 989 Market Street, San Francisco, CA 94103-1741, USA

Wiley-VCH Verlag GmbH, Boschstr. 12, D-69469 Weinheim, Germany

John Wiley & Sons Australia Ltd, 42, McDougall Street, Milton, Queensland 4064, Australia

John Wiley & Sons (Asia) Pte Ltd, 2 Clementi Loop #02-01, Jin Xing Distripark, Singapore 129809

John Wiley & Sons Canada Ltd, 22 Worcester Road, Etobicoke, Ontario, Canada M9W 1L1

Wiley also publishes its books in a variety of electronic formats. Some content that appears in print may not be available in electronic books.

Library of Congress Cataloging-in-Publication Data

Fisher, Peter F., 1995-
Re-presenting GIS / Peter Fisher, David Unwin.
p. cm.
Includes bibliographical references and index.
ISBN-13: 978-0-470-84847-0 (cloth: alk. paper)
ISBN-10: 0-470-84847-2 (cloth alk. paper)
1. Geographic information systems. I. Unwin, D. (David John) II. Title.
G70.212F57 2005
910′. 285–dc22 2005003238

British Library Cataloguing in Publication Data

A catalogue record for this book is available from the British Library

ISBN-13 978-0-470-84847-0 (HB)
ISBN-10 0-470-84847-2 (HB)

Typeset in 10/12pt Times by Thomson Press (India) Limited, New Delhi
Printed and bound in Great Britain by Antony Rowe Ltd., Chippenham, Wiltshire
This book is printed on acid-free paper responsibly manufactured from sustainable forestry in which at least two trees are planted for each one used for paper production.

Contents

PART II NOT JUST SPACE

PART III TIME AS WELL

PART IV NOT ‘THERE’ YET?

List of Contributors

Ola Ahlqvist Department of Geography, The Pennsylvania State University, University Park, PA 16802, USA. Email: oka1@psu.edu

Michael Batty Centre for Advanced Spatial Analysis, University College London, 1-19 Torrington Place, Gower Street, London, WC1E 6BT, UK.
Email: mbatty@geog.ucl.ac.uk

Peter Bibby Department of Town & Regional Planning, The University of Sheffield, Geography and Planning Building, Winter Street, Sheffield, S3 7ND, UK.
Email: p.r.bibby@shef.ac.uk

Tao Cheng Advanced Centre for Spatial Information Technology, Department of Land Surveying and GeoInformatics, The Hong Kong Polytechnic University, Kowloon, Hong Kong. Email: tao.cheng@polyu.edu.hk

Alexis (Lex) J. Comber ADAS Woodthorne, Wergs Road, Wolverhampton, WV6 8TQ, UK. Email: alexis.comber@adas.co.uk

Matt Duckham Department of Geomatics, The University of Melbourne, Victoria 3010, Australia. Email: mduckham@unimelb.edu.au

Peter Fisher Department of Information Science, City University, Northampton Square, London, EC1V 0HB, UK. Email: pff1@city.ac.uk

Antony Galton School of Engineering, Computer Science and Mathematics, University of Exeter, Exeter, EX4 4QF, UK. Email: a.p.galton@exeter.ac.uk

Subhrajit Guhathakurta School of Planning and Landscape Architecture, College of Architecture and Environmental Design, Arizona State University, Tempe, Arizona 85287-2005, USA. Email: subhro.guha@asu.edu

Francis Harvey Department of Geography, Univeristy of Minnesota, Minneapolis, MN 55455, USA. Email: fharvey@geog.umn.edu

Marcos Llobera Department of Anthropology, University of Washington, Seattle, WA 98195-3100, USA. Email: mllobera@u.washington.edu

Harvey J. Miller Department of Geography, University of Utah, Salt Lake City, Utah 84112-9155, USA. Email: harvey.miller@geog.utah.edu

Robert Muetzelfeldt Simulistics Ltd, Edinburgh Technology Transfer Centre, King's Buildings, Edinburgh, EH9 3JL, UK. Email: r.muetzelfeldt@ed.ac.uk

Jonathan Raper Department of Information Science, City University, Northampton Square, London, EC1V 0HB, UK. Email: raper@soi.city.ac.uk

Nadine Schuurman Department of Geography, Simon Fraser University, Burnaby, BC, Canada, V5A 1S6. Email: schuurman@sfu.ca

Joanne Sharp Department of Geography and Geomatics, University of Glasgow, Glasgow, G12 8QQ, UK. Email: jsharp@geog.gla.ac.uk

David J. Unwin School of Geography, Birkbeck College, Malet Street, London, WC1E 7HX, UK. Email: d.unwin@geography.bbk.ac.uk and Department of Geomatics, University College London, Gower Street, London, WC1E 6BT, UK.

Richard Wadsworth Centre for Ecology and Hydrology, Monks Wood, Abbots Ripton, Huntingdon, Cambridgeshire, PE28 2LS, UK. Email: rawad@ceh.ac.uk

Preface

Re-presenting GIS in the third in a series of edited volumes, stretching back to 1994, that share a common format and approach. *Visualization in Geographical Information Systems* (Hearnshaw and Unwin, 1994) established the style and this was followed in 2002 by *Virtual Reality in Geography* (Fisher and Unwin, 2002). These volumes are alike in three respects. First, all have a content that sits, according to taste, at the interface between academic geography and computer science or in the now well-established and mature field of geographic information science. Second, although all have been edited collections, the temptation to produce yet another set of vaguely connected solicited conference papers has in each case been avoided by use of a more collaborative approach. This involved the use of a residential seminar/workshop extending over several days at which all the individual chapter authors were present to discuss each other's contributions and, critically, to work as teams to produce the scene-setting introductions to each section of the final book. Individual chapters were also revised in the light of the formal and informal discussions. Third, this approach must have some initial support. For *Visualization* this came from the then fledgling UK *Association for Geographic Information* and the eventual publishers. For *Virtual Reality* support come from the *Advisory Group on Computer Graphics*. The present volume was produced with the generous assistance of the UK *Economic and Social Research Council*, under its Research Seminar series, Grant number R451265140. As convenors and editors, and on behalf of all the participants, it is a pleasure to acknowledge our gratitude.

Peter Fisher and David J. Unwin
14th February 2005

References

Fisher, P. and Unwin, D. (eds), 2002, *Virtual Reality in Geography*, London and New York: Taylor and Francis.

Hearnshaw, H.M. and Unwin, D.J. (eds), 1994, *Visualization in Geographical Information Systems*, New York and Chichester: John Wiley and Sons Ltd/Inc.

Acknowledgements

The authors, editors and publishers would like to thank the following for permission to reproduce previously published figures: The University of Toronto Press for permission to reproduce Figures 3.1, 3.2 and 3.3; Taylor and Francis for agreement to reproduce Figures 7.1, 7.4, 7.5, 7.6, 15.1, 16.3, 16.4; Kluwer Academic Publishers for permission to reproduce 15.2, 15.3, and 16.5. In addition, spatial data which is in the copyright of the Ordnance Survey and made available through the Edina-Digimap JISC service, and the Office of the Deputy Prime Minister, were used in the construction of Figures 5.1 and 5.2.

1

Re-presenting Geographical Information Systems

Peter Fisher and David J. Unwin

1.1 Why 'Re-Present' GIS?

The increasingly widespread use of Geographical Information Systems (GISystems, widely known as GIS) has meant that a version of 'geography' has been exported to many other disciplines and walks of life where this technology has been found to be useful. As an undergraduate, one of us read *Applied Geography* by Dudley Stamp (Stamp, 1960). It is full of ideas and examples of the applications of geography in the real world, but it has taken the better part of half a century for much of Stamp's vision to become reality; the rest of the world – including many cognate sciences – have discovered the power of something they call 'geography'. It is wise to treat the word with caution, however, since there are at least three ways in which 'geography' is used. First, and at its simplest, geography is the places and spaces on our planet. Second, in the analysis of geographic information, 'geography' is often used as a short hand for the spaces and distances used to explain or model some phenomenon. Third, there is the Geography of our schools, colleges and research institutes; the academic study of the previous two readings. Typically, it is the usefulness of the second usage, in turn almost entirely a consequence of the phenomenon of spatial autocorrelation, which explains the evident popularity of GISystems.

In his essay in *Political Geography Quarterly*, Peter Taylor (1990) referred to this process as the 'imperialism of the new geography'. In the vanguard of this imperialism has been the technology of GISystems, which Taylor paints as the villain of the piece.

Re-presenting GIS Edited by P. Fisher and D. Unwin

GISystems have been the subject of much academic boosterism – verging on evangelism – from both geographers and others. A necessary condition for the widespread use of GIS has been the availability of relevant locational information and suitable computing environments, and recent decades have seen these both become commonplace. The imperialism of GISystems has, however, not been a case of geographers conquering the territory of other disciplines but rather of those disciplines generating an internal demand for GISystems fuelled by the discovery that 'geography', defined loosely as 'where things are', actually 'matters' (see, for example, Hall, 1992).

So, large parts of the world have discovered geography through the use of GISystems and this is only likely to increase with the recent introduction of Location Based Services (LBS) on the back of mobile phone technology. Arguably, GISystems technology has at least hastened, if not caused, the resurgence of 'Geography' as a discipline in US universities and schools, a resurgence that does not seem to be happening in the UK. Together with this widespread adoption of GISystems, however, can come the impression amongst users that GISystems *are* geography and, worse still, that the representations of geographical phenomena stored within GISystems are unproblematic in academic Geography. Nothing could be further from the truth. Since 1990 or so, there has been a geographers' version of the so-called 'science wars' in which cultural and social geographers have attacked something they call 'GIS'.

The first shot in this geographical war was probably that fired by Peter Taylor in his 1990 editorial about what he called GKS (Geographical Knowledge Systems). A few years later, John Pickles (1995) edited *Ground Truth*, a set of essays that led to conference debates and a collection of papers in *Environment and Planning A* (Flowerdew, 1998; Clark, 1998). More recently still, we have seen publication of Michael Curry's *Digital Places* (1998) and a typically forthright attack on these critiques from Stan Openshaw (1998).

A summary of the internecine war in geography over GIS is outlined by Taylor and Johnston in an essay on 'GIS and Geography' in *Ground Truth* (Taylor and Johnston, 1995). Their story is roughly as follows. During the 1960s something called quantitative geography grew up and was for a few years the dominant research paradigm. There are difficulties and contradictions in this approach when it is applied to social studies, and so in the last twenty years leading human geographers have moved on to find explanations grounded in critical social theory; dialectics, rather than data, have become the main research tool. A more recent articulation of a similar nature can be found in Hamnett's account of contemporary human geography (Hamnett, 2003). A spirited response from Johnston *et al.* (2003) points out that quantitative analysis in Geography has not gone away, but this rejoinder has little to say about the GIS phenomenon, one of the principal tools being used in the analysis of large social science datasets. So, while the academic debate has proceeded, some geographers, the 'unreconstructed quantifiers' of Taylor's account, together with spatial statisticians, cartographers and computer scientists, have contributed to the creation and use of GISystems. In turn, this has reinvigorated empirical analysis, often of a strongly applied nature, within the discipline. A large part of this group of academic geographers would nowadays almost certainly regard themselves as practitioners in a discipline they call geographic information science (Goodchild, 1992), with a content and concerns drawn in part from academic Geography but also widened to include geomatics, cartography and parts of computer and information sciences.

The need for GIS to be *Re*-presented to the wider world in an accessible form is grounded in the problem, both within the discipline and the wider world, that GIS may be seen as an unproblematic encapsulation of Geography, and that researchers developing or using GIS may not be aware of the distinction. This book is intended to give such a Re-presentation, dwelling primarily on representation which is at the heart of the issues raised by the critical geographers and also the roots of the possibly simplistic views of users of GISystems. In this introduction we note the separation of the GISystems and the GIScience that underlies it. Then, we discuss the basis of this representation issue and introduce the parts of this book that attempt to address it.

1.2 Separation of GISystem and GIScience

1.2.1 The GI Continuum

As noted above, those involved in the Geographical Information project have argued for a separation between the Systems and the Science (Goodchild, 1992). This separation seems to have been ignored by some contributors to the GI wars. To clarify the concepts, Wright *et al.* (1997) used an email discussion list to examine the relationship between GISystems and GIScience. Arising from contributions to the discussion, they proposed a continuum with Geographical Information Science (GIScience) at one pole, and Geographical Information Systems at the other. To them, GIScience includes all areas of interest to those engaging with the theory of the methods used, and GISystems is use of the software itself. Between the poles, they placed GIS as 'toolmaking', mediating between the science and the system. This model is, however, theoretically problematic, as suggested by Pickles (1997) and Fisher (1998). It results in polarisation of the subject material. Among other things, it implies that:

- users of a GISystem cannot be indulging in a valid scientific endeavour unless it is in the scientific domain of their subject (archaeologists using GISystems can be doing archaeology but not valid spatial science); and
- the unthinking use of a system by a soil scientist may be doing good soil science, but is never doing more than using spatial science.

This seems extremely unsatisfactory because the uncritical use of a GISystem can never be good science in any sense. Furthermore, this model has no explanatory power as to how either GIScience or GISystems develop.

It would appear that this continuum model has more to do with how people theorise (science), develop (toolmaking) and employ (system) GIS (in both the Science and the System sense). Neither the user nor the theorist of spatial information is necessarily an exclusive specialist. Many individuals develop new and interesting spatial theory using existing GISystems, or they may develop original computer programs because no GI System can be used for implementation of the concepts. The labelling of Application Sciences and Spatial Sciences is intended to illustrate the idea that soil scientists and demographers may add to spatial science in the course of their own domain research. Therefore, people do not see themselves at the poles but on the continuum from pure theory development to extensive system use. Indeed, the primary evidence for the model of Wright *et al.* (1997) is personal statements made on the email list. Pickles (1997) points out, however, that such statements are most commonly informed by opinion, rather than by critical reading or understanding of the issues in the philosophy of science. The continuum concept describes people well, but it fails to theorise the relationship of either the science or the system.

An alternative basis for exploring the interactions of GI Systems and GI Science is proposed by Fisher (1998) (see Figure 1.1). This cyclic metaphor recognises that theory of some kind underlies any tool and is implemented in the production of that tool. Thus the Tool, a particular GISystem, is no more than the realisation at a particular time of some of the theoretical spatial concepts of GIScience. This interpretation is based on how almost all scientific instruments work from, a Seismograph to an Atomic Absorption Spectrometer, which would not exist without scientific theory relating to transmission of shock waves through the earth in the first case and the resonance of atoms in response to electromagnetic radiation in the second.

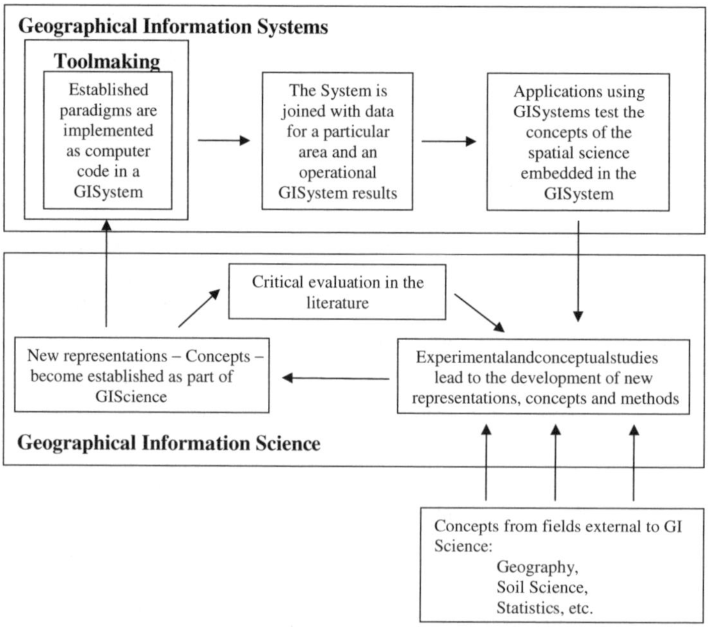

Figure 1.1 *The GI Science–System cycle (after Fisher, 1998)*

Many people will recognise the GIScience cycle of Figure 1.1. Here we can see the development of concepts resulting from the critical evaluation of existing ideas through publication and critique in the scientific literature, and resulting in the development of new concepts. This is not a closed system, but rather can take as input ideas from other spatial sciences including mainstream geography. This is similar to the advances in other sciences, but can be seen as the essence of spatial science and spatial theory, criticised though it has been as disembodied and denatured (Tilley, 1994; Pickles, 1995). The criticism seems to be that any method should be specific to data, and can only have any meaning in a particular context.

As well as the lower conceptual loop, Figure 1.1 shows an upper loop, the implementation of the GISystem itself. A GISystem is a realisation of a particular set of spatial concepts at a particular time, and so this loop shows that users of GISystems make use of the concepts embedded in that system in their applications. Therefore they are in a position to evaluate critically those concepts in the context of their application. Since the concepts at any one time are only a state of understanding, they may or may not work well for a particular application. If they fit well, the user of the system can provide valuable feedback and supporting critical valuation of the embedded concepts. If the concepts do not fit well, then the critical evaluation is equally valid and helpful to spatial theorists and system developers in revising either the spatial theories or the GISystem.

Some concepts may be used legitimately within a particular GISystem or indeed many GISystems for a considerable period of time. There is the risk that the majority of practitioners will not critically evaluate these concepts, and they may be considered paradigms in the sense of Kuhn (1970). Fisher (1996) has listed a number of GIScience concepts that might be considered in this way including the paradigms of the Boolean map, the layer-based raster or vector models, Euclidean metrics of space, production cartography, and the image. As with all paradigms, a concept that reaches such 'heights' does not have to be correct; it is merely a convenient and agreed way to conceptualise and interpret the world at a particular time. Critical thinking should always be evaluating such paradigms.

Arguably those using a particular GISystem may be so removed from the general theory of Spatial Information that they can only think in terms of the implementation with which they are familiar (anecdotal and empirical evidence of this abounds; Medyckyj-Scott and Hearnshaw, 1993; Davies and Medyckyj-Scott, 1994). Indeed, it follows from this that Taylor and Johnson (1995) and others among the GIS critics are correct, in that particular GISystems can be criticised for over simplifying spatial concepts and for (possibly unknowingly) imposing particular approaches on the data and ultimately on the people about whom decisions are being made. This does not mean that there are not more appropriate methods in the larger toolkit of spatial theory. Thus, either theoretically or practically, the inherited constraints in the implementation need not constrain the developer, but their level of knowledge of Spatial Information Theory in general (GIScience), and their ability to implement those concepts with the toolkit they have available, are constraints which may result in the inappropriate use of technology. Not every possible conceptualisation of spatial information in the totality of GIScience is implemented in some GISystem. If that were the case, there would be little research remaining in GIScience, which is certainly not the case! Rather, active researchers in the theory of spatial information are attempting to expand the concepts specifically to relax these constraints.

1.3 Why Representation?

Taylor and Johnston's (1995) argument that 'in GIS data are usually treated unproblematically except for technical concerns about errors' is only true in the application of GISystems where it is a necessary evil in the use of such systems. However, it is anathema to GIScience researchers; in GIScience there is enormous concern to produce models that better match our existing domain knowledge and approaches. Object orientation, fuzzy boundaries and object classes, three and more dimensional models, multiple representations, spatial languages, and multi-media may be absurdly primitive, but they are all steps along a difficult road to revise continuously the representations of the world that can be said to be at the heart of the GISystems.

In most critiques of GIS, one issue that repeatedly arises is the tendency of GIS to promote or privilege certain types of representation at the expense of alternatives (Aitken and Mitchel, 1995; Martin, 1999; Raper, 1999; Sheppard, 1995; Sheppard *et al.*, 1999). This 'problem' is neither unique to geography, nor is it particularly new (see, for example, Tuan, 1957; Lowenthal, 1961). All disciplines have to address it and it is not clear that the issue of digital representation is either any more complex or more problematic than analogue or other representations. Some of the representational issues that arise in GIScience relate to:

- *Space vs. place.* GI theory articulates the idea of absolute Euclidean spaces quite well, but the socially-produced and continually changing notion of place has to date proved elusive to digital description except, perhaps, through photography and film.
- *Entitation.* What are the objects of interest and is it legitimate to 'objectify' individuals?
- *Description.* Objects of interest, be they areas or people, may well themselves be incapable of crisp description. In a technical sense they are 'fuzzy' and, if they are regions of the earth's surface like a town centre, they have uncertain boundaries.
- *Temporality and change.* Digital geographies are usually static descriptions and the technology has found the representation of change in time extremely hard to operationalise.
- *Creating space and time.* The continued, and tightly coupled, creation of space and time conflates all the above issues into a single representational problem.

To some degree each of these issues is addressed in this volume, and in the remainder of this introduction we examine some of the progress that has been made and what remains to advance them further.

As in geography generally, scale is a fundamental aspect of the representation of geographical information. Treatments of this problem are varied, but this is not part of the agenda for this book. The interested reader is referred to the recent treatments of this issue in the GIScience literature (Quattrochi and Goodchild, 1997; Tate and Atkinson, 2001).

1.3.1 'Not Just Objects'

If a relational view of space is to be of any use, we need to represent objects that create and represent the 'place' in the space. Most GISystem implementations assume that objects of interest are uncontroversial, and have definite, fixed boundaries that can be represented in a digital world. This certain world of the GISystem is, of course, mostly a fiction and this has been recognised through more than two decades of work on error and uncertainty (Fisher, 1999). Does the unobserved or un-modelled detail matter? At least two sources of imprecision are of concern in GIScience: what we define as objects and how we delineate them.

Defining objects has a social dimension and so it is not a value- or culture-free operation. This matters, for example, in attempts to make databases interoperable, or in transforming geographic information for differing purposes when the context in which that information was created is ignored. For example, work by Bibby and Shepherd (2000) shows that even the apparently unproblematic idea of land use cannot readily be captured in a digital representation. Importantly, they show that the apparently abstract theoretical considerations raised by entitation have had important implications for a range of policy related research. Similarly, Thurstain-Goodwin and Unwin (2000) have attempted to define town centres, a project that immediately poses difficult problems of definition and delineation since the notion of a town centre can be viewed as an archetypal example of a fuzzy object with an uncertain boundary. Formal approaches have been proposed to this problem using fuzzy and rough set theory. In this book, Chapters 3–6 by Schuurman, Harvey, Bibby and Fisher *et al.*, address a selection of issues related to the social context of geographical objects and information.

In GIScience, these types of concern have usually been articulated by the notions of 'error' and 'uncertainty' in the ways in which we chose to model objects (Unwin, 1995). Uncertainty arises because information about 'geography' is always *imperfect*, being often either *imprecise*, *inaccurate*, *vague,* or some combination of all three. In Chapters 7 and 8, first Ahlqvist then Duckham and Sharpe address these issues. The latter recognise four different formal models that are being used to manage and describe uncertainty in geographic information:

- *Stochastic models*, using well-established ideas from both classic and geo-statistics (e.g. Ehlers and Shi, 1996; Heuvelink, 1998; Leung and Yan, 1998; Shi, 1998);
- *Fuzzy set theory*. This has been used successfully to describe the inaccuracy of land cover classifications (Woodcock and Gopal, 2000), boundary imprecision (Leung, 1987) and so on. The assignment of fuzzy membership values is the Achilles heel of this approach, and is still not clearly understood, but the approach is an attractive one.
- *Three valued logic*, most often using rough set theory in which elements are either 'in' 'out' or 'neither in nor out' of a set (Worboys 1998; Ahlqvist *et al.*, 2000).
- *A variety of alternative logic models*. Three value logics are themselves an example of the more general multi-valued logics that can also be applied to spatial data.

These models attempt to create a consistent representation of an inconsistent reality and, according to circumstances, all are valid approaches but none are well-handled by traditional relational database technology which sets great store by consistency.

1.3.2 'Not Just Space'

One of the great achievements of western science has been the notion of Euclidean space as an infinitely extended and infinitely sub-divisible continuum in which each point can be specified by means of a tuple of numerical co-ordinates. This physical conception of space has proved to be enormously useful in virtually all physical and natural science.

Whether implemented in a field or an object data model, this common sense view of space makes a number of assumptions:

- Space is a 'given' and is usually conceptualised in an absolute sense as providing a fixed frame of reference in which to locate objects Note that it is assumed to exist independently of the objects themselves.
- Entities are uncontroversial and are in some sense externally defined.
- Space is more important than time. If it is considered at all, time is seen as an attribute of objects.

This does not mean that the language is adequate for all needs (Mark and Frank, 1991), and it does not mean that these assumptions are a necessary part of space as people experience it. In fact, even in its own terms, it can give problems when the representation is created in a digital environment. In what Goodchild (1995) called the 'absurdly primitive world of the digital computer' we represent locations using (x, y) tuples in which the values are expressed in fixed, finite increments. This is fine – sooner or later we always have to do this – but outside of the computer we change the word length to suit the problem ('no. of significant figures'). To argue that 32 bits is not enough for pure location would be silly, it almost always is, but:

- In calculations, all the problems listed by David Douglas (1974) in his classic paper 'It makes me so cross' are related to the inability to geo-reference objects with infinite precision. Similarly, many authors have pointed out and illustrated the computational problems of using a finite and fixed numerical precision (Unwin, 1975, for example), and Worboys (1995, p. 188) discusses the 'problems arising from discretization and the Green-Yao algorithm'.
- The issues tend to be discussed in conversion of vector to raster data (or other tessellation-based data structures) and so cannot be avoided. Much of the technical literature is about the 'errors' rather than the necessary 'artefacts' introduced into derived fields (such as gradient) or objects (such a view and water sheds) from this discretisation.

More importantly, as Lefebvre (1991) points out, this representation of space has problems when attempts are made to use it in the study of the social world that people experience, whether it is in describing personal interaction or the physical environment. The space is not an infinitely empty space, it is a populated space. Indeed, mathematical and computational geometry are the realm of the study of that empty space, and surveying is its application to the world. At the heart of geography is the description of what populates the space on the surface of the earth. Geographical knowledge is far more than the specification of positions by means of co-ordinates, and, reasonably, to some it has nothing to do with that co-ordinate position. This has been recognised, at least implicitly, throughout the development of geography. Nonetheless, the traditional mathematical conception of space is widely perceived as constituting the dominant influence in providing a theoretical basis for GIScience, yet it falls dramatically short of the kind of rich and highly structured conceptions of space that are required to do justice to all the concerns of either the natural or social sciences (Egenhofer *et al.*, 1999; Yuan, 2001).

In this book, the essays in Part II address how we conceptualise space, examining so-called 'qualitative' (Galton), 'network' (Batty) and 'perceived' (Llobera) spaces.

1.3.3 'Time as Well'

In representation, time is seen by many as the 'other'. As a concern, it lies beyond objects and space, but for representational completeness time should be explored as well (Raper, 2000). One of the major difficulties in using GIS to support science is still the poverty of the mechanisms we have for integrating time into our representations. In 1960, the integration of spatial and temporal description was taken by Henry Clifford Darby as the theme for his IBG Presidential address on 'The problem of geographical description' (Darby, 1960). First, Darby recognised the necessity to treat space and time *together* in the same representation and, second, the six literary and cartographic strategies he recognised as solutions to his 'problem' map very closely into what has been attempted in GIS.

There has been a range of different strategies and approaches to the representation of the spatio-temporal, including:

- transformations from 4D to 2D, 2D + time, or 1D plus time (Langran, 1992);
- addition of dynamic behaviour to spatial representations (Wesseling *et al.*, 1996)
- visualisation and animation of change (Hearnshaw and Unwin, 1994);
- use of concepts from the 'time' geography of Hagerstrand (1970) (Miller and Wu, 2000; Miller, Chapter 16, this volume);
- formalisation of qualitative spatio-temporal change.

The papers by Massey (1999, 2001; see also Raper and Livingstone, 1995, 2001) in which she explores the communalities between physical and human geography in conceptualising space, time, and space-time, address general issues in how we represent space and time in digital 'geographies'. Massey directs attention to some emerging similarities of concern that have not as yet been articulated but which have both theoretical and practical implications for spatial science. Incorporation of time will not be easy, however, for three reasons:

- Emergent phenomena. In the context of a coastal spit, Raper and Livingstone (1995) devised a data model that enabled them to handle phenomena that emerge as time passes, common in all process studies. In this volume, Chapter 14 by Raper explores this issue.
- Different perceptions of time. In Chapter 15, Cheng looks at some different spatio-temporal structures.
- Simulations of temporal processes. Muetzelfeldt and Duckham (Chapter 17) introduce the interactive and object based Simile environment for simulation with particular emphasis on its spatial processing abilities.
- Personal descriptions of events. Finally, in this volume in Chapter 18, Guhathakurta describes an approach using narrative emerging from use of GIS as a way of investigating an area.

Of course, objects, space and time interact and a crucial task for GIScience is to further the integration of the three, in particular accommodating the challenges of each as articulated here, as well as challenges which have been set elsewhere and those which have not yet been devised.

1.4 The Uses and Ethics of GIS

Finally, two issues that greatly exercised some of the contributors to *Ground Truth* were the democratisation of access to GIS technology and the ethical implications of its use. We believe that it is in these areas that those of us who work in GIS have a great deal to learn from our social and cultural geographer colleagues.

In *Ground Truth*, Harris *et al.* (1995) (see also Weiner *et al.*, 1995) pointed out that GIS can be used to pursue social goals through what they called participatory GIS (PGIS, also known as public participation GIS, PPGIS). This area is rapidly being developed, at least in the USA (see Rundstrom, 1995; Sieber, 2000; Talen, 2000), but perhaps less so elsewhere. It is tempting to suggest that the lack of development elsewhere can be explained by perhaps two major factors: first, the contrast in data supply policy between easy and essentially free access to 'framework' data in the US compared with notions of copyright and tradable information, with its associated high costs, in most other countries; and second, it might be that centralised government disempowers local people and hence provides little incentive for development of PPGIS.

A particularly taxing essay in *Ground Truth* was that by Michael Curry (1995a) in which he highlighted what he saw as inevitable ethical problems in the application of the technology, a theme he returned to later (Curry, 1995b; Curry, 1998). His argument seems to revolve around the idea that users of the technology are not worried about any ethical implications. In fact, there has been a steady flow of papers that have shown concerns for such issues, including Openshaw (1993), Rix and Markham (1994), Dale (1994), Obermeyer (1995), Onsrud (1995) and Crampton (1995), to name a few which appeared before or very soon after Curry's paper. It would have been a good conclusion to this volume to have extended its scope to cover the issues that arise here but, to our regret, this has not proved to be possible.

1.5 Conclusion

All the papers in this book challenge the current paradigmatic models of space represented within current GISystems. None leaves the current models unquestioned. This book will have been a success if the reader finishes with the impression that the representations we use within current GISystems are problematic for the GIScience community. It will be even more successful if some of the representational issues raised in the chapters have an influence on the development of future GISystems, something only time will tell.

References

Ahlqvist, O., Keukelaar, J. and Oukbir, K., 2000, Rough classification and accuracy assessment, *International Journal of Geographical Information Science*, **14**(5), 475–496.

Aitken, S.C. and Mitchel, S.M., 1995, Who contrives the 'real' in GIS? Geographic information, planning and critical theory, *Cartography and Geographic Information Systems*, **22**, 17–29.

Bibby, P. and Shepherd, J., 2000, GIS, land use and representation, *Environment and Planning B: Planning and Design*, **27**, 583–598.

Clark, M.J., 1998, GIS – Democracy or delusion, *Environment and Planning A*, **30**, 303–316.

Crampton, J., 1995, The ethics of GIS, *Cartography and Geographic Information Systems*, **22**, 84–89.

Curry, M.R., 1995a, GIS and the inevitability of ethical inconsistency. In: Pickles, J. (ed.), *Ground Truth*, New York: Guilford, pp. 68–87.

Curry, M.R., 1995b, Rethinking rights and responsibilities in geographic information systems: beyond the power of the image, *Cartography and Geographic Information Systems*, **22**, 58–69.

Curry, M., 1998, *Digital Places: Living with Geographic Information Technologies*, London: Routledge.

Dale, P., 1994, *Professionalism and Ethics in GIS*, London: Association for Geographic Information, available at www.agi.org.uk.

Darby, H.C., 1960, The problem of geographical description, *Transactions of the Institute of British Geographers*, **30**, 1–14.

Davies, C. and Medyckyj-Scott, D., 1994, GIS usability: recommendations based on the user's view, *International Journal of Geographical Information Systems*, **8**, 175–189.

Douglas, D.H., 1974, It makes me so cross. In: Peuquet, D.J. and Marble, D.F. (eds), *Basic Readings in Geographic Information Systems*, London: Taylor and Francis, pp. 303–307.

Egenhofer, M.J., Glasgow, J., Gunther, O., Herring, J.R. and Peuquet, D.J., 1999, Progress in computational methods for representing geographical concepts, *International Journal of Geographic Information Science*, **13**(8), 775–796.

Ehlers, M. and Shi W., 1996, Error modelling for integrated GIS, *Cartographica*, **33**(1), 11–21.

Fisher, P.F., 1996, Concepts and paradigms of spatial data. In: Craglia M. and Couclelis H. (eds), *Geographic Information Research: Bridging the Atlantic*, Taylor and Francis, London, pp. 297–307.

Fisher, P.F., 1998, Geographic Information Systems and Geographic Information Science: A Case of the Wrong Metaphor. In: *Proceedings of the 8th International Symposium on Spatial Data Handling*. Simon Fraser University, Vancouver, pp. 321–330.

Fisher, P.F., 1999, Models of uncertainty in spatial data. In: Longley, P.A., Goodchild, M.F., Maguire, D.J. and Rhind, D.W. (eds), *Geographical Information Systems, Volume 1 Principles and Technical Issues*, Wiley: Chichester, pp. 191–205.

Flowerdew, R., 1998, Reacting to *Ground Truth, Environment and Planning A*, **30**, 289–301.

Goodchild, M.F., 1992, Geographical information science, *International Journal of Geographical Information Systems*, **6**, 31–45.

Goodchild, M.F., 1995, Geographic information systems and geographic research. In Pickles, J. (ed.), *Ground Truth*, New York: Guilford, pp. 31–50.

Hagerstrand, T., 1970, What about people in regional science? *Papers of the Regional Science Association*, **24**, 7–21.

Hall, S., 1992, *Mapping the Next Millennium*, New York: Random House.

Hamnett, C., 2003, Contemporary human geography: fiddling while Rome burns? *Geoforum*, **34**, 1–3.

Harris, T.M., Weiner, D., Warner, T.A. and Levin, R., 1995, Pursing social goals through participatory geographic information systems. In: Pickles, J. (ed.), *Ground Truth*, New York: Guilford, pp. 196–222.

Hearnshaw, H. and Unwin, D., 1994, *Visualization in Geographical Information Systems*, London: Wiley.

Heuvelink, G., 1998, *Error Propagation in Environmental Modelling with GIS*, London: Taylor and Francis.

Holloway, S.L., Rice, S.P. and Valentine, G. (eds), 2003, *Key Concepts in Geography*, London: Sage.

Johnston, R., Hepple, L., Hoare, T., Jones, K. and Plummer, P., 2003, Contemporary fiddling in human geography while Rome burns: has quantitative analysis been largely abandoned – and should it be? *Geoforum*, **34**, 157–161.
Kuhn, T.S., 1970, *The Structure of Scientific Revolutions*, 2nd edition, Chicago: The University Press.
Langran, G., 1992, *Time in Geographic Information Systems*, London: Taylor and Francis.
Lefebvre, H., 1991, *The Production of Space*, Translated by Nicholson-Smith, D., Oxford: Blackwells.
Leung, Y., 1987, On the imprecision of boundaries, *Geographical Analysis*, **19**(2), 125–151.
Leung, Y. and Yan, J., 1998, A locational error model for spatial features, *International Journal of Geographical Information Systems*, **12**(6), 607–620.
Lowenthal, D., 1961, Geography, experience and imagination: towards a geographical epistemology. *Annals of the Association of American Geographers*, **51**, 241–260.
Mark, D. M. and Frank. A.U. (eds), 1991, *Cognitive and Linguistic Aspects of Geographic Space*, Dordrecht: Kluwer.
Martin, D., 1999, Spatial representation: the social scientists perspective. In Longley, P.A., Goodchild, M.F., Maguire, D.J. and Rhind, D.W. (eds), *Geographical Information Systems, Volume 1 Principles and Technical Issues*, Wiley: Chichester, pp. 71–80.
Massey, D., 1999, Space-time, 'science' and the relationship between physical geography and human geography, *Transactions, Inst. Brit. Geographers*, NS **24**, 261–276.
Massey, D., 2001, Talking of space-time, *Transactions, Inst. Brit. Geographers*, NS **26**, 257–261.
Medyckyj-Scott, D. and Hearnshaw, H.M., 1993, *Human Factors in Geographical Information Systems*, London: Belhaven Press.
Miller, H.J. and Wu, Y.-H., 2000, GIS software for measuring space-time accessibility in transportation planning and analysis, *GeoInformatica*, **4**, 141–159.
Obermeyer, N.J., 1995, The hidden GIS technocracy, *Cartography and Geographic Information Systems*, **22**, 78–83.
Onsrud, H., 1995, Identifying unethical conduct in the use of GIS, *Cartography and Geographic Information Systems*, **22**, 90–97.
Openshaw, S., 1993, Commentary: GIS 'crime' and GIS 'criminality', *Environment and Planning A*, **25**, 451–458.
Openshaw, S., 1998, The truth about *Ground Truth. Transactions in GIS*, **2**, 7–24.
Pickles, J., 1995, *Ground Truth: the Social Implications of Geographic Information Systems*, New York: Guilford.
Pickles, J., 1997, Tool or science? GIS, technoscience, and the theoretical turn, *Annals of the Association of American Geographers*, **87**, 363–372.
Quattrochi, D.A. and Goodchild, M.F., 1997, *Scale in Remote Sensing and GIS*, Boca Raton: Lewis.
Raper, J.F., 1999, Spatial representation: the scientist's perspective. In: Longley, P.A., Goodchild, M.F., Maguire, D.J. and Rhind, D.W. (eds), *Geographical Information Systems, Volume 1 Principles and Technical Issues*, Wiley: Chichester, pp. 61–70.
Raper, J., 2000, *Multidimensional Geographic Information Science*, London: Taylor and Francis.
Raper, J. and Livingstone, D., 1995, Development of a geomorphological spatial model using object-oriented design, *International Journal of Geographical Information Systems*, **9**, 359–383.
Raper, J. and Livingstone, D., 2001, Let's get real: spatio-temporal identity and geographic entities, *Transactions, Inst. Brit. Geographers*, NS **26**, 237–242.
Rix, D. and Markham, R., 1994, GIS certification, ethics and professionalism, *Proceedings AGI'94*, 6.2.1–6.2.5.
Rundstrom, R.A., 1995, GIS, indigenous peoples, and epistemological diversity, *Cartography and Geographic Information Systems*, **22**, 45–57.
Sheppard, E., 1995, GIS and society: towards a research agenda, *Cartography and Geographic Information Systems*, **22**, 5–16.
Sheppard, E., Couclelis, H., Graham, S., Harrington, J. and Onsrud, H., 1999, Geographies of the information society, *International Journal of Geographic Information Science*, **13**(8), 797–823.
Shi, W., 1998, A generic statistical approach for modelling error of geometric features in GIS, *International Journal of Geographical Information Systems*, **12**(2), 131–143.

Sieber, R.E., 2000, Conforming (to) the opposition: the social construction of geographical information systems in social movements, *International Journal of Geographical Information Science*, **14**, 775–793.

Stamp, D., 1960, *Applied Geography*, Harmondsworth: Penguin.

Talen, E., 2000, Bottom-up GIS: a new tool for individual and group expression in participatory planning, *Journal of the American Planning Association*, **66**(3), 279–294.

Tate, N.J., and Atkinson, P.M. (eds), (2001), *Modelling Scale in Geographical Information Science*, Chichester: Wiley.

Taylor, P.J., 1990, 'GKS', *Political Geography Quarterly*, **9**, 211–212.

Taylor, P.J. and Johnston, R., 1995, GIS and Geography. In: Pickles, J. (ed.), *Ground Truth*, NewYork: Guilford, pp. 51-67.

Thurstain-Goodwin, M. and Unwin, D., 2000, Defining and delimiting the central areas of towns for statistical monitoring using continuous surface representations, *Transactions in GIS*, **4**, 305–317.

Tilley, C., 1994, *A Phenomenology of Landscape: Places, Paths and Monuments*, Oxford: Berg.

Tuan, Yi-Fu., 1957, The use of simile and metaphor in geographical description, *The Professional Geographer*, **9**, 8–11.

Unwin, D.J., 1975, Numerical error in a familiar technique: a case study of polynomial trend surface analysis, *Geographical Analysis*, **7**, 197–203.

Unwin, D.J., 1995, Geographical information systems and the problem of error and uncertainty, *Progress in Human Geography*, **19**(4), 549–558.

Weiner, D., Waner, T., Harris, T.M. and Levin, R.M., 1995, Apartheid representations in a digital landscape: GIS, remote sensing and local knowledge in Kiepersol, South Africa, *Cartography and Geographic Information Systems*, **22**, 30–44.

Wesseling, C.G., Karssenberg, D.-J., Burrough, P.A. and Van Deursen, W.P.A., 1996, Integrating dynamic environmental models in GIS: the development of a dynamic modelling language, *Transactions in GIS*, **1**, 40–48.

Woodcock, C. and Gopal, S., 2000, Fuzzy set theory and thematic maps: accuracy assessment and estimation, *International Journal of Geographic Information Science*, **14**(2), 153–172.

Worboys, M., 1995, *GIS: A Computing Perspective*, London: Taylor and Francis.

Worboys, M., 1998, Computation with imprecise geospatial data, *Computers, Environment and Urban Systems*, **22**(2), 85–106.

Wright, D.J., Goodchild, M.F. and Proctor, J.D., 1997, GIS: Tool or Science? Demystifying the persistent ambiguity of GIS as Tool versus Science, *Annals of the Association of American Geographers*, **87**, 346–362.

Yuan, M., 2001, Representing complex geographic phenomena in GIS, *Cartography and Geographic Information Science*, **28**, 83–96.

Part I

Not Just Objects

2

Not Just Objects: Reconstructing Objects

Ola Ahlqvist, Peter Bibby, Matt Duckham, Peter Fisher, Francis Harvey and Nadine Schuurman

One of the dominant discourses in GIScience (GISc) literature stresses the importance of objects as the rational basis for describing geographical phenomena, encoding their behaviours, and archiving their relations (Buttenfield, 1995; Frank and Raubal, 1998; Jackson, 1994; Leung *et al.*, 1999; Voisard and Schweppe, 1998). While authors from a variety of disciplines have raised concerns about discourses involving objects (Frank, 1996; Henderson, 1995; Mack, 1990; Smith and Varzi, 1997), objects are commonly viewed within GISc as uncontroversial or neutral, with a utilitarian function. Furthermore, the term 'object' is often used without a clear indication of what exactly is being referred to. Objects embody multiple meanings in different contexts.

Unpacking objects and examining the processes that lead to their construction reveals that many objects, if not all, are vague in both semantic and spatial definition, ambiguous and imprecise. The nebulous character of objects conceals the contentious technological and institutional processes through which every object is constructed. However, certain areas of research within GISc do exhibit an understanding of the limits and capabilities of objects distinguished. GISc research often recognizes objects as digital incarnations rather than realist reflections. The philosophical premises of class, and its close relation category, are increasingly questioned, and the relationship between scale and object is receiving scrutiny. A number of researchers have questioned the ontological merit of the objects permitted by current data models, while research into uncertainty attempts to address imperfection in the location and attribution of objects. Cognitive science has assumed a large presence in GISc and undertaken studies of object perception. Work on

Re-presenting GIS Edited by P. Fisher and D. Unwin

automated generalization has developed ways of extending cartographic objects to include scale and location-dependent behaviour. The chapters in Part I of this book are an attempt to provide a snap-shot of some of the diverse work in this complex area. In order to situate these contributions, we begin by reviewing some of the research literature in this area.

2.1 What are Objects, and How Do We Talk About Them?

Cognitive research emphasizes a view of objects as elements of a naturalized human cognition (Campari, 1991; Fabrikant and Buttenfield, 2001; Mark, 1997; 1999; Peterson, 1994). Objects are often defined, in part, with reference to their dichotomous relationship to fields (Couclelis, 1992; Goodchild, 1989). Objects are seen as natural categories at a large geographical scale, with fields believed to dominate human cognition at smaller scales (Couclelis, 1992; Frank, 1999; Kemp, 1997; Mark, 1997). Many researchers have argued that some geographic entities are closer to fields than objects (Couclelis, 1992; Goodchild, 1992). For example, elevation, temperature, and soil moisture arguably behave in field-like ways (NCGIA 1999). Some GISc authors have preferred to reserve the term 'objects' to refer exclusively to things in a digital database, using instead the term 'entities' to refer to things 'out there' (Mark, 1999).

However, the difficulties posed by objects extend much further than relating digital to material geography. Objects are never solely a construct used to discuss our geographic environment. Rather, they are part of discourses that shape our thinking about space, time, and spatial relations (Massey, 1999). Objects encourage a compartmentalization of related concepts that often seems so 'natural' that it is easy to regard them as essential and forget that they were ever constructed from other objects and concepts. As these packages become integrated into a discourse, the forces and processes that led to their inception are collectively discarded. To illustrate, we concentrate on four of the major discourses about objects over recent years: the object's relationship to cartography, category, uncertainty and ontology.

2.1.1 Cartography and Objects

Much of our understanding of digital objects can be traced back to the traditions of cartographic representations (Fisher, 1998). Over millennia, cartography has constituted our primary representation of geography and imbued Western geographical imaginations (Cosgrove, 2001; Gregory, 1994). Throughout its history, GIS has been overly concerned with cartography and cartographic products and the problems associated with digitizing cartographic materials, produced using pen, ink, and lithography, have left their mark on vector-based GIS (Hohl, 1998). The points, lines, and areas commonly used to locate and delineate objects in vector-based GIS have their antecedents in cartographic primitives. This focus on recreating cartography and cartographic objects without questioning the concepts embedded within cartography inevitably puts some significant restrictions upon such a GIS. A range of technical and institutional issues influence the creation of digital objects, but the retention of the map's division of space is fundamental. Digital objects created in absolute space can be moved, stacked, and manipulated just like bricks or books, in ignorance of the complex interpreted relationships to which the original map product referred. Both cartographic and GIS objects refer to entities that are part of an experiential relational space that the interpretation of objects partially ascertains (Muehrcke and Muehrcke, 1992). This social space of experiences is simply not commensurate with geometrical space (Schatzki, 1991). Recent efforts to develop digital generalization procedures need to tackle these issues. While generalization is of utmost importance in cartography, it is still not widely understood (McMaster and Shea, 1992; Muller *et al.*, 1995). Apart from cartographic generalization research, increased attention is now put on issues of object generalization concerned with abstraction processes and categorization of real world phenomena, and model generalization partly dealing with transformation between different categorization levels (Weibel, 1995; Weibel and Dutton, 1999).

2.1.2 Categories and Objects

GISc discourse has stressed the need to formalize objects and processes, and rightly so given the need for computational implementation. But this very process of formalization can often obscure the meaning of objects, their relationships to other objects, and the contexts in which they are developed and used. The computational environment has reinforced certain visions of geography in which objects can play a key role (Schuurman, 1999). In a computational sense, objects are instances of classes (or categories). However, these categories do not exist in the world, they exist in the mind and are therefore subject to cross-cultural differences in interpretation and implementation (Lakoff, 1987). Despite wide recognition of this variation in meaning, object-orientation (OO) requires that we identify entities and categorize them without regard for different interpretations. Classes and systems of inheritance can obscure the relationship between natural language and interpretation. Categories (which become classes in an OO environment) reflect social and political values, but these values are often made implicit in the implementation of a system. The categories used by institutions, either implicitly or explicitly, inevitably reflect and facilitate the activities of the institution, be it governmental or private. Their instantiation as objects makes them complex referents for multiple meanings associated with that category. Indeed, sometimes the complexity is so great that although multiple groups utilize the same category and object they may not be referring to the same things. 'Agreeing to disagree' when it comes to objects is what social scientists refer to as 'boundary objects' (Star and Griesemer, 1989), an idea that has been used to study difference in wetland classifications (Harvey and Chrisman, 1998).

2.1.3 Uncertainty and Objects

The process of categorizing objects can produce changes in cognition so that categories begin to be seen to constitute reality, a process known as convergence (Bowker and Star, 2000). Even when we differentiate digital objects from the material entities that populate the earth's surface, once displayed on the screen they assume a materiality that often supplements the actual situation. The roads, bridges, tidal flats, and urban spaces that are delineated on the screen gain a fierce materiality. Research into uncertainty in geographic information has begun to address some of these issues. Classic approaches to uncertainty in GISc have tackled the uncertainties associated with location (Blakemore, 1984), boundary (Leung, 1987), and attribution (Goodchild and Dubuc, 1986) of objects. However, most approaches to uncertainty are still restricted. Despite techniques for representing the uncertainty associated with the location and boundary of an object, presenting the uncertainty associated with the existence of an object at all is more problematic. In an information system, uncertainties associated with a category are also difficult to represent. Categories such as 'pond' or 'mountain' vary in meaning. Within one country such as Canada, pond can mean a large water body covering several square kilometers or a small, semi-annual pool (Mark, 1993); a mountain in Wales becomes a small hill when subjected to the effects of context and social structures (Fisher and Wood, 1998). Recent research into uncertainty has begun to address problems like definition (Duckham *et al.*, 2001; Fisher, 2000) and granularity (Guesgen and Albrecht, 2000; Worboys, 1998) of objects and categories. Rather than being uncertain, some concepts are in clear competition (Bibby, Chapter 5; Fisher *et al.*, Chapter 6); uncertainties based on ambiguity and discord are epitomized by political disagreement over ownership of tracts of land between states. So far, very little work in GISc has addressed any such problems, although Ahlqvist *et al.* (2000) and Ahlqvist (Chapter 7) demonstrate one approach using fuzzy and rough set theory to represent vagueness and ambiguity in vegetation map reclassification.

2.1.4 Ontology and Objects

Ontology is the branch of metaphysics that deals with the nature of being. The GIScience community has begun to address some of the problems posed by categories through a study of ontology (Guarino and Giaretta, 1995; Mark and Egenhofer, 2001; Smith and Varzi, 1997). A range of different approaches exist, characterized at one extreme by a positivist approach to identifying a single Ontology, that underlies all phenomena, and at the other extreme by an approach closely related to computer science which works with multiple ontologies that each represent a unique understanding of the world. As yet there exists no clear direction in GISc research literature over the precise role and importance of ontology. A key problem in all work in this area is that people express ontologies in different ways. The differences between expression of an ontology in natural language and programming language are not just curious artifacts, but indicate some of the substantial ontological issues between different representations of phenomena that still need to be addressed. However, it remains the case that we see a relative sparsity of ontologies of space in GIS; the object and field views of space continue to dominate GISc.

2.2 Re-presenting Objects

The discussion so far has attempted to indicate some of the ways objects may play an active role in discourse. The processes that lead to the promotion of particular objects and categories may have clear social, economic and political dimensions. Interdependencies between the objects and interactions we recognize, correspond to interdependencies between the classifications and the spaces within a GIS. For example, census data commonly conflates people and the surface of the earth. While there is a connection between the two, describing the socio-economic characteristics of a region can never adequately characterize the people who are closely connected with that region. Understanding the objects in a GIS means understanding the negotiations that have occurred to arrive at these objects and the active part objects have to play in these negotiations.

Despite the naive representations of objects within GISc in the past, it does not follow that GISc must be straightjacketed by these limitations in the future. Increased dialogue between social theorists and GISc researchers is enlivening and broadening the subject. At the same time, GISc remains closely connected with questions of information and computation. An important aspect of theory in GISc is its practical application within a computational setting; its codability (Schuurman, 1999). 'Not just objects' recognizes this axiom, and several of the papers are concerned with implementing emerging concerns about the role of objects in GISc. The six chapters in this section are founded on an understanding of objects as constructions, and attempt to show how this understanding can affect the development of information systems.

Increased attention to the role of data models in enabling specific ontologies is commensurate with recognition that defining objects is a social process. To date, the emphasis on ontological determinants has been largely technical. Schuurman (Chapter 3) argues that perception and deployment of objects varies across communities of interest, and it is only by investigating institutional culture that these differences can be documented, and analysed. The chapter begins by reviewing the development of the discourse of ontologies and objects in the GISc community in the context of a growing philosophical and cognitive awareness. The present framework is then extended by illustrating the power of social and institutional influences to affect object definition, and suggesting vehicles to incorporate their study in GISc.

Recent studies of semantics offers inroads to study the production and communication of geographic knowledge. Language is the key arbitrator in these processes. Drawing on post-structuralist studies of semantics and deconstructionist work, Harvey (Chapter 4) presents an approach, called social linguistics, to studying the situatedness and relational characteristics of geographic knowledge. This approach is closely tied to recent work by Doreen Massey and others who argue for a relational understanding of space-time. Applied to semantic interoperability, this conceptual framework extends formalized knowledge domains through the consideration of informal knowledges. Concurrently, this approach provides a way to consider the nebulous character of objects and processes of constructing and representing objects.

Objects may, however, be viewed relationally; as the product first of the attention of a reflective observer, second of the matter and processes observed, and third of the representational devices deployed. Collectively we overlook this, allowing particular objects to become *naturalized* in the discourse of our daily lives, so that they come to appear as the inevitable constituents of our world. In this manner, a world is continually reproduced (largely unthinkingly), both at the level of representations, and in material terms, guided by these representations. Bibby's perspective (Chapter 5) treats GIS use as being embedded in this process, and sees familiar GIS problems (such as MAUP and ecological fallacy) as arising from the relationship between different representational systems in a particular discursive setting. He argues that the relation between GIS and natural language is particularly important. Adopting an approach resting ultimately on the work of Quine (1960), Jubien (1993), and particularly Goodman (1978), he considers the way that representational systems are used in constructing objects. From this perspective, both natural language and GIS can be used to configure different objects from the same matter. These objects may be coextensive, such as the *wood* that is also a *nature reserve*, or the *flower-rich meadow* that might also be a *housing site*. Alternatively the same elements of matter may contribute to objects that overlap but are not coextensive, such as a town centre and a bus corridor. Language is used by particular interest groups to bind particular bundles of matter, thereby creating objects meaningful to them, involving qualitative differences that cannot be reduced to degraded quantitative ones. Bibby therefore treats GIS use as a computational variant of what Goodman (1978) terms *worldmaking*. Linking GIS and logic programming, he attempts both to investigate the repertoire of objects deployed by others and to exploit more actively the creative potential of GIS use by, for example, constructing land-use maps from natural language texts, computational interpretation of Ordnance Survey basic scale maps, and building a gazetteer of rural settlements from postal addresses.

Fisher, Comber and Wadsworth (Chapter 6) examine definitions for Land Use and Land Cover mapping from remotely sensed imagery. Such mapping is now very widespread with the term being used in many agencies both national and international. From a re-reading of some of the classic texts in this area they remind us that, although typically identified as a single class of information, Land Use and Land Cover are actually very different. Surprisingly, the grouping of the two as one is a product of the consideration of cartographic information density. The confusion, however, is now endemic in this area, and may be a problem for many applications using this information.

Ahlqvist (Chapter 7) demonstrates a formal implementation to integrate uncertain object information. Bringing together ideas of crisp, fuzzy, and rough classifications, the rough fuzzy formalization demonstrates the possibility of representing and merging different views on a certain conceptualization, each carrying different forms of uncertainties or limitations. This brings vagueness explicitly into the analysis and the resulting set of alternative maps indicates a potential of the suggested approach for group decision processes. Use of a spatial frame to provide common points of reference enable crisp, fuzzy, and rough representations to be used together to articulate similarities and differences between different views as well as efforts trying to formalize the process of reaching a negotiated agreement.

Finally, Duckham and Sharpe (Chapter 8) argue that significant trends in current research into uncertainty and imperfection in geographic information are already beginning to converge on a much more diverse, pluralistic view of objects. The widening variety of formal models of imperfect objects available to GISc researchers is often highly suitable for use in a computational environment. At the same time, adoption of these models can provide room to challenge the apparent materiality of on-screen digital objects and unpack some of the processes that led to their construction. By integrating multiple formal models of imperfection within one information system, it should be possible to enable users to switch dynamically between very different representations of objects, selecting the representations that are most appropriate for a particular situation.

References

Ahlqvist, O., Keukelaar, J. and Oukbir, K., 2000, Rough classification and accuracy assessment, *International Journal of Geographical Information Science*, **14**(5), 475–496.

Blakemore, M., 1984, Generalisation and error in spatial databases, *Cartographica*, **21**(2/3), 131–139.

Bowker, G. and Star, S.L., 2000, *Sorting Things Out: Classification and its Consequences*, Cambridge, MA: MIT Press.

Buttenfield, B., 1995, Object-oriented map generalization: modelling and cartographic considerations. In *GIS and Generalization: Methodology and Practice*, edited by Muller, J.C., Lagrange, J.P. and Weibel, R., Bristol, PA: Taylor and Francis, pp. 91–105.

Campari, I., 1991, Some notes on geographic information systems: the relationship between their practical application and their theoretical evolution. In: *Cognitive and Linguistic Aspects of Geographic Space*, edited by Mark, D.M. and Frank, A.U., London: Kluwer Academic Publishers, pp. 35–44.

Cosgrove, D., 2001, *Apollo's Eye. A Cartographic Genealogy of the Earth in the Western Imagination*, Baltimore: John Hopkins University Press.

Couclelis, H., 1992, People manipulate objects but cultivate fields: beyond the raster-vector debate in GIS. In: *Theories and Methods of Spatial-Temporal Reasoning in Geographic Space*, edited by Frank, A.U., Campari, I. and Formentini, U., Berlin: Springer-Verlag, pp. 65–77.

Duckham, M., Mason, K., Stell, J. and Worboys, M., 2001, A formal approach to imperfection in geographic information, *Computers, Environment and Urban Systems*, **25**, 89–103.

Fabrikant, S.I. and Buttenfield, B., 2001, Formalizing semantic spaces for information access, *Annals of the Association of American Geographers*, **91**(2), 263–280.

Fisher, P.F., 1998, Is GIS hidebound by the legacy of cartography? *Cartographic Journal*, **35**(1), 5–9.

Fisher, P.F., 2000, Sorites paradox and vague geographies, *Fuzzy Sets and Systems*, **113**(1), 7–18.

Fisher, P.F. and Wood, J., 1998, What is a mountain? Or the Englishman who went up a Boolean geographical concept but realised it was fuzzy, *Geography*, **83**(3), 247–256.

Frank, A.U., 1996, The prevalence of objects with sharp boundaries in GIS. In: *Geographic Objects with Indeterminate Boundaries*, edited by Burrough, P.A. and Frank, A.U., London: Taylor and Francis, pp. 29–43.

Frank, A.U., 1999, Ontology of Fields: Reflections on a Varenius Conference. Paper read at *Association of American Geographers 95th Annual Meeting*, March 23, 1999, at Honolulu, Hawaii.

Frank, A.U. and Raubal, M., 1998, Specifications for interoperability: formalizing image schemata for geographic space. In: *Proceedings from the 8th International Symposium on Spatial Data Handling*, International Geographical Union, Burnaby, Canada, edited by Poiker, T.K. and Chrisman, N., pp. 331–348.

Goodchild, M.F., 1989, Modeling error in objects and fields. In: *Accuracy of Spatial Databases*, edited by Goodchild, M. and Gopal, S., London: Taylor and Francis, pp. 107–113.

Goodchild, M.F., 1992, Geographical information science, *International Journal of Geographical Information Systems*, **6**(1), 31–45.

Goodchild, M.F. and Dubuc, O., 1986, A model of error for chloropleth maps with applications to geographic information systems. In: *Proceedings Auto-Carto 8*, ACSM, Bethesda, MD, pp. 165–174.

Goodman, N., 1978, *Ways of Worldmaking*, Hassocks: Harvester.

Gregory, D., 1994, *Geographical Imaginations*, Cambridge: Blackwell.

Guarino, N. and Giaretta, P., 1995, Ontologies and knowledge bases: towards a terminological clarification. In: *Towards Very Large Knowledge Bases*, edited by Mars, N.J.I., Amsterdam: IOS Press, pp. 25–32.

Guesgen, H.W. and Albrecht, J., 2000, Imprecise reasoning in geographic information systems, *Fuzzy Sets and Systems*, **113**: 121–131.

Harvey, F. and Chrisman, N.R., 1998, Boundary objects and the social construction of GIS technology, *Environment and Planning A*, **30**(9), 1683–1694.

Henderson, K., 1995, The visual culture of engineers. In: *The Cultures of Computing* edited by Star, S.L., Oxford: Blackwell Publishers/The Sociological Review, pp. 196–217.

Hohl, P., (Ed.), 1998, *GIS Data Conversion: Strategies, Techniques, Management*, Santa Fe: OnWord Press.

Jackson, R.W., 1994, Object-oriented modeling in regional science: an advocacy view, *Papers in Regional Science*, **73**(4), 347–367.

Jubien, M., 1993, *Ontology, Modality and the Fallacy of Reference*, Cambridge: Cambridge University Press.

Kemp, K., 1997, Fields as a framework for integrating GIS and environmental process models. Part 1; Representing spatial continuity, *Transactions in GIS*, **1**(3), 219–233.

Lakoff, G., 1987, *Women, Fire, and Dangerous Things: What Categories Reveal about the Mind*, Chicago: University of Chicago Press.

Leung, Y., 1987, On the imprecision of boundaries, *Geographical Analysis*, **19**(2), 125–151.

Leung, L., Leung, K.S. and Zhong, H.J., 1999, A generic concept-based object-oriented geographical information system, *International Journal of Geographical Information Science*, **13**(5), 475–498.

Mack, P.E., 1990, *Viewing the Earth. The Social Construction of the Landsat Satellite System*, Cambridge, MA: MIT Press.

Mark, D.M., 1993, Towards a theoretical framework for geographic entity types. In: *Spatial Information Theory: A Theoretical Basis for GIS*, edited by Frank, A.U. and Campari, I., Berlin: Springer-Verlag, pp. 270–283.

Mark, D.M., 1997, Cognitive perspectives on spatial and spatio-temporal reasoning. In: *Geographic Information Research: Bridging the Atlantic*, edited by Couclelis, H. and Craglia, M., Bristol, PA: Taylor and Francis.

Mark, D.M., 1999, Spatial representation: a cognitive view. In: *Geographical Information Systems: Principles, Techniques, Management and Applications*, edited by Longley, P. A.,Goodchild, M.F., Maguire, D. J. and Rhind, D. W., New York: Wiley.

Mark, D.M. and Egenhofer, M.J., 2001, Research on Ontology and Ontologies. Paper presented at the *UCGIS Summer Assembly*, June 2001, Buffalo, NY.

Massey, D., 1999, Space-time, 'science' and the relationship between physical geography and human geography, *Transactions of the Institute of British Geographers*, **24**, 262–276.

McMaster, R.B. and Shea, K.S., 1992, *Generalization in Digital Cartography*, Washington: Association of American Geographers.

Muehrcke, P. and Muehrcke, J., 1992, *Map Use. Reading, Analysis, Interpretation*, 3rd edition, Madison: JP Publications.

Muller, J.C., Weibel, R. Lagrange, J.P. and Salge, F., 1995, Generalization: state of the art and issues. In: *GIS and Generalization: Methodology and Practice*, edited by Muller, J.C., Lagrange, J.P., Weibel, R., Bristol: Taylor and Francis, pp. 3–18.

NCGIA, 1999, The NCGIA Core Curriculum in GIScience. http://www.ncgia.ucsb.edu/education/curricula/giscc/cc_outline.html Updated August 13, 2000.

Peterson, M.P., 1994, Cognitive issues in cartographic visualization. In: *Visualization in Modern Cartography*, edited by MacEachren, A.M. and Taylor, D.R.F., Tarrytown, NY: Elsevier Science Inc (Pergamon).

Quine, W.V., 1960, *Word and Object*, Cambridge, MA: M.I.T. Press.

Schatzki, T., 1991, Spatial ontology and explanation, *Annals of the Association of American Geographers*, **81**, 650–670.

Schuurman, N., 1999, Critical GIS: theorizing an emerging discipline, *Cartographica*, **36**(4), 1–108.

Smith, B. and Varzi, A.C., 1997, Fiat and bona fide boundaries: towards an ontology of spatially extended objects. In: *Spatial Information Theory. A Theoretical Basis for GIS*, edited by Hirtle, S., and Frank, A.U. Lecture Notes in Computer Science 1329, Berlin: Springer-Verlag, pp. 103–119.

Star, S. L. and Griesemer, J.R., 1989, Institutional ecology, 'translations', and boundary objects: amateurs and professionals in Berkeley's Museum of Vertebrate Zoology, *Social Studies of Science*, **19**, 387–420.

Voisard, A. and Schweppe, H., 1998, Abstraction and decomposition in interoperable GIS, *Geographical Information Science*, **12**(4), 315–333.

Weibel, R., 1995, Three essential building blocks for automated generalisation. In: *GIS and Generalisation: Methodological and Practical Issues*, edited by Muller, J.C., Lagrange, J.P. and Weibel, R., London: Taylor and Francis, pp. 56–69.

Weibel, R. and Dutton, G., 1999, Generalising spatial data and dealing with multiple representations. In: *Geographical Information Systems: Principles, Techniques, Management and Applications*, edited by Longley, P.A., Goodchild, M.F., Maguire, D.J. and Rhind, D.W., New York: Wiley, pp. 125–155.

Worboys, M.F., 1998, Computation with imprecise geospatial data, *Computers, Environment and Urban Systems*, **22**(2), 85–106.

3

Social Dimensions of Object Definition in GIS

Nadine Schuurman

3.1 Introduction: Ontologies and Objects

How we define objects really depends on what objects are; and what they are is closely related to their ontologies, which depends on how we define objects. A fine mess. The gist is that object definition is closely bound to ontologies. This paper will attempt to sort out this relationship, not in a definitive way, but as a means of moving beyond the technical parameters of the issue to a preliminary examination of *social dimensions* of object definition. It is only possible at this time to move beyond these initial questions because so much careful work has been done in the last decade to clarify these issues (Burrough, 1996; Frank, 1996, 2001; Kuhn, 2001; Mark, 1991; NCGIA, 1998; Smith and Mark, 1998, 2001; Winter, 2001). Before treading upon social territory, it is salutary to examine the growing body of research concerning geographic objects and their ontologies in order to clarify the following questions: what is object definition? what is the relationship of object definition to ontologies? and how have the two been treated in GIS?

Object definition – or entitation – has increased in importance during the past decade of GIS research. Previously, researchers did not ignore this important element of representation in GIS as much as they framed it in different terms. Early on, the complexities of object definition were confined to debate over the relative merits of data structures and models. Physical structuring devices were regarded as the primary influences on entitation. Attention to cognitive and philosophical implications of encoding features using different data models in the early 1990s led to examination of epistemologies and ontologies. With the 2001 publication of a special issue of the

Re-presenting GIS Edited by P. Fisher and D. Unwin

International Journal of Geographic Information Science (IJGIS) on ontologies, the importance of this area was firmly ensconced, yet, with few exceptions (Harvey, Chapter 4), the social and institutional processes that structure ontologies have not been documented.

Ontologies are difficult to agree upon at the best of times because they imply a foundationalism that is itself contentious. In GIS, these discussions are further complicated as the meaning of ontology remains vague, depending on the context in which geographic information is being modelled. To a philosopher, the word implies the essence of being, an ultimate and stable reality (Gregory, 1994). To a computer scientist, ontology refers to an internally consistent formal system in which all elements are precisely described and relations between entities are coherent (Gruber, 1995). Even within a circumscribed information science community such as GIS, understandings of what ontologies are or imply vary widely. Winter (2001) claims that ontology is a new term for what were formerly data models, formal specifications, and semantics. Smith and Mark (2001) use a concept-driven explanation that centers on the ways in which geographic objects are understood and conceptualized within a discipline and externally. In this scenario, categorizations of human subjects are naturalized, and ontologies are based on how people envisage phenomena. Another approach to defining ontologies is offered by Andrew Frank (2001) who focuses on the ways that human beings interact with the world. This approach recognizes five tiers of ontology that range from independent reality to subjective knowledge, and recognizes that the properties of objects depend on context. Werner Kuhn (2001) prefers a language-based ontology. For him, it makes more sense to study texts in order to derive ontologies based on the *use* of language. Like Winter, Kuhn regards 'ontologies' as old actors wearing new hats: '[t]hey have informally been around for decades in the form of feature-attribute catalogues' (Kuhn, 2001, p. 613).

There is some truth in this view, but it fails to recognize that GIS researchers have incorporated new layers of abstraction in recognizing that: (i) language and categories are naturalized with use; and (ii) that conceptualizations influence the results of queries and research generally. Recognition of the role of ontologies was accompanied by new respect for epistemology – or the way that phenomena are studied and the presuppositions that are embedded in the realization of knowledge. For the purposes of this discussion, I use Raper's (1999) explanation of ontology which marries these multiple perspectives by linking ontology to object identity in a GIS context, thus acknowledging the influence of both social and technical influences in establishing identity.

Which leaves us with the question of what objects are. Like ontologies, there are many ways of looking at objects in GIS. The simplest approach is to say that geographical entities are those features that exist on the earth, while objects are their incarnation in a GIS (Mark 1993). Since the advent of object-oriented data models, objects have come to refer to entities described distinctly from the field in which they are situated. Objects *interrupt* the vision of the world as a series of locationally registered layers, each representing a single attribute. Rather than focus on location, object-oriented GIS defines geographical phenomena, such as telephone poles or streets, as objects. Location becomes one of many other attributes associated with a particular object. Objects can be points, lines, areas or volumes with 3 dimensions. Confusion abounds in the discipline about what exactly object data models are. Vector data models were historically referred to as objects as they are constructed from points, lines and areas – objects – and also have clear boundaries. With the inception of object-oriented computing science, vector data models have been pushed back into the field universe. Legitimate objects (as opposed to vectors) carry their own confusing baggage of terms and implications. They can refer to object-oriented programming languages (OOPL) – one of the clearer designations – or to database structures or to programming that uses a conceptual model based on objects.

Every object in an object-oriented GIS database is not defined individually; that would be a painstaking task. Rather, groups of like objects are organized into classes. Vehicles used for transportation might comprise one class; sub-classes would include cars, trains, trucks, etc. Classes and sub-classes have attributes that apply to the entire group. In addition, operations or methods that describe possible pertinent actions, can be defined with reference to the class. Attributes and procedures are bequeathed through a hierarchical system of inheritance (Figure 3.1). Using a hierarchical system allows

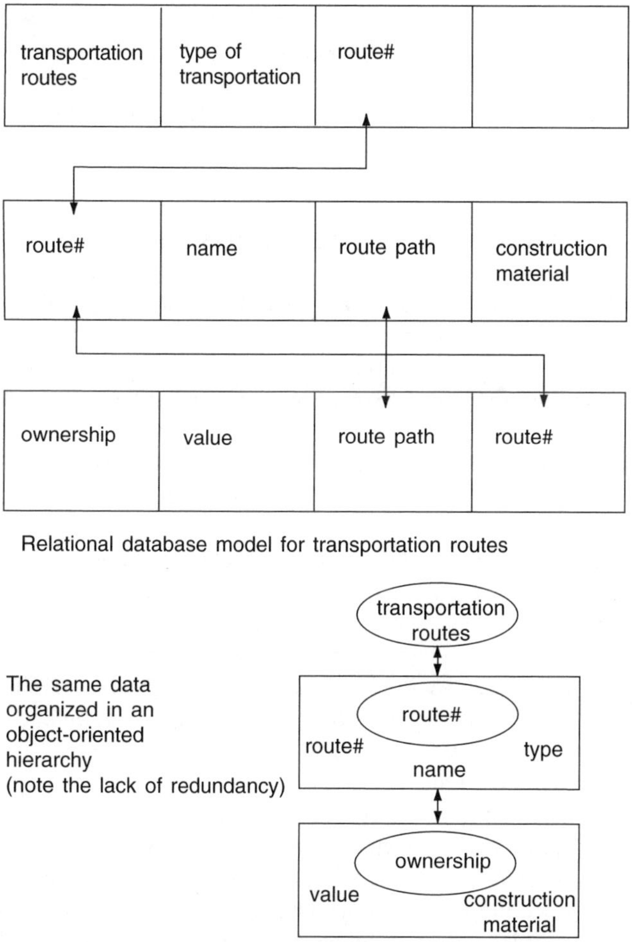

Figure 3.1 *The difference between relational and object-oriented databases (Reproduced from Schuurman, N. 1999. Critical GIS: theorizing an emerging discipline. Cartographica,* ***36****(4))*

rapid updating of general characteristics. That is not, however, the chief conceptual attraction of object-oriented data models. Rather they are believed, by some, to parallel more closely human conceptualizations than location-based field models. Another reputed advantage of object data models over fields is that entities can be defined by function rather than by name (Kuhn, 1994). This assumes, however that functions can be precisely delimited.

These, like many other assumptions about data models, merit critical inquiry for field and object models rely ultimately on a Newtonian view of the world coupled with Euclidean geometry; neutral space is always assumed (Couclelis, 1999). Neither model allows the characterization of complex-interrelated geographic entities (Burrough, 1996; Goodchild, 1992). They are each simplistic characterizations of a complex geographical reality. For the purposes of this paper, and general discussion, however, it is safe to assume that objects are geographical entities represented in a digital context.

The following section contains a review of the development of disciplinary attention to object definition and ontologies since the 1960s and, in the process, distinguishes between the two. In Section 3.3, the impact of social and institutional context on the process of defining objects is illustrated in an effort to show the effects of social practice on GIS objects. Finally, in Section 3.4, an appeal is made to encourage the growth of GIS that supports multiple epistemologies and ontologies.

3.2 Fields and Objects as Structuring Devices, 1960–2001

In the 1960s and 1970s, objects had not yet entered the GIS imagination. Discussions of feature representation were limited to debates over the relative merits of raster and vector data structures. Conference participants wore buttons pledging their allegiance to one or the other. A sample pin proclaimed that 'raster is faster but vector is better' (T. Poiker, personal communication). When objects migrated onto the GIS scene from computer science in the late 1980s, they cast a new light on fields (Goodchild, 1995). Goodchild (1992) described six types of fields, but these can be broadly described as covering the extent of the study area such that an attribute of each layer will be associated with each set of geographical coordinates. Typically, fields are used to describe scientific phenomena such as surface temperature, elevation, and soil type. Objects, on the other hand, exist independently of each other. They are not required to cover a study area. Moreover, they can overlap. Objects are believed to better represent geographical phenomena as humans perceive them (Couclelis, 1992; Mark, 1999); they are also more suitable than fields for applications that focus on particular, limited entities such as pipelines or fiber optic networks (Figure 3.2).

Both object and field data models can be represented by either raster or vector data structures. Those data structures, in turn, are implemented using *structuring devices* at a computational level. Rasters, for instance, are held by arrays of 1 to *n* dimensions, and vector data structures are held by lists. Ultimately, each of these structuring devices are linked to registers which literally stack the addresses of bits and bytes. This information is so fundamental that it is common knowledge for high school students. Its significance in GIS, however, is underrated. These *physical* structuring devices – from the data model to data structures to lists to the lowly register – were initially considered the chief

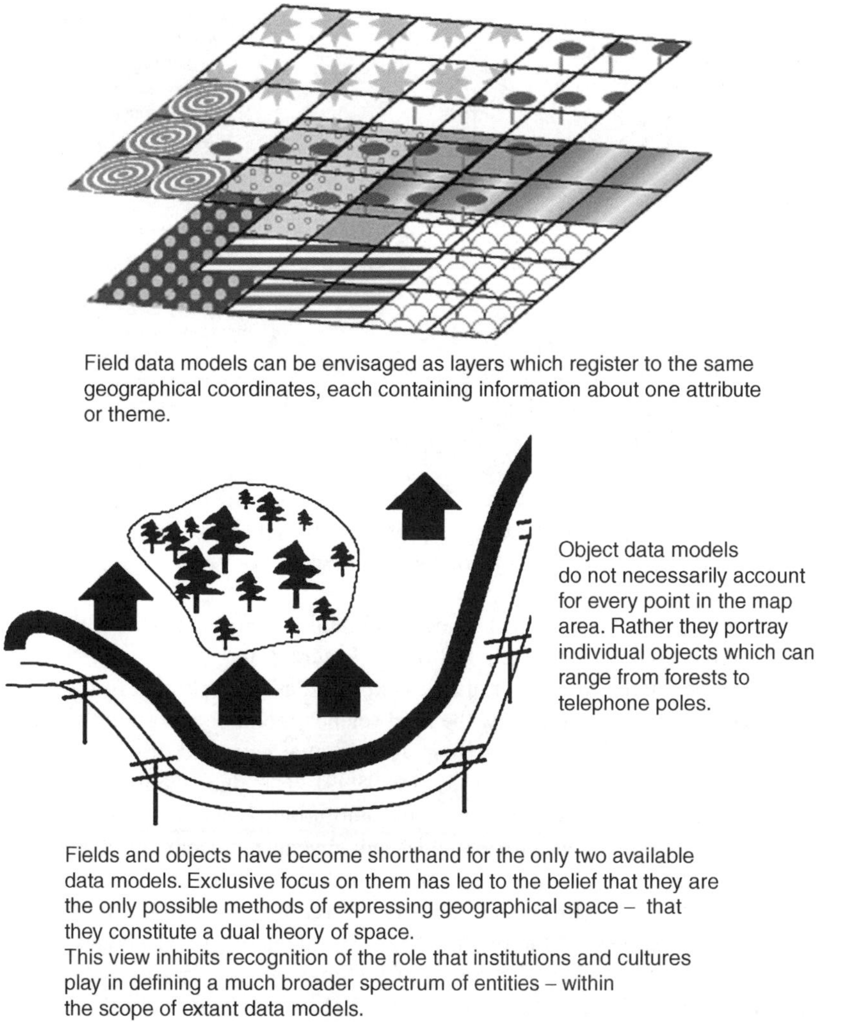

Figure 3.2 *Field and object data models (Reproduced from Schuurman, N., 1999. Critical GIS: theorizing an emerging discipline. Cartographica,* ***36****(4))*

constraints to object definition. A great number of articles exploring the implications of employing one data model, with its attendant structuring devices, over another have reinforced their stature as the primary parameters affecting digital objects (Brown, 1999; Frank, 1996; Kemp, 1997; Mark, 1984; Nunes, 1991; Peuquet, 1984, 1988; Raper, 1999). A series of scientific developments in the GIS community contributed to a binary of fields and objects with its implications for research in object definition.

In the late 1980s and 1990s, preferences over data structures were re-framed as between debates over fields versus objects. This distinction coincided with efforts to better understand geographical space. Objects and fields were recast as the two sides of spatial representation. Just as light is considered by physicists to have a dual wave/particle

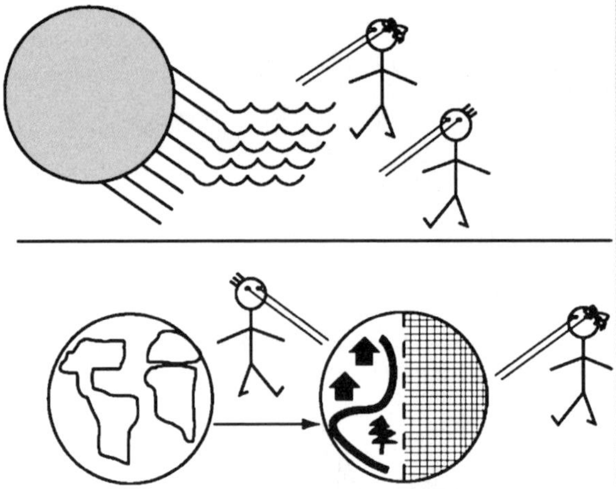

Figure 3.3 *The dual theory of geographic space and its role in entitation (Reproduced from Schuurman, N. 1999. Critical GIS: theorizing an emerging discipline. Cartographica,* ***36****(4))*

nature, researchers hypothesized that objects and fields constitute a duality that describes space and spatial relations (Peuquet, 1988) (see Figure 3.3).

This is a very seductive view as it justifies existing data models, and follows a physics ('hard science') precedent. Ironically, the hard science model is increasingly discredited, but retains a hold on disciplines such as geography that have had difficulty establishing foundational objects and principles (Massey, 1999). The dual theory of geographical space is a quick fix for representation – and entitation. It assumes that existing data models with their correspondence to fields and objects represent the extent to which geographical features can be digitally represented. I have argued elsewhere that the continued binary between objects and fields is self-reinforcing, and limits the likelihood of developing alternatives (Schuurman, 1999). The dual theory of geographical space represents an extreme version of the sort of technicism that critics of GIS have been critical of in the past (Lake, 1993; Pickles, 1993, 1995; Sheppard, 1993). Moreover, it fails to do GIS scholars justice, most of whom recognize social and philosophical influences and conflicts associated with their research (Burrough, 1996; Couclelis, 1996; Smith and Mark, 2001).

Coincident with debates over objects and fields in the 1990s, there has been a philosophical turn in GIS research. Researchers are increasingly recognizing that there are epistemological and ontological repercussions to the choice of data models and structures. Raper (1999) highlights a changed discipline when he notes that science always carries philosophical implications – even if they are unacknowledged. This recognition amongst GIS researchers has been manifest in two ways. First, there is increasing acceptance that epistemology – the study methods that we recognize and propound within our discipline – has an influence on the choice of data models. Second, geographers are beginning to document ontological implications of how spatial data is structured inside the computer (Kemp and Vckovski, 1998; NCGIA, 1998; Schuurman, 1999; Smith and Mark, 1998).

Field and object data models are products of different ways of viewing the world. Debates over their relative merits generated a compendium of agreed-upon knowledge about ontologies. Objects privilege the formal boundaries of entities while field models emphasize location above all else (Couclelis, 1992). Fields are flexible enough to accommodate changing relations between geographic entities while objects, once inscribed, are more immutable. Definition of objects is ill-justified by the transience of geographical relations and many entities. But, *both* fields and objects imply concrete boundaries, and despite preliminary efforts, few GIS applications can accommodate fuzzy objects (Burrough and Frank, 1996). The fact remains that phenomena are modeled in one data model or another, and this affects their range of description and analysis. Object-based relations, for instance, describe taxonomy and membership while locational relationships lend themselves to areal generalization, overlay and neighborhood functions (Peuquet, 1988). Related objects call up an entirely different set of similarities than related locations (Kemp, 1997). Certainly GIS scholars do not claim that either is a perfect model: 'The only perfect model is the phenomena itself, and we are probably always going to be faced with going from one type of representation to another' (Peuquet, 1984, p. 84).

Objects, fields and ontologies gave rise to a new discourse in GIS research, but one that was concerned primarily with the virtues and limitations of data models. More recently, researchers have acknowledged the roles of conceptualization and categorization in understanding ontologies (Kuhn, 2001; Winter, 2001). Smith and Mark (2001) argue that ontologies require empirical support in the form of robust classification systems. Moreover, they acknowledge that classification systems become 'naturalized' over time. Digital ontologies act as 'surrogate created worlds' (Smith and Mark, 2001). As such, their power to influence science must be recognized (Bowker and Star, 2000).

Acknowledgment of the power of data models to shape ontologies and analyses has been a profound advance, and has provided the basis for recognition of the role of social practice to affect objects and their representation. To date, social dimensions of object creation have received considerably less attention than technical considerations. Only recently did Bibby and Shepherd (2000) demonstrate that object definition is shaped by social purpose. Perception and deployment of objects varies across communities of interest, and by investigating institutional culture these differences can be documented, and analyzed. Recognizing and documenting the role of social processes will extend rather than contract geographical understanding of entitation.

3.3 The Social Construction of Objects

Emphasis on data models and classification as the primary influences on object definition – and by extension representation – generates three problems. First, methodologies that can be used to discern points of social influence have not been ratified by the discipline: ethnography is one example. Second, there has been a reluctance to acknowledge that *cultures* of data production are a primary influences on object definition. Third, the potential effects of interoperability and standardization, as enforcers of object definition, has been given little attention. These problems point to a need for increased recognition of strong social influences on GIS. For instance, the scale at which researchers choose to conduct a given study is socially influenced, and has practical and intellectual repercussions (Marston, 2000). Likewise, there is no ontological basis for the current practice of defining discrete cartographic scales at 1:10 000 or 1:250 000 (Goodchild and Proctor, 1997). Clearly, social influences affect not only the levels of abstraction at which we view objects, but their very definition.

3.3.1 The Uses of Ethnography

In order to discern social impacts on entitation, one has to study the ways that objects have been created in different institutional settings. Latour (1987) in a ground-breaking study of scientific laboratories, *Science in Action*, illustrated that scientific objects such as proteins or microbes slowly become recognized entities through a process of reification. They start as noted effects (for example, the differentiation of cell contents in a centrifuge), and through a process of observation *and* academic citation slowly acquire object status. Later, new objects call upon older, institutionalized objects 'in their reified form' to establish their own identity (Latour, 1987, p. 92). Latour thus drew scholarly attention to the ways in which entities in physics and chemistry are constructed through a complex interaction of social and scientific effects. The mechanics of entity creation in GIS are, ironically, even more complex than this because our objects are often presumed 'natural' and, therefore, uncontroversial. In order to better understand social dimensions of their creation, we need to follow Latour's example, and study the processes by which we encode geographical objects.

Ethnographic methods are essential to this exercise. One explanation for low recognition and acceptance of social influences on entitation is that our current and accepted GIS methodologies exclude them from purview. Steve Herbert (2000) has outlined the traditional resistance of geographers to ethnographic method – an undervalued technique in geography. Three criticisms of ethnography prevail: (i) it relies on interpretation and so is not 'scientific'; (ii) an intense focus on a single or limited number of situations make it difficult to generalize; (iii) representation is clouded by the failure of researchers to problematize their methods. Herbert points out that all science relies on interpretation, and often on empirical evidence. Thus, the problem of interpretation is not unique to social science. Likewise, data are not naive, and are themselves discerned and interpreted through social processes. The problem of generalization can also be addressed by noting that studies of single institutions have provided insights into their workings. *Science in Action*, for instance, provides one of the first documentations of the processes, social and scientific, that guide scientific research. Another is provided by John Law's ethnography of Daresbury Nuclear laboratory in the UK (Law, 1994). Finally, the 'problem of representation' is broader and plagues not just ethnography but all representations. It is sufficient to say that representation is always limited, but nevertheless necessary, if natural and social processes are to be understood. The strength of ethnography is that it can be used to uncover the ways that systems of meaning are generated and reinforced.

3.3.2 Cultures of Data Production and the Development of Scientific Practice

Raper (1999) noted that systems of meaning can only be understood by persons within a shared 'cognitive environment'. Meaning is forged by interactions among people and agents within such an environment. In the context of GIS object definition, such environments constitute cultures of data production. Ethnography is a key tool for understanding how particular objects are generated in data production environments, and what limitations they have.

Understanding interactions between people, location and processes of data collection can shed light on the value and limitations of entities. These processes are partly consolidated through institutional culture. Interviews and site studies are a key component of 'studying-up', or examining the institutions that influence social interactions. For instance, refugees have long been studied by anthropologists and geographers. 'Studying-up' has gained ground among both anthropologists and geographers interested in the ways in which power is inscribed, in institutional settings (Abu-Lughod, 1991; Pred and Watts, 1992). In reality, these institutions influence refugees' experience from the length of time that they spend in camps to their chances for resettlement (Hyndman, 2000). This tradition of focusing on the particular at the expense of the institutional ('studying-down'), while valuable, has eclipsed significant aspects of social process. In GIS, it has reinforced a focus on data structures. Even when social processes are acknowledged, they are assumed to be local rather than institutional.

The role of technology in creating objects is shaped by institutions. An example is provided by the transformation of water well data into aquifers and aquitards (impediments to the flow of groundwater). In the province of Ontario, legislation passed in 1946 requires private drillers to report well-log data to the provincial government, including material, depth and lithological description (Russell *et al.*, 1998). Few of the drillers have any geological training, and the water well reports were designed primarily to protect well owners rather than provide data for the provincial government. These are, however, the chief source of data from which sub-surface models are developed. Higher quality data provided by continuous cored boreholes and geotechnical reports are either not publicly available or too sparse to allow interpolated sub-surface models for the area (Logan *et al.*, 2001). Such models are potentially useful in guiding waste management decisions, development policy, and determining groundwater flow direction among other things (Kenny *et al.*, 1996; Russell *et al.*, 1996). The Geological Survey of Canada (GSC) – a federal body – undertook the standardization of a portion of these data, in the Oak Ridges Moraine, in order to develop better tools with which to understand groundwater issues affecting the Greater Toronto Area (GTA). Eighty-two possible material types were simplified to eight major sediment types using truncation as well as inference based on the drillers' ability to identify lithology (Russell *et al.*, 1998). Control drilling data were used to define limits of accuracy and provide context for driller descriptions of sedimentology. This standardization effort resulted in a preliminary model of aquifers and aquitards in the GTA – a vital tool for a large metropolitan area.

From the perspective of entitation, the effects of this local standardization effort extend beyond a single dataset. This project became the basis for subsequent standardization in the provincial Ministry of the Environment in Ontario, and for a similar study in British Columbia. Furthermore, the rule-based standardization procedures developed during the project have been integrated into other standardization programs at the provincial level. Semantic standardization is entirely necessary if data from disparate sources (such as private drillers) are to be used to develop models of sub-surface groundwater. The GSC contributed substantially to the ability of the province to delineate and protect permeable areas. In fact, models from the Oak Ridges dataset were recently used in a legal enquiry over E-coli pollution in the groundwater that killed seven people in Walkerton, a town in Southern Ontario. Institutionalization of these local standardization practices will, however, have a lasting impact on entitation of sub-surface data in Canada.

The rule-based classification procedure developed for the sub-surface sedimentology is based on the parameters of data quality associated with the Oak Ridges Moraine data. The water well data from Ontario are based on reports from water well drillers who have little if any geological training. Frequently, they use locational data based on estimated distances from semi-permanent structures or highway junctions. Moreover, the reports are filled in (by law) after the fact, so the scenario of drillers assembled in a doughnut shop, filling in 'clay' for most categories is not impossible. Indeed, clay is the most common classification used for 40% of all fields. By contrast, log data from continuous boreholes used for ground-truthing reveal that clay constitutes 2% of the material in the area (Russell *et al.*, 1998). As one geologist put it: 'the drillers can't tell clay from muck, but they know bedrock when they hit it.' As a result of low accuracy, the standardized data are designed to allow only primitive modeling of aquifers and aquitards – or porous material from impermeable material. More reliable, complex data would allow greater confidence to be placed on the modeling of porosity and flow direction. The standardization system, while well adapted to disparate data from multiple sources, is unable to handle more complex descriptions associated with higher quality data.

This solution reflects a particular cognitive environment shaped by emphasis on integrating poor quality data for modeling geological phenomena. The group of scientists at the GSC share a particular set of valid assumptions about the problem and the technology that is required to solve it. Jonathan Allen (2000) notes that social or scientific groups harness particular visions and interpretations of reality 'to form complex networks of practice which create and sustain sociotechnical systems' (Allen, 2000). These networks of practice greatly influence the direction that future innovation takes. The problem framing or cognitive environment in which a group of scientists works can explain how a technological innovation develops over time (Allen, 2000; Latour, 1988). From a sociological perspective, the interesting and important point is that this is becoming the basis for other standardization projects in the country. Indeed, the author has developed a project to standardize British Columbia (BC) water well data using the set of procedures developed by the GSC. In BC, this approach is warranted. Like Ontario, BC relies on water well data from drillers that is generally of a low quality. In Newfoundland, by contrast, water well drillers must be licensed, and all are required to attend a rigorous training program and use fixed lithological classifications. The rule-based standardization procedures used in Ontario are not as appropriate in such a data environment for two reasons. First, the rules are designed for a geological environment dominated by unconsolidated materials. Second, and more significantly, drillers in Newfoundland work from a common category list, and drillers are trained to recognize the lithological classifications. Rules to collapse a heterogeneous classification system are superfluous in this situation, and would reduce the complexity of the data.

In the case of the Oak Ridges Moraine sub-surface data, the parameters used to frame the problem (i.e. the need for a regional stratigraphic framework coupled with low quality data) led to a solution that permitted scientists, especially hydrogeologists, a preliminary basis for creating models of groundwater. A particular set of practices developed within a particular cognitive environment, and a problem-framing shaped by local factors, may well become the basis for national standards. This is illustrative of the extent to which cultures of data production affect entitation far beyond their native jurisdiction. There are instances in which richer – or more limited classification systems – are suitable depending on data quality and the range of objects permitted for analysis.

The more important point is that the complexity of any classification system determines the complexity of subsequent analysis; rich semantics permit richer analysis. This dictum is illustrated in the case of the Oak Ridges data which are constrained by low limits of confidence in the sedimentological reporting, but reasonable accuracy associated with the differentiation between bedrock and permeable material. Degrees of porosity cannot be modelled while true aquitards can. This distinction is the basis for the classification system used, and similar ones being developed in Ontario and British Columbia. Object definition is affected by particular sets of social and scientific circumstances that are later institutionally consolidated.

Technical limitations are also frequently reinforced at an institutional level through institutional cultures. For instance, the current aquifer classification system used at the BC Ministry of Water, Land and Air Protection categorizes known aquifers based on their level of development, use and vulnerability to contamination (Kreye *et al.*, n.d.). This classification system limits future flexibility as aquifers change ranking in terms of productivity and vulnerability as a result of environmental changes and better data. The Ministry currently uses a relational database, but a shift to an object-oriented data structure (such as the proposed North American Geological Data Model (NADM)) would allow them to define classes of aquifers as well as individual ones, and then associate degrees of vulnerability and productivity to each. This shift in structuring would permit greater flexibility because aquifers could be classified according to a number of different, dynamic characteristics. Categories could then be constructed on the fly based on varying combinations of criteria. Moreover, the sub-surface sedimentological classification discussed above could be greatly extended in an object oriented environment depending on the sophistication of data available for each borehole. Unfortunately, the politics of data structures may intervene.

There is a push on the part of some research scientists to introduce object oriented data structures such as NADM to geological surveys in Canada and the US, and Environment Ministries in Canada, but there is attendant resistance. NADM would enable greater ease of interoperability as it allows for the creation of meta-categories based on conceptual frameworks. Feature-based representation is supplemented with ontological context that will allow easier reconciliation of the meaning of data categories (Brodaric and Hastings, 2002). NADM integrates recognition that some categories may depend on the context that the individual researcher is working within, and also that ontologies are created by utilizing diverse types of knowledge, including situational knowledge (Brodaric and Gahegan, 2001). NADM allows the separation of facts from conceptualization, thus allowing multiple conceptualizations to emerge from the same set of facts.

Ministries in Canada are under financial stress, however, and some perceive NADM as just another top-down initiative for which they might receive little financial support. The Ministry of Water, Land and Air Protection in British Columbia, for instance, is reluctant to invest hastily in any shifts associated with interoperability frameworks for fear that these specification will be later be superceded (Glen, 2000). As this example illustrates, institutional cultures are not linear influences on the adaptation of technology; rather, they are tangled up with policy, people, and technology. The value of ethnography is in unravelling this web. It in no way supercedes the physical parameters of object definition, but enhances understanding of social and institutional practices that contribute to the development of ontologies and objects in GIS.

3.3.3 Standardization as a Meta-influence on Entitation

Cultures of data production are not isolated fiefdoms. Government agencies that use GIS are linked by common software, and policy regimes. Moreover, they are guided by internationally and federally endorsed meta-institutions such as the Open GIS Consortium (OGC), the Spatial Data Transfer Standard (SDTS) and the Federal Geographic Data Committee (FGDC) in the US. In Canada, the equivalent body is GeoConnections charged with developing interoperability standards for geospatial data. Each of these institutions influences standardization and interoperability, and in the process determines the parameters of object creation. Standardization governs how data can be described and used. In many cases standardization renders raw data usable in the first instance, but nevertheless it constitutes a little recognized institutional influence on entitation.

Indeed, one cannot talk about object definition without talking about standardization and its close relation classification. Both are social and political processes that implicitly acknowledge and endorse certain points of view (Bowker and Star, 2000). Current efforts to develop and endorse standard semantics within domains (Kottman, 1999) are based on reigning systems of classification within domains. This trend carries with it two implications. First, classification for information systems tends to emphasize low-level convergence rather than high-level semantic classification. Second, every classification system changes and perpetuates perception of geographical objects such that the systems categories are seen to constitute reality (Bowker, 2000). This is a recursive process in which scientific classification reflects social structures in its organization.

Returning to the example of standardization of sub-surface strata in the Greater Toronto Area, we see both of these implications at work. Low-level convergence necessarily became the basis for standardization of terms, and more significantly, the binary of aquifers and aquitards became the basis of stratigraphic description. Clearly, practice is a social factor that needs to be factored into the development of standardization/classification systems, and the role of institutional influence must be acknowledged. This will become increasingly true as the dream of interoperability between systems as well as datasets is advanced through national and international convention. This is not to discount these very valuable efforts towards data integration, and improved range of spatial analyses, but to note that systems approaches to standardization are fraught with social influence and implications.

3.4 Making a Space for Multiple Objects and Ontologies

This chapter has emphasized the power of the social and institutional to influence digital entitation of geographical objects. It is also imperative, however, to recognize that a wider scope for object definition will ultimately be achieved through technical solutions.

I have stressed elsewhere that only through technical understanding can social influences be properly differentiated, and constructively integrated into research (Schuurman, 2000). In this instance, recognizing the complexity of influences – social and technical – that shape entitation will enable the adaptation of solutions that support multiple objects and ontologies.

While, historically, entitation has been considered in GIS research as a process entirely dependent on technical structuring devices, I have illustrated that this view elides the social contexts that frame the process. Object definition and interpretation in GIS is not a sterile scientific process, but one that is influenced by, and influences, social and institutional cultures. The adoption of ethnographic methods will better allow researchers to discern the ways that cultures of data production influence object definition. Likewise, a recognition that interoperability and standardization are paramount institutional influences on the range of possible object definition, will alert researchers to how knowledge representation is influenced. Objects are also influenced by the perceptual framework in which that are viewed. Geologists and housing developers interpret the earth's sub-surface very differently, and there are no neutral means of portraying prototypes in GIS. Objects reflect the agenda and value of their users.

References

Abu-Lughod, L., 1991, Writing against culture. In: *Recapturing Anthropology*, edited by Fox, R.G., Sante Fe: School for American Research.

Allen, J.P., 2000, Information systems as technological innovation, *Information, Technology and People*, **13**(3), 210–221.

Bibby, P. and Shepherd, J., 2000, GIS, land use, and representation, *Environment and planning B: Planning and Design*, **27**, 583–598.

Bowker, G., 2000, Mapping biodiversity, *International Journal of Geographical Information Science*, **14**(8), 739–754.

Bowker, G. and Star, S.L., 2000, *Sorting Things Out: Classification and its Consequences*. Cambridge, MA: MIT Press.

Brodaric, B. and Gahegan, M., 2001, Learning geoscience categories *in situ:* implications for geographic knowledge representation. In: *Proceedings of the Ninth International Symposium on Advances in GIS*, Atlanta, GA, Nov. 9–10, 2001 edited by Aref G.W. ACM Press, New York, pp. 130–135.

Brodaric, B. and Hastings, J., 2002, An object model for geologic map information, In: *Advances in Data Spatial Handling. The Proceedings of the 10th International Symposium on Spatial Data Handling*, Berlin, Springer-Verlag, pp. 55–68.

Brown, L., 1999, Change, continuity, and the pursuit of geographic understanding, *Annals of the Association of American Geographers*, **89**(1), 1–25.

Burrough, P.A., 1996, Natural objects with indeterminate boundaries. In: *Geographic Objects with Indeterminate Boundaries*, edited by Burrough, P.A. and Frank, A.U., London: Taylor and Francis.

Burrough, P.A, and Frank, A.U. Eds, 1996, *Geographic Objects with Indeterminate Boundaries*. London: Taylor and Francis.

Couclelis, H., 1992, People manipulate objects but cultivate fields: Beyond the raster-vector debate in GIS. In: *Theories and Methods of Spatial-Temporal Reasoning in Geographic Space*, edited by Frank, A.U., Campari, I. and Formentini, U., Berlin: Springer-Verlag.

Couclelis, H., 1996, Towards an operational typology of geographic entities with ill-defined boundaries. In: *Geographic Objects with Indeterminate Boundaries*, edited by Burrough, P.A. and Frank, A.U., London: Taylor and Francis.

Couclelis, H., 1999, Space, time, geography. In: *Geographical Information Systems: Principles, Techniques, Management and Applications*, edited by Longley, P.A., Goodchild, M.F., Maguire, D.J. and Rhind, D.W., New York: John Wiley and Sons, Inc.

Frank, A.U., 1996, The prevalence of objects with sharp boundaries in GIS. In: *Geographic Objects with Indeterminate Boundaries*, edited by Burrough, P.A., and Frank, A.U., London: Taylor and Francis.

Frank, A.U., 2001, Tiers of ontology and consistency constraints in geographical information systems, *International Journal of Geographical Information Science*, **15**(7), 667–678.

Glen, A., 2000, Towards Interoperable Geographic Information Systems at the British Columbia Ministry of Environment, Lands, and Parks. Unpublished Honours Thesis, Simon Fraser University.

Goodchild, M.F., 1992, Geographical information science, *International Journal of Geographical Information Systems*, **6**(1), 31–45.

Goodchild, M.F., 1995, Geographic Systems Information and Research. In: *Ground Truth*, edited by Pickles, J., New York: Guildford Press.

Goodchild, M.F. and Proctor, J., 1997, Scale in a Digital Geographic World, *Geographical and Environmental Modelling*, **1**(1), 5–23.

Gregory, D. 1994, Ontology. In: *The Dictionary of Human Geography*, edited by Johnston, R.J., Gregory, D. and Smith, D.M., Oxford: Blackwell.

Gruber, T., 1995, Toward principles for the design of ontologies used for knowledge sharing, *International Journal of Human-Computer Studies*, **43**, 907–928.

Herbert, S., 2000, For ethnography, *Progress in Human Geography*, **24**(4), 550–568.

Hyndman, J., 2000, *Managing Displacement: Refugees and the Politics of Humanitarianism*, Minneapolis: University of Minnesota Press.

Kemp, K., 1997, Fields as a framework for integrating GIS and environmental process models. Part 1; Representing spatial continuity, *Transactions in GIS*, **1**(3), 219–233.

Kemp, K. and Vckovski, A., 1998, Towards an ontology of fields. Paper read at The Third International Conference on GeoComputation, at Bristol, UK, September 17–19.

Kenny, F.M., Russell, H.A.J., Hinton, M.J. and Brennand, T.A., 1996, Digital elevation models in environmental geoscience, Oak Ridges Moraine, southern Ontario. In: *Current Research 1996-E*: Geological Survey of Canada, pp. 201–208.

Kottman, C., 1999, The Open GIS Consortium and Progress Toward Interoperability in GIS. In: *Interoperating Geographic Information Systems*, edited by Goodchild, M., Egenhofer, M., Fegeas, R. and Kottman, C., Boston: Kluwer Academic Publishers.

Kreye, R., Ronneseth, K. and Wei, M., n.d, *An Aquifer Classification System for Groundwater Management in British Columbia*. British Columbia Ministry of Environment, Lands and Parks 1994 [accessed April 2, 2005]. Available from http://www.env.gov.bc.ca/wat/aquifers/Aq_Classification/Aq_Class.html.

Kuhn, W., 1994, Defining semantics for spatial sata transfers. Paper read at The Sixth International Symposium on Spatial Data Handling (SDH), at Edinburgh, Scotland.

Kuhn, W., 2001, Ontologies in support of activities in geographical space, *International Journal of Geographical Information Science*, **15**(7), 613–631.

Lake, R.W. 1993, Planning and applied geography: positivism, ethics, and geographic information systems, *Progress in Human Geography*, **17**, 404–413.

Latour, B., 1987, *Science in Action*, Cambridge MA: Harvard University Press.

Latour, B., 1988, *The Pasteurization of France*, Cambridge MA: Harvard University Press.

Law, J.,1994, *Organizing Modernity*, Oxford: Blackwell.

Logan, C., Russell, H.A.J. and Sharpe, D.R., 2001, Regional three-dimensional stratigraphic modelling of the Oak Ridges Moraine areas, southern Ontario, *Geological Survey of Canada Current Research*, 2001-D1.

Mark, D.M., 1984, Digital Elevation Models and Data Structuring Issues. Paper read at Computer Assisted Cartography and Geographic Information Processing: Hope and Realism, at Ottawa.

Mark, D.M., 1991, Object modeling and phenomenon-based generalization. In: *Map Generalization*, edited by Buttenfield, B. and McMaster, R., Essex, UK: Longman Scientific and Technical.

Mark, D.M., 1993, Towards a theoretical framework for geographic entity types. In: *Spatial Information Theory: A Theoretical Basis for GIS*, edited by Frank, A.U. and Campari, I., Berlin: Springer-Verlag.

Mark, D.M., 1999, Spatial representation: a cognitive view. In: *Geographical Information Systems: Principles, Techniques, Management and Applications*, edited by Longley, P.A., Goodchild, M.F., Maguire, D.J. and Rhind, D.W., New York: Wiley, Inc.

Marston, S.A., 2000, The social construction of scale, *Progress in Human Geography*, **24**(2), 219–242.

Massey, D., 1999, Space-time, 'science' and the relationship between physical geography and human geography, *Transactions Institute of British Geographers*, **24**, 261–276.

NCGIA, 1998, The Ontology of Fields: Report of a Specialist Meeting held under the auspices of the Varenius Project. National Center for Geologic Information Analysis, Santa Barbara, CA

Nunes, J., 1991, Geographic space as a set of concrete geographical entities. In: *Cognitive and Linguistic Aspects of Geographic Space*, edited by Mark, D.M. and Frank, A.U., London: Kluwer Academic Publishers.

Peuquet, D., 1984, A conceptual framework and comparison of spatial data models, *Cartographica*, **21**(4), 66–113.

Peuquet, D.J., 1988, Representations of geographic space: Toward a conceptual synthesis, *Annals of the Association of American Geographers*, **78**, 375–394.

Pickles, J., 1993, Discourse on method and the history of discipline: Reflections on Jerome Dobson's 1993 'Automated geography', *Professional Geographer*, **45**(4), 451–455.

Pickles, J., 1995, Representations in an Electronic Age: Geography, GIS, and Democracy. In: *Ground Truth*, edited by Pickles, J., New York: Guildford Press.

Pred, A. and Watts, M., 1992, *Reworking Modernity*, New Jersey: Rutgers University Press.

Raper, J., 1999, Spatial representation: the scientist's perspective. In: *Geographical Information Systems: Principles, Techniques, Management and Applications*, edited by Longley, P.A., Goodchild, M.F., Maguire, D.J. and Rhind, D.W., New York: Wiley, Inc.

Russell, H.A.J., Logan, C., Brennand, T.A., Sharpe, D.R. and Hinton, M.J., 1996, Regional geoscience database for the Oak Ridges Moraine project (southern Ontario). In: *Current Research 1998-E*: Geological Survey of Canada, pp. 191–200.

Russell, H.A.J., Brennand, T.A., Logan, C. and Sharpe, D.R., 1998, Standardization and assessment of geological descriptions from water well records, Greater Toronto and Oak Ridges Moraine areas, southern Ontario. In: *Current Research 1998-E*: Geological Survey of Canada, pp. 89–102.

Schuurman, N., 1999, Critical GIS: theorizing an emerging discipline, *Cartographica*, **36**(4), 1–108.

Schuurman, N., 2000, Trouble in the Heartland: GIS and its critics in the 1990s, *Progress in Human Geography*, **24**(4), 569–590.

Sheppard, E., 1993, Automated Geography: What kind of geography for what kind of society? *Professional Geographer*, **45**, 457–460.

Smith, B., and Mark, D.M., 1998, Ontology and geographic kinds. In: *Proceedings from the 8th International Symposium on Spatial Data Handling*, edited by Poiker, T.K. and Chrisman, N., International Geographical Union, Burnaby, Canada, pp. 308–320.

Smith, B., and Mark, D.M., 2001, Geographical categories: an ontological investigation, *International Journal of Geographical Information Science*, **15**(7), 591–612.

Winter, S., 2001, Ontology: buzzword or paradigm shift in GI science? *International Journal of Geographical Information Science*, **15**(7), 587–590.

4

The Linguistic Trading Zones of Semantic Interoperability

Francis Harvey

Men content themselves with the same words as other people use, as if the very sound necessarily carried the same meaning.

John Locke

4.1 Introduction

In this chapter, I will examine semantic interoperability as a process of communication inseparable from language. To examine this particular aspect of geographic knowledge production, I draw on Galison's trading zone concept. This is a concept from socio-linguistics that articulates both the process of communication and comprehension between people lacking a common language and the place it occurs. The term originates in anthropological studies of culture and trade between Pacific Islanders. In these studies, each culture is largely self-contained, but at distinct intervals would travel, or greet travelers, and engage in trade. Since the other people did not speak the same language, a trading-zone 'language', or minimal contact language, would enter into use to facilitate the exchanges between people and overcome semantic disagreements. If the contact language is utilized frequently enough it could evolve into what linguists refer to as a 'pidgin' language. This concept has been extended to studies of exchanges between scientists and engineers (Galison, 1997). Galison's study sees the creation of trading zone language as a critical underpinning for interactions between different professions. I apply the trading zone concept in this chapter to study the possible resolution of semantic disagreements in geographic information interoperability.

Re-presenting GIS Edited by P. Fisher and D. Unwin

This chapter starts by engaging the question of how people produce and represent meaning through the vehicle of language. To begin, I assert that the codes used by computers to represent geographic information are rooted in human language. Understanding the meaning of codes for land use, vegetation, zoning, cadastre, etc. involves assessing their sociolinguistic roots. Wittgenstein's work on language games (Stroud, 1996; Wittgenstein, 1958a, 1958b) provides a widely accepted basis for this study. Like other language games, semantic interoperability is profoundly social, full of changing meanings; it is more than finding, creating, applying, and translating definitions. Many times, the meaning of attributes, while apparently concretized in the database and specification, are compromises laden with ambiguity. Much as a beaker carved and decorated for South Sea islanders' annual trade and exchange meeting, geographic information codes are boundary objects (Star and Griesemer, 1989) between different groups. Codes exhibit dual characters of binding and separation providing relative semantic stability for both flexible interactions and discrepant meanings at the same time. For example: two groups may both use the term wetlands; they may both agree that wetlands should be protected, but may disagree that wetlands are consistently inundated by water (Harvey, 1998).

Before turning to the theoretical foundations and examples of this approach that largely draw on literature outside of geography, I should make mention of the rich body of literature that exists on language and the role of geography in forming place (Buttimer, 1993; Tuan, 1971) which responded to geography's mid-twentieth century emphasis on structuralism. Following a presentation of the theoretical underpinnings in Section 4.2, I turn in Section 4.3 to consider the possibilities for drawing on linguistic insights for improving semantic interoperability.

4.2 Knowledge and Language

To discuss trading zones of semantic interoperability, we must begin with an introspective look into the concepts and theories of sociolinguistics. Wittgenstein's concept of language game is the foundation for many studies of sociolinguistic activity and concepts of multiple perspectives on reality, but it differs distinctively from theoretical studies of language (i.e. Chomsky's universal grammar) and the realist assertion that a word contains the reality of its meaning (Malinowski, 1953). To understand the trading zones concept, we need first to examine dominant assumptions underpinning modern scientific thought, geography, and computer science.

Realist assertions about the veridical relationship between words and reality have been the currency of most modern thought and geography. This assumption restrains understanding of the role of language in articulating knowledge because it assumes there is one underlying unique and universal valid definition for each term and word. The fundamental tenet of logical positivism, the most out-spoken school of realism in the modern period, as expressed by Rudolf Carnap early in the twentieth century was: 'We assumed that there was a certain rock bottom of knowledge, the knowledge of the immediately given, which was indubitable' (cited in Galison, 1997, p. 784). While numerous geographers would cite the influence of logical positivism, few would claim to be strict adherents to its tenets, drawing more widely from philosophy.

In the Western tradition, people often assume a direct connection between thought and representation, a dangerous oversimplification of Cartesian and Neo-Kantian philosophy. Geographers often focus on Descartes' work on geometry with its fundamental roles for scientific practices. Cartographic projections, spherical coordinate systems, and Global Positioning Systems (to name a few) rely on Cartesian innovations in geometry and mathematics. While the preeminence of Cartesian thought in Western civilization is seen by many geographers to be primarily a question of representing objects on an orthogonal plane by pairs of coordinates, I will argue, following the tradition of Husserl, Heidegger, Wittgenstein, Winch, Ryle, Schutz and others, that the Cartesian separation of mind and body, or *res cogitans* and *res extensa*, that accompanies the arrival of modern epistemology and mechanical thinking in the Renaissance (Lynch and Collins, 1998), exerts a more important influence on geography and other sciences.

The mind separated from body is also separated from the world and is reduced to the role of observer. The relevance of the separation of mind and body goes hand-in-hand with the development of linear perspectivalism. Albrecht Dürer's (1471–1528) advancement of linear perspectival drawing helped lay the foundation for modern ways of geographical knowledge involving detachment, distance, and limited perspective. As twentieth century analyses of the shift in the illustrative technology of the sixteenth century point out, culture thus gained a representational technique that mimics sight, but constrains the breadth of our vision to a focus on portions of objects as constructed by the viewer (Romanyshyn, 1989). This problem results from the loss of breadth and instrumentalization of a single perspective in observing and thinking and underlies the disembodiment of scientific studies to which feminists point (Domosh and Seager, 2001).

The ascendancy of techniques based on linear perspective representation and mechanistic thinking has been further enhanced by developments in mathematics, culminating in the development of Turing's computing machine (Sheppard, 1995). The unparalleled resources awarded to mathematics in the last four hundred years have led to a refinement of techniques and the development of symbolic languages that are uniquely capable of describing the world and any extra-terrestrial observations in terms of linear perspective and mechanical causality. Coupled with the shift in thinking that accompanied perspectival representation and thinking, Western civilization has established an unparalleled technology to study, describe, and control nature in line with Galileo's philosophical views regarding the role and prominence of mathematics:

> *The book of nature is written in mathematical language, using triangles, circles, and other geometrical figures as alphabet; and without these tools it would be impossible for humans to but understand even one word of this language.*

Strengthened by mathematical advances over four centuries, ascendency of scientific ideals, and metaphysical millenarism, most Westerners (and indeed through Western colonization, domination, and orientation of media, most people of the earth) have become so accustomed to this way of thinking that our language implicitly reflects the preeminence of sight in gaining and communicating knowledge (Wertheim, 1999).

While the suggestion of 'minds in vats' seems wildly exaggerated and distortive, the production of knowledge in the modern epoch implicitly requires putting our minds not only at a distant point to observe, but simultaneously using this point as the fulcrum to gain leverage for our mathematical tools. Mechanistic cognitive explanations complement perspectivalism and the turn to realism transcends the artist's subjective intervention. In geography, perspectivalism has arguably led to an emphasis on detachment between observer and observer, and on rigorous methods to distinguish the object from the subject. The language of geographic science and its codes can becomes oddly detached from the social act of communication. Resting on quantitative advances language becomes information and the vehicle for transferring observations (Goodchild, 2000).

This approach to language is less than fifty years old, but has become for many the unquestioned staple of how we conceive the role of language. Based on Shannon and Weaver's theory of communication in their *Mathematical Theory of Communication* (Shannon, 1949), it focuses on syntax and eliminates semantics. This is a development of realist approaches that fail to articulate what constitutes meaning. Disciplines and fields that were influenced by computer science developed systems approaches to study and understanding. The systems approach instrumentalizes many implicit aspects of realist language and perspectivalism. The realist visual metaphor is quintessential to geography as Ron Abler's analogy of the relative importance of GIS in geography to the microscope in biology (Abler, 1991) underscores.

4.2.1 Non-Realist Concepts of Language and Knowledge

Until recently, a heated inner-disciplinary debate raged between proponents of modernist, realist approaches and post-structuralist studies (Dobson, 1983, 1993; Pickles, 1995). These debates addressed the underlying ontology and epistemology of geography and the legitimization of GIS as a set of conventions for observation, analysis, and representation. I want to draw on John Pickles engagement with these questions to present the relevance of linguistics to non-realist studies of semantic interoperability, In particular, John Pickles takes up Walter Benjamin's work to examine critically the exuberance with which geographers embraced cartography for colonization and later in the construction of digital earth (Pickles, 1999). Complementary to Doreen Massey's arguments about geography as a process of flows and relationships (Massey, 1993), Pickles' archaeology of geography's visual representations points out that the attempt to create a permanent world on exhibition by any means ranging from ordinance survey mapping to geostationary satellite is limited when compared to the millions of individual experiences. We have perhaps the technology to observe the world 24 hours a day from a single satellite, but our languages do not seem up to the task of making sense of all the information; it is a common malady of the modern person. In fact, information overload was the very problem of verbal communication, for tasks that needed to be precisely coordinated against the background of loud machines, that led Shannon and Weaver to develop their realist mathematical theory of information to distinguish information, and data, from noise.

The meaning/background differentiation problem that the mathematical information theory addresses only accounts for syntax and not semantics. We can describe an image but we cannot explain it. This limitation was recognized following the Second World War by Weaver, Mead, and Bateson in continuing meetings of the Macy Foundation. These reflections, developing from discontent with realism's fallacies, have grown and diversified to fields ranging from mathematics (Hofstadter, 1979), logic (Devlin, 1997), systems design (Coyne, 1995), and philosophy (Feenberg, 1995). Integral to these critical works has been a marked break with modernist scientism and its assumptions about the relationships between language and meaning grounded in realism and logical positivism. This body of work has clearly strongly influenced geography.

For this chapter I wish to focus on the problems of claiming that a geographical term has one 'true' meaning. Take, for example, the term 'wetlands'. While widely used, without quotation marks, in the US and elsewhere, in environmental legislation that regulates or limits the activities of property owners, there is no administrative, political, or even scientific agreement as to what defines a wetland (Harvey and Chrisman, 1998). With no agreement as to what defines a wetland, even after many millions of dollars spent on research and litigation, this is a case that illustrates the limits of realist approaches in finding a single unified definition of any term. The structuralist attempt to circumvent the logical positivist dead-end was to declare that different definitions are the results of different contexts. Researchers deployed code-word analysis and other techniques to find commonalities between different contexts with some intriguing results, but never with clear enough findings to claim that a indubitable common meaning could be ascertained (Mark and Frank, 1996).

4.2.2 Language Games and Semantic Interoperability

Wittgenstein's work on language games provides a basis for productively engaging these issues and readdressing semantic interoperability barriers. Replacing the realist understanding that meaning lies unequivocally in a single definition of the term, Wittgenstein's influential work makes clear that, in fact, there is no single meaning to any term but, actually, many different meanings that different people and groups associate with a term in question. Understanding a sentence is like understanding a language and being able to master the technique of acting and responding linguistically in appropriate ways (Stroud, 1996). Take, for example, the term 'game'. To some people this may refer to team sports such as football. Others may take it to indicate a child's entertainment, yet others will think of board games. The term could also mean board-game, card-game, ball-game, school-game.

Are there any commonalities to these definitions of games? Are they all pleasurable or competitive, calling for skill or merely chance? Wittgenstein shows through his detailed arguments that what we believe to be commonalities dissolve into complex networks of similarities between different clusters of characteristics that an individual or group may associate with the term (Wittgenstein, 1958a, 1958b). Communicating becomes knowing, applying, and acting according to the community's use of the term. This means that there are no necessary and sufficient criteria that pick out all and only the things we know as games. Kuhn referred to this as 'meaning incommensurability' or 'the inability of one language and its referential structure to translate fully into another language system' (cited by Galison, 1997, p. 795). Kuhn also recognized that 'there is no "protocol language" that would serve as a common referent for the two languages' (cited by Galison, 1997, p. 795).

Acceptance of the multiplicity of meanings and meaning incommensurability does not lead to terminological relativism or sound the death knell for interoperability. Turning from Wittgenstein's example to geographic information, we uncover that phenomena such as parcels, buildings, wetlands, forests – to name a few common geographic phenomena – certainly have no necessary and sufficient distinguishing criteria. In other words, geographic meaning is not canonical, but the signs of relationships (Latour, 1997). Given the diversity of meaning and the incommensurability of terms, standards serve as linkage points between different groups. Like treaties, they are crucial to interactions, including computer system interoperability, between these groups who otherwise may have little in common and much disagreement (Harvey, 2001). Terms and codes invoked in standards are the boundary objects that, like boundary markers, provide common points of reference and distinction for people who otherwise use different languages (Star and Griesemer, 1989). In the trading zones of daily work places, people develop boundary objects that have shared meanings (i.e. wetlands should be protected) but which differ in other ways (i.e. wetlands are always inundated for some, while for others a temporally limited, but regular, immersion is sufficient). Instead of meaning translation, interoperability becomes part of a dynamic process of communication in a trading zone.

4.2.3 Trading Zones and Interoperability

The concept of the trading zone helps conceptualize the process through which different communities transcend differences to develop semantic interoperability. If we accept that computer codes and representations are intrinsically related to our languages, then the linguistic concept of trading zones can help illuminate and understand the processes of interoperability. The concept of trading zones draws primarily on work by Harvard historian Peter Galison, who has developed Wittgenstein's concept of language games into a historiographic approach for studying the rash growth of applied physics in the Second World War.

4.2.4 The Trading Zone and its Linguistic Roots

> *Within a certain cultural arena [the trading zone]... two dissimilar groups can find common ground. They can exchange fish for baskets, enforcing subtle equations of correspondence between quantity, quality, and type, and yet utterly disagree on the broader (global) significance of the items exchanged. Similarly, between the scientific subcultures of theory and experiment, or even between different traditions of instrument making or different cultures of theorizing, there can be exchanges (coordinations), worked out in exquisite detail without global agreement.*
>
> (Galison, 1997, p. 46)

The trading zone concept is intricately related to the notion of pidgin and creole languages. A pidgin language is used to make communication possible between people with different native languages. 'A pidgin is a relatively unstable variety [of language] developed in marketplaces, plantations, and similar environments for limited communication among native speakers of different languages...' (Heath, 2001, p. 441). A creole language is a pidgin language that becomes the native language for a new generation and is characterized by a more stable grammar (Heath, 2001). Linguistics focus in their studies on changes in grammar, the lexical relationship to the original lexifier language, and diglossic relationship. Pidgin languages can be understood as examples of the linguistic concepts of borrowing or code-switching. Borrowing is the use of historically transferred form in a target language and code-switching is the switch from one language to another (Heath, 2001). For example, a German discussing the results of work with a GIS analytical operation may use the German verb to discuss the process (*Verschneiden*, meaning to cut or to clip) but speak of the outputs as *die Overlays*. This is likely to be an example of borrowing, but more contextual information would be called for. A sociolinguist would also add that with more context, this example could possibly be revealed as a case of linguistic borrowing: the actual copying of the forms into the target language. If the German speaker spoke the word overlay with an English accent the case would be clear that this is linguistic transfer.

Galison applies these linguistic terms to a study of epistemological diversity, contradictions, and resolution between the cultures of theoretical physics, experimental physics, and engineering. Since pidgin languages arise from the need to establish trade or exchange, Galison asserts, as an example, the need for theoretical physics to exchange their computational models and results with the engineers building the first microwave antennas. Based on an exhaustive study of physics during the first fifty years of the twentieth century, the key point epistemologically for Galison is: 'The physicists and engineers of Room 4-133 are not engaging in translation as they piece together their microwave circuits, and they are not producing "neutral" observation sentences: they are working out a powerful, locally understood language to coordinate their actions' (Galison, 1997, p. 833). The importance of this insight comes in a re-conceptualization of science as 'an intercalated set of subcultures bound together through a complex of hard-won locally shared meanings...' (Galison, 1997, p. 840) that is neither relativist nor realist. Galison calls this view 'historicized neo-Kantianism' (Galison, 1997, p. 840). The ontologies and epistemologies of scientists and engineers are empirically realistic, not transcendentally realistic. Empirical realism is the process of coordinating different symbolic and material actions by people that creates a binding culture of science.

How would we distinguish the practices of interoperability from Galison's study of physicists' and engineers' interactions? While there are indubitably differences in the subject, the cultures of computer scientists and of domain experts and administrators, and the environment in which they work, differences of principle seem negligible. I would assert that because of the similarities of linguistic and epistemic issues faced in interoperability, Galison's trading zone concept can be applied extremely effectively to study the diversity of semantic interoperability. The OpenGIS concept of 'community' underlies the cultural dimension of interoperability and provides an accepted inroad to study the trading zones of interoperability.

4.3 Semantics and Interoperability

Computers and computer networks change the process of producing and communicating maps and geographic information. The representation of geographic information on a map is somewhat static and bound in large institutions because of the costs involved. In post-Fordian economic relations where the economies of scale in new market segments have become so large that centralized collection, preparation, publication, and distribution is rapidly breaking down (Rhind, 1997) and computer networking technology makes it possible to almost instantly exchange data, the different semantics of geographic information become more relevant.

Since our knowledges are always incomplete, semantic interoperability must provide support explicitly for interactive knowledge discovery and informal knowledge representation, i.e. narrative and pictures. Working with geographic information requires an understanding of these different forms of knowledge. Semantic interoperability and knowledge discovery provides a supportive framework for producing multiple geographic knowledges in 'trading zones'. Using multiple ontologies, people working with geospatial information can develop richer analytical tools and higher levels of interoperability and integration (Goodchild *et al.*, 1999). Engineers, environmentalists, government staff, and researchers, have become profoundly aware of the substantial discrepancies that lie between different information sources (Goodchild *et al.*, 1999). The scientific and engineering area in which work on addressing these issues occurs is know as interoperability because of its emphasis on improving the exchange of information in networked computer settings. Through substantial industry efforts, public specifications for exchanging various types of data have been developed or are under development (see, www.opengis.org).

4.3.1 Rethinking (Semantic) Interoperability

In this section, I want to build on the trading zones concept and post-realist computer science scholarship to articulate a theoretical approach to resolving semantic differences in interoperable environments.

The problems of geographic information interoperability are not unique; like any other techno-science, it involves trading zones. A substantial amount of literature influenced by Wittgenstein and other post-realist scholars in computer science and related fields provides a theoretical basis for considering heterogeneous semantic interoperability. Reconceptualizing meaning and knowledge as processes intricately interwoven in intersubjective discourse has become a common way of thinking in many disciplines ranging from mathematics (Restivo, 1990), physics (Collins and Pinch, 1998), computer science (Suchman, 1987; Winograd, 1995; Suchman *et al.*, 1999), and literary studies, and has strongly influenced recent geographic studies (Massey, 1993; Thrift, 1996).

Considerable work on semantic interoperability has developed approaches to define a common ontology for translating between a defined number of information sources and more recently has extended this model to support interactive construction of ontologies through knowledge discovery. Most of this work has occurred in the field of computer science on federated database systems, distributed information systems, and multimedia data management (Sheth, 1996; Kashyap and Sheth, 1997; Bishr, 1998). The three most prevalent semantic integration approaches explored are attribute equivalence, context and domain definitions, and shared ontology. The first two approaches model data in a database to compare domain, constraints, and operation or combine different ontologies in a single database to evaluate semantic equivalence, or more generally, semantic distance or proximity (Larson *et al.*, 1989; Sheth and Kashyap, 1992; Ouksel and Naiman, 1994). The work on shared ontologies relied on term definitions and interrelations. Semantic similarity studies use a shared ontology or a global ontology. These realist-based approaches are successful for constrained environments, e.g. an airline ticket price comparison web utility, but fail when confronted with semantics or syntax from outside their defined domains.

Extending the realist approach to a neo-Kantian framework, a key issue is the resolution of semantic similarity measures. Semantics for Amit Sheth (Sheth and Gala, 1989; Kashyap and Sheth, 1996, 1997; Sheth, 1996, 1997) need to be assessed in terms of the context. The concept of semantic proximity refers to an abstraction or mapping between the domains of two objects. Establishing similarities calls for comparing the intensional (contextual) descriptions of the two objects, described in a description logic language that links the semantic and schematic level. Conceptually, semantic integration in this approach consists of two phases. In the first phase objects are identified in different databases that are conceptually similar. In the second phase, the semantic differences are resolved between semantically related objects (Kashyap and Sheth, 1996).

Semantic proximity, in contrast to semantic similarity, uses a declarative language to define objects *a priori*, and strong ontological definitions that involve vocabulary, content and structure (Sheth, 1996). Semantic proximity refers to the similarities between objects, relationships, and context. To resolve semantic disagreements, linguistic studies should extend semantic proximity to support the dynamic development and redefinition of ontologies and epistemologies. People working with this system have access to their communities' own ontologies and through the 'trading zone' can enter into multiple social discourses. Meaning is dynamic, but the semantic interoperable information system can aid in identifying and developing different meanings.

Agent-based technology plays a crucial part in implementing linguistically-aware semantic interoperable solutions. The agents should become part of the communicative discourses. Recent developments allow knowledge discovery through interactions with different information communities and the generation of ontologies in a knowledge base. Associating behavior with the properties and behavior of information allows for specialized and generalized reasoning that exploits specific characteristics of the knowledge base (Wickler, 1998) and facilitates a more efficient use of information resources. It deploys computational techniques to assist in knowledge discovery and automates tasks that require access to multiple data repositories. Based on a dynamic archive, called the knowledge base, finding out information held in a federated database system is no longer a process of querying, but of learning the different types of information, their representation, uses, limits, etc. The process is interactive and relies on a rich multimedia environment to assist people from a variety of backgrounds not only to find what they seek, but understand what it means by embedding data and meta-data in the interface.

The discursive dimension that supports 'trading zones' comes into its own in the user interface and interaction support. Sheth's InformationSpace (Sheth, 1996) supports user interaction, knowledge discovery, and learning. It is the interface to incorporate procedural knowledge that the user articulates during interaction and the information that the user brings to the system. 'Democratizing' GIS through participatory techniques calls for extensions of the InformationSpace to incorporate narrative format knowledge and interface techniques to support human needs and capabilities. Using Harvey's (1997) participatory design methodology to resolve semantic disagreements and develop ontological commitments, semantic interoperability will be more extensible and support multiple knowledge representations.

4.4 Concluding Thoughts

This chapter presents preliminary research into non-realist linguistics and semantic interoperability. This brief overview of work in linguistics and related fields provides insights into the multiple meanings with which we associate words and ways to extend semantic interoperability to include the critical linguistic components of communicative discourse.

I would like to conclude with some reflections that summarize the procedural way of thinking about the relationship between knowledge and language. Derrida believes language to be all-encompassing. Language is like a swirling vortex with no starting points, ending points, or boundaries. Any concepts passed off as first principles '. . . may always be "deconstructed": they can be shown to be a product of a particular system of meaning, rather than what props it [that meaning] up on the outside' (cited in Eagleton, 1983, p. 132). This means that any concept held up as a first principle, universal, or separate from language, is actually part of language. The multiplicity of geographic terms is not relativistic, but evidence of multiple process. Language produces the world and simultaneous to the production of language is a production of geospatial arrangements.

Any GIS is part of this 'swirling'. Semantic disagreements are friction points. Project orientated systems clearly focus on a limited purpose in comparison to land information systems. The differences can be categorized in different ways, i.e. organizationally, but this type of distinction holds in itself an indicator for other discourses in which the organizations are engaged. We can think of this as a set of levels, but there is no hierarchy here, just an ecological flow of knowledge through the ongoing process of learning. Language and technology are mutually integral to the flow and production of geographic knowledge. Semantic interoperability will always be a trading zone.

Acknowledgments

Portions of this research were supported with funding from the National Science Foundation (EIA-0091534). I also wish to thank participants in the ESRC seminar for their comments on an earlier version of this paper.

References

Abler, R.F., 1991, *Something old, something new . . . GIS, Geography, and the Social Sciences*, Paper presented at the NCGIA conference on GIS in the Social Sciences, Toronto.

Bishr, Y., 1998, Overcoming the semantic and other barriers to GIS interoperability, *International Journal of Geographic Information Science*, **12**(4), 299–314.

Buttimer, A., 1993, *Geography and the Human Spirit*, Baltimore: The John Hopkins University Press.

Collins, H., and Pinch, T., 1998, *The Golem. What you should know about Science* (Second Edition), Cambridge: Cambridge University Press.

Coyne, R., 1995, *Designing Information Technology in the Postmodern Age. From Method to Metaphor*, Cambridge, MA: The MIT Press.

Devlin, K., 1997, *Goodbye, Descartes. The End of Logic and the Search for a New Cosmology of the Mind*, New York: John Wiley and Sons, Inc.

Dobson, J., 1983, Automated Geography, *The Professional Geographer*, **35**(2), 135–143.

Dobson, J., 1993, The geographic revolution: A retrospective on the age of automated geography, *Professional Geographer*, **45**(4), 431–439.

Domosh, M. and Seager, J., 2001, *Putting Women in Place. Feminist Geographers Make Sense of the World*, New York: Guilford.

Eagleton, T., 1983, *Literary Theory. An Introduction*, Minneapolis: University of Minnesota Press.

Feenberg, A., 1995, Subversive rationalization: technology, power, and democracy. In: *Technology and the Politics of Knowledge*, edited by Feenberg, A. and Hannay, A. Bloomington, IN: Indiana University Press, pp. 3–22.

Galison, P., 1997, *Image and Logic. A Material Culture of Microphysics*, Chicago: University of Chicago Press.

Goodchild, M.F., 2000, Communicating geographic information in a digital age, *Annals of the American Association of Geographers*, **90**(2), 344–355.

Goodchild, M.F., Egenhofer, M.J., Fegeas, R. and Kottman, C.A. Eds, 1999, *Interoperating Geographic Information Systems*, Boston: Kluwer Academic Publishing.

Harvey, F., 1997, Improving multi-purpose GIS design: participative design. In: *Spatial Information Theory. A Theoretical Basis for GIS*, edited by Hirtle, S. and Frank, A.U., Berlin: Springer Verlag, Vol. 1329, pp 313–328.

Harvey, F., 1998, *Interoperability and Integration: Finding Semantic Agreement*, Paper presented at the Interoperating GISs: Report of the Specialist Meeting, 5–6 December 1997, Santa Barbara, CA.

Harvey, F., 2001, Constructing GIS: actor networks of collaboration, *URISA Journal*, **13**(1), 29–37.

Harvey, F. and Chrisman, N.R., 1998, Boundary objects and the social construction of GIS technology, *Environment and Planning A*, **30**(9), 1683–1694.

Heath, J., 2001, Borrowing. In: *Concise Encyclopedia of Sociolinguistics*, edited by Mesthrie, R., Amsterdam: Elsevier, pp. 432–442.

Hofstadter, D., 1979, *Gödel, Escher, Bach: an eternal golden braid*, New York: Basic Books.

Kashyap, V. and Sheth, A., 1996, Semantic and schematic similarities between database objects: a context based approach, *VLDB Journal*, **5**, 276–304.

Kashyap, V. and Sheth, A., 1997, Semantic heterogeneity in Global Information Systems: The role of metadata, context, and ontologies. In: *Cooperative Information Systems: Current Trends and Directions*, edited by Papzoglou, M. and Schlageter, G., New York: Academic Press, pp. 1–30.

Larson, J., Navathe, S. and Elmasri, S., 1989, A theory of attribute equivalence in databases with application to schema integration, *IEEE Transactions on Software Engineering*, **15**(4), 449–463.

Latour, B., 1997, *On recalling ANT*, Paper presented at the CSTT Workshop 'Actor Network and After', Keele, UK, 10–11 July 1997.

Lynch, M. and Collins, H.M., 1998, Introduction: humans, animals, and machines, *Science, Technology, and Human Values*, **23**(4), 371–383.

Malinowski, B., 1953, The problem of meaning in primitive languages. In: *The Meaning of Meaning*, Fifth Edition edited by Ogden, C.K. and Richards, I.A., New York: Harcourt, Brace and Company, pp. 296–336.

Mark, D.M. and Frank, A.U., 1996, Experiential and formal models of geographic space, *Environment and Planning B*, **232**, 3–24.

Massey, D., 1993, Power-geometry and a progressive sense of place. In: *Mapping the Futures. Local Cultures, Global Change*, edited by Bird, J., Curtis, B., Putnam, T., Robertson, G. and Tickner, L., New York: Routledge.

Ouksel, A.M. and Naiman, C.F., 1994, Coordinating Context Building in Heterogeneous Information Systems, *Journal of Intelligent Information Systems (JIIS)*, **3**(2), 151–183.

Pickles, J., 1999, *Cartography, Digital Transitions, and Questions of History*, Paper presented at the 19th International Cartographic Conference, Ottawa, Canada, 14–21 August 1999.

Pickles, J. Ed., 1995, *Ground Truth. The Social Implications of Geographic Information Systems*, New York: Guilford Press.

Restivo, S., 1990, The social roots of pure mathematics. In: *Theories of Science and Society*, edited by Cozzens, S.E. and Gieryn, T.F., Bloomington, IN: University of Indiana Press, pp. 120–143.

Rhind, D. Ed., 1997, *Framework for the World*, Cambridge: GeoInformation International.

Romanyshyn, R.D., 1989, *Technology as Symptom and Dream*, London: Routledge.

Shannon, C.E., 1949, *The Mathematical Theory of Communication*, Urbana, IL: The University of Illinois Press.

Sheppard, E., 1995, GIS and society: towards a research agenda, *Cartography and Geographical Information Systems*, **22**, 5–16.

Sheth, A., 1996, *Data Semantics: What, Where and How?* Paper presented at the 6th IFIP Working Conference on Data Semantics (DS-6), Atlanta, GA.

Sheth, A., 1997, *Semantic Interoperability in Infocosm: Beyond Infrastructural and Data Interoperability in Federated Information Systems*, Paper presented at the International Conference on Interoperating Geographic Systems (Interop'97), Santa Barbara, 3–4 December 1997.

Sheth, A. and Kashyap, V., 1992, *So Far (Schematically) yet So Near (Semantically)*. Paper presented at the DS-5 Semantics of Interoperable Database Systems, Lorne, Australia.

Sheth, A.P. and Gala, S.K., 1989, *Attribute Relationships: An Impediment in Automating Schema Integration*. Paper presented at the Workshop on Heterogeneous Database Systems, Chicago, 11–13 December, 1989.

Star, S.L. and Griesemer, J.R., 1989, Institutional Ecology, 'Translations', and Boundary Objects: Amateurs and Professionals in Berkeley's Museum of Vertebrate Zoology, *Social Studies of Science*, **19**, 387–420.

Stroud, B., 1996, Mind, meaning, and practice. In Sluga, H. and Stern, D.G. Eds, *The Cambridge Companion to Wittgenstein*, Cambridge: The Cambridge University Press, pp. 296–319.

Suchman, L., 1987, *Plans and Situated Actions. The Problem of Human-Machine Communication*. Cambridge: Cambridge University Press.

Suchman, L., Blomberg, J., Orr, J.E. and Trigg, R., 1999, Reconstructing technologies as social practice, *American Behavioral Scientist*, **43**(3), 392–408.

Thrift, N., 1996, *Spatial Formations*, London: Sage Publications.

Tuan, Y.-F., 1971, Geography, phenomenology and the study of human nature, *Canadian Geographer*, **15**, 181–192.

Wertheim, M., 1999, *The Pearly Gates of Cyberspace. A History of Space from Dante to the Internet*, New York: W.M. Norton and Company.

Wickler, G., 1998, On the integration of specialized and general reasoning. In: *Towards Very Large Knowledge Bases*, edited by Mars, N.J.I., Amsterdam: IOS Press, pp. 165–172.

Winograd, T., 1995, Heidegger and the design of computer systems. In: *Technology and the Politics of Knowledge*, edited by Feenberg, A. and Hannay, A., Bloomington, IN: Indiana University Press, pp. 108–144.

Wittgenstein, L., 1958a, *The Blue and Brown Books. Preliminary Studies for the 'Philosophical Investigations'*, New York: Harper Torchbooks.

Wittgenstein, L., 1958b, *Philosophical Investigations*, Second edition, Oxford: Blackwell.

5

GIS, Worldmaking and Natural Language

Peter Bibby

5.1 Introduction: Some old Curiosa Made Manifest

Alongside many other effects, the widespread use of GIS has stimulated awareness of some old 'curiosa' of quantitative geography. Chief among these is the 'ecological fallacy', which arises when correlations between phenomena evident at the level of (say) census tracts are imputed to individual persons (where they may be absent, implausible or impossible). It is not unusual, for example, for the proportion of elderly households in a census tract to increase as the proportion of very young households increases (reflecting the concentration of smaller dwellings). To infer on this basis that increasing age implies increasing youth is, however, to fall victim to the fallacy. Wider use of GIS has, moreover, prompted wider awareness of the related Modifiable Areal Unit Problem (MAUP) and substantially heightened its practical relevance. Concern over MAUP (identified by Openshaw (1984, 1996) originates in the finding that entirely different statistical relations between properties (such as rainfall and wheat yield) pertain for different sets of areal objects.

From the perspective of this chapter, such familiar difficulties in analysing relationships between abstracted properties appear to derive from fundamental confusions concerning the *objects* deemed to possess them. The modifiable areal unit problem lies in the fact that none of the competing sets of regions over which relations between properties might be estimated merits any particular priority. In other words, the scale and configuration of the regions used has no substantive significance or intuitive meaning in the context of the relationship being investigated. Underlying these two curiosa is an

Re-presenting GIS Edited by P. Fisher and D. Unwin

implicit tension between approaches that conceive of geographic objects as coherent 'things' (at an extreme, with internal structures and causal powers), and approaches that treat geographic objects simply as more or less arbitrary divisions of a continuous surface. In the limit, these latter approaches may entirely dispense with 'things' allowing only space-time-attribute triples on a regular grid to stand in the place of objects. More generally, hidden in the specification of geographic objects lie a welter of implicit conceptualizations of the relationships between parts and wholes, between people and the tracts of land they occupy, between groups and their members and, of course, between people themselves. This chapter is concerned to open up some of these issues.

Perhaps less obviously, the definition of objects also rests upon and hides relationships between words and things, and more generally between 'signs' and things. Signs in this sense include words, point symbols, and digital representations of boundaries. Later sections of this chapter attempt to illustrate the manner in which sign systems are instrumental in the construction of more or less arbitrary divisions of the earth's surface, and in the recognition of prior divisions. (The term 'sign systems' is intended here to encompass so-called 'natural languages' (such as English), formal languages (such as algebras, geometries and the predicate calculus) and (less tractably) pictures and ultimately the whole domain of semiotics.) GIS may be thought of as a box of representational tools inheriting from several of these varying systems subsuming them in digital form. It is often observed that GIS embraces three technologies (database, computer mapping and spatial analysis). As a corollary, it embraces three groups of sign systems, and thus embodies three batteries of what might be termed 'structural metaphors'. Metaphors are frequently thought of as embellishments that might be applied to adorn a literal text. The idea of a structural metaphor following Lakoff and Johnson (1980) is a far more fundamental device – a frame for organizing thought. A Euclidean plane might be thought of as a structural metaphor which can be used to capture certain properties of the surface of the earth. Different sign systems are bound to their own structural metaphors (as discussed later in this chapter). Thus, there is a possibility that the objects that we represent may depend to some degree on representational tools.

This chapter attempts to explore the manner in which geographic objects are defined in a number of steps. First, it offers a very simple definition of an object for the purposes of this paper (Section 5.2). Then, adding to the familiar curiosa, the chapter provides other examples of oddities that arise out of the relationship between words and things (Section 5.3). It then sets out a more systematic account of the relationship between words and things based on the work of the philosopher Michael Jubien, demonstrating that natural language is implicated in the definition of objects (Section 5.4). With some caution, this argument may be set in a broader context stressing the socially embedded nature of language, and seeing the construction of objects as an example of what Nelson Goodman termed 'worldmaking' (Goodman, 1978). Then, in Section 5.5, GIS is presented as an operational constructional system and examples of worldmaking practices are demonstrated by means of the discussion of a particular GIS-based study. Section 5.6 raises the contentious issue of whether worldmaking practices deform, and Section 5.7 attempts to open up less usual but far more specifically articulated ways of conceptualizing social and geographic objects – milieu-behaviour synomorphs and actor networks.

5.2 Objects are What We Talk About

For the purpose of this chapter, an object is defined simply as something that can be expressed in speech or writing as a noun (phrase). An object is thus taken to be something 'signified', i.e. something that may be imagined by a sentient subject. An important aspect of this definition is that it is grounded explicitly in discourse. Later sections, rather than abstracting from this discursive origin, attempt to explore whether in fact the use of GIS may be enriched by taking explicit account of it. Although a discursive perspective is potentially very widely applicable (Woolgar's (1988) work on hard science, for example), this chapter will focus on the discourse of urban planning with its repertoire of objects such as Green Belts, developers, local planning authorities and so on. Some of these may have material existence (e.g. the Royal Albert Hall), but others do not such as 'Heathrow Terminal 5' (a proposed development), or the 'Ashton-under-Lyne Business Improvement Quarter' (a plausible fiction). Others may have a contested existence, such as 'rural England', a commonplace denied by academics such as Hoggart (1990).

This is obviously a liberal definition. The key restriction of the concept outlined is that it must be possible to name an object of discourse or the classes to which it belongs; an object need not exist but cannot be ineffable. It need not be a dateable placeable part of the physical world, that is to say it need not have *extension.* The definition is thus *intensional*, i.e. it depends on a description or a set of properties. These might be thought of as fitting some more or less well-defined idealized cognitive model (or ICM) – in a sense relating to that of Lakoff (1987). The distinction between *intension* and *extension* is important and critical to this chapter.

A conception of this breadth is necessary not only to allow discussion of objects that may be brought into being, but also to create a 'space' in which varying ontological positions and representational tropes might be discussed. This definition of objects admits anything we may wish to discuss including, of course, such analytically unattractive objects as Heller's Maryalice: 'the object that exactly fills the region that we would ordinarily describe as including the front half of my car and all its contents from noon until two' (Heller, 1990, p. 55). Crucially this definition allows distinctions between objects that are more or less analytically useful and invites exploration of the objects that GIS users in particular social contexts have chosen to identify. Much of the chapter is devoted to exploration of this space.

It might be objected that while this broad conception may be useful, it is inappropriate to label it 'object'. Only two candidate terms, however, suggest themselves: 'object' or 'thing'. It has sometimes been suggested in the GIS literature that it would be preferable to reserve the term 'object' for items represented using a particular trope (i.e. in an object-oriented system). In this chapter, where necessary, these are referred to as OO-objects, aligning the simple term 'object' with a much longer and broader tradition. Deprived of this use of the term, it would be very difficult to discuss non-existent objects (Parsons, 1980), or to distinguish in the manner of Heller (1990) 'conventional' objects and 'non-conventional' objects. Significantly too, the term 'object' implies some relation to an observer (the subject), while the term 'thing' not only suggests clear separation, but also conveys a sense of solidity inappropriate to social objects (such as 'community'). The term 'object' seems most appropriate to describe items that are regarded neither as a pure creation of the subject nor as some Kantian 'thing in itself', inaccessible to the subject. The term 'thing' will be reserved here for apparently solid self-evident entities supposed to constitute the everyday world, but no definition will be offered. Rather the term 'thing' is assumed to allude to objects matching the predispositions of the GIS user.

5.3 Words and Things: Some more Curiosa

A range of other problems, less frequently discussed than MAUP and the ecological fallacy, arise out of the relationships between signs and things and emerge in GIS use. These include problems that arise where elements of different sign systems – specifically the name and geographic limits – compete as essential properties of an object or where the same region of matter or space-time is treated as part of many different objects. Before considering these relationships further in general terms, it may be worth considering some illustrations.

If an object has a property *essentially*, then it ceases to exist if that property is changed. An object may sometimes be deemed to have its boundary *essentially*. More frequently in social practice, an object is considered to have its *name* essentially. Thus in describing the growth of urban populations, an object is typically regarded as persisting as long as its name (say 'Tamworth') continues, with tabulations showing discontinuities when municipal boundaries change. Figure 5.1(a) represents two urban areas that are about to coalesce (Tamworth and Fazeley). Will the number of urban areas be reduced by one when this occurs? Will Tamworth be treated as having expanded, 'annexing' the population of Fazeley. The issues illustrated in Figure 5.1(a) are endemic within analyses of urban population change, and rather than meriting note even as curiosa are shrugged off or forgotten. At the same time, the question of the nature of the allegedly persistent 'thing' corresponding to the sign 'Tamworth' from 757AD when Offa named it capital of the Kingdom of Mercia and in 2004, goes unasked and probably remains unanswerable.

The possibility that natural language terms may, by facilitating the identification of objects, prejudice our appreciation of things is far more vividly illustrated by Figure 5.1(b). This example is drawn from a GIS-based reconstruction of field patterns. In this instance, the spatial extent of an object identified by name is first expanded, but later reduced in such a manner that no part of its ultimate configuration overlaps its initial configuration. Treating names as 'essential' properties of an object would in this case lead to the implausible conclusion that the field known as 'Mary Hyde's Little Meadow' 'moves'. The riddle may be resolved by debunking the name, treating it simply as an attribute of matter, consistent with a general principle discussed in Section 5.4. It will be argued that proper names provide only the illusion of persistence or rather persistence by convention that secures both 'Manchester' and 'Mary Hyde's Little Meadow'. In practice, of course, familiarity desensitizes us to the former instance, but we are less well prepared for the latter. Analogous problems may occur whenever objects are identified through both a naming process and a bounding process. Such situations are frequent in GIS use.

Moreover, naming is just one of a whole class of linguistic actions potentially underwriting geographic objects (which will be considered a little more systematically below). Where a public body commits itself to exercise its legal powers in a particular

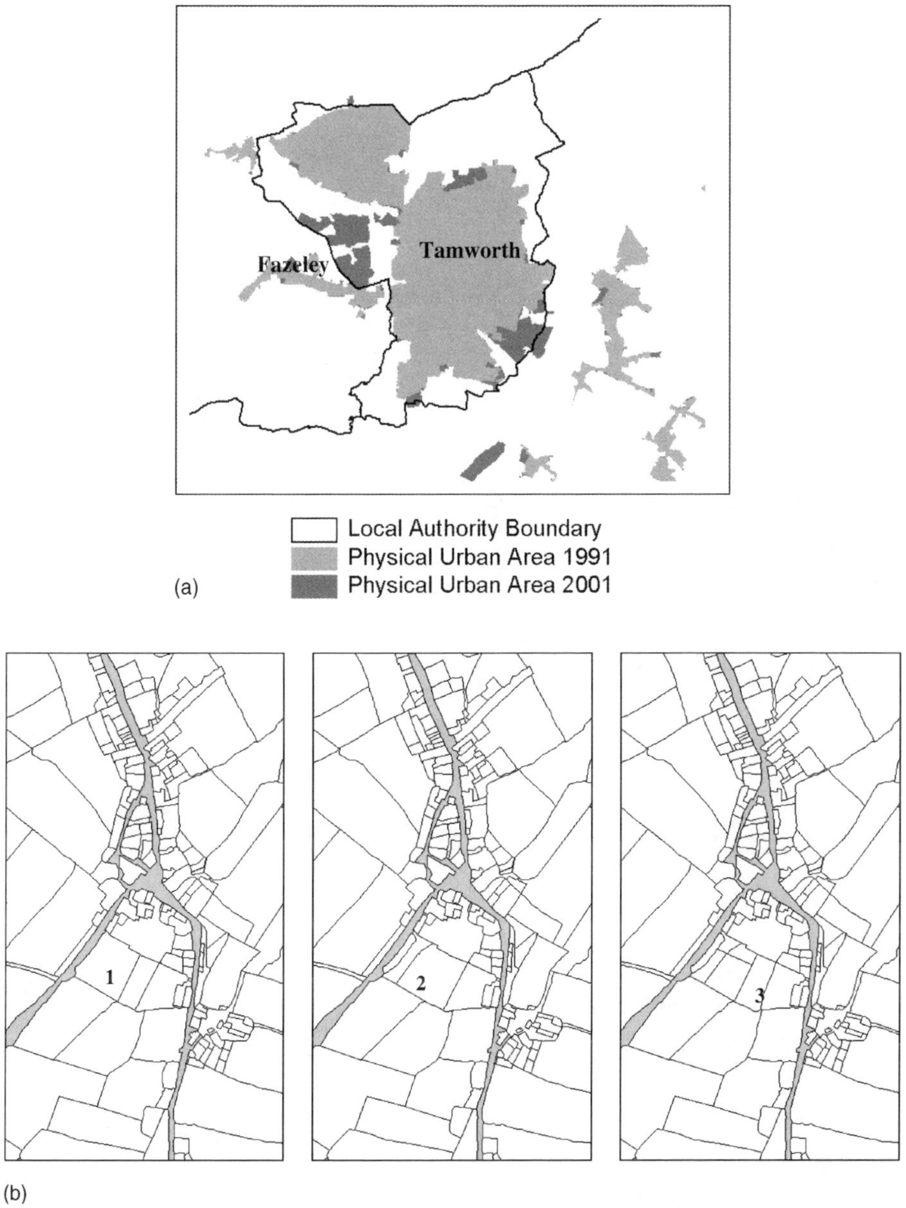

Figure 5.1 *Definitional curiosa in naming and bounding: illustrations from (a) Tamworth, Staffordshire, UK; and (b) Mottram-in-Longdendale, Greater Manchester, UK (Urban area boundaries in S. Lancashire appear with permission of the Office of the Deputy Prime Minister)*

way, it might define what are sometimes termed 'policy areas' such as Green Belts in the UK (where there is strong presumption against development). Both the boundaries and the attributes of such objects are defined by fiat (Smith, 1995). Any region, however small, may be subject to a multiplicity of such commitments. As illustrated

Figure 5.2 *Detail of local plan covering Newton-le-Willows, Merseyside, UK: (a) existing settlements and areas to which Green Belt (GB1213) and certain major development policies apply; (b) extent of other selected environmental policies; (c) policy polygons derived from the union of all policy 'extents'*

in Figure 5.2, these objects may overlay and intersect each other in complex ways so that the same region of matter or space-time may be part of many such objects. This proliferation of linguistically defined geographic objects fits uncomfortably into familiar GIS representational tropes. Intuitively, it may seem more attractive to treat commitments simply as attributes of geographic objects, rather than the essential properties defining them, but this leaves the problem of just what objects these commitments might be regarded as being attributes *of* (c.f. the problem underlying the MAUP). It is, of course, possible to treat each commitment as defining a GIS layer and to treat the union of all these layers as defining a series of resultant objects, each associated with a variable number of commitments. Such objects can persist only until further cross-cutting commitment is made. Representing each commitment as a grid or raster perhaps best satisfies the intuition that they are more appropriately treated as attributes rather than things.

The above examples should serve to illustrate that the role of language is not limited to describing the same objects in different ways, but rather in defining *different* objects. Sometimes social scientists engage in forms of critical discourse analysis problematizing the manner in which an individual might be alternately dubbed a 'terrorist' or a 'freedom fighter' (Fairclough, 1989) (or perhaps a parcel of land might be dubbed 'open space' or 'scrubland'). This, however important, is not enough. As the above examples show, when portions of the earth's surface are considered, the role of language is not limited to applying different labels to coincident tracts. Different objects with different boundaries can be drawn out for different purposes.

5.4 Matter, Words and Things: Towards a more Systematic Account

All of these curiosa arise out of particular relationships between signs and things. The rest of this chapter explores some aspects of the nature and implications of these relationships. This section considers the relationship between words and things in a more systematic though sketchy manner. Section 5.5 both generalizes this discussion and engages with the specifics of GIS.

For the purpose of exposition, this section begins by abstracting as far as possible from the intensional objects that are this chapter's prime concern. It attempts to identify what might be termed *extensional individuals* in the physical world rather than objects of discourse. The term *individuals* will be used to distinguish them from objects in the sense of this paper. The relationship between individuals and objects is explored later in the section.

To identify extensional individuals, a simple constructional system is deployed (Goodman, 1977). This system assumes a world of matter, simply undifferentiated physical stuff, and a space-time coordinate system. It is recourse to the coordinate system that allows elements of matter to be identified extensionally, avoiding some of the complications associated with the use of natural language descriptions. The smallest element of matter that might be addressed will be called a coordinate system atom or simply a c-atom. The size of a c-atom, that is the granularity of the system, depends on the coordinate system. If the coordinate system is expressed as real numbers, the smallest element is in principle infinitesimal. It is not assumed that a c-atom corresponds to an atom in the sense of the physical sciences. It should also be emphasized that the present concern is with the elements of *matter* that can be identified by reference to the coordinate system, *not* with elements of absolute space-time. This concern with matter forms the starting point for the theory of physical things set out by Jubien (1997, Chapter 9; see also Jubien, 1993). Variant perspectives based on elements of absolute space-time will be touched upon in Section 5.5.

From a particular perspective, that of Lesniewskian mereology, an individual might be thought of as one of these c-atoms or *any* combination of c-atoms defined in this system. Mereology is the formal theory of part-whole relations (Goodman, 1977; Simons, 1987) and forms the foundation for the present definition of extensional individuals. This particular perspective allows any two individuals to possess a *mereological sum* or simply a *sum*, even though they may be disjoint, widely separated in space and time and of different kinds. Any of these sums may be thought of as an individual and so individuals are allowed to proliferate. Given three atoms (X,Y,Z) there are thus seven potential individuals (Simons, 1987, p. 17; Searle, 1995, p. 162), as follows: {X}, {Y}, {Z}, {X + Y}, {X + Z}, {Y + Z}, {X + Y + Z}. More generally, if there are 'n' atoms, then there are $2^n - 1$ *individuals*. It should be immediately obvious that these sums (e.g. {X + Y} and {X + Z}) may overlap, and so the corresponding individuals may also overlap.

The notion of an individual allowed by mereological sums is extremely liberal, admitting both overlapping and scattered individuals and making minimal requirements of them. If *n* c-atoms were to be visualized as grains of matter, their sum would be identical whether those grains were widely scattered, heaped in a cone or compressed into a cube. Intuitively, therefore, mereological sums are likely to provide individuals far more numerous and far less structured than our everyday purposes typically require. GIS users may be particularly disinclined to identify scattered individuals, but whether an individual is regarded as scattered depends upon the scale at which it is observed (a lump of anything having spaces between its individual atoms). Moreover, at everyday scales, some scattered objects (such as a jigsaw, a dismantled bicycle or collectives such as a flock of sheep) seem 'natural' (Jubien, 1997, p. 157) and are likely to be important, as will be illustrated below.

Most mereological sums will define individuals that appear to be of no particular interest. This need not be a problem, however, in that sums that are of no interest may be discounted (Quine, 1960). Particular mereological sums will correspond to objects in the sense of this chapter, identified by names such as 'Manchester' or 'Maryalice'. Seemingly 'natural' sums might be compared with all other possibilities, perhaps exposing something of our predispositions about analytic objects. It might be said, moreover, that at an instant, an object of discourse might comprise a particular mereological sum. The term 'comprise' seems apt as it suggests 'holding together', while being neutral with respect to how or in what form c-atoms are held together. The intensional criteria (or ICMs) for particular classes of 'thing' will usually demand that the atoms comprising particular mereological sums possess other properties (such as topology, form, organic structure, or social structure). These further criteria might be thought of as making demands at a series of higher ontological levels (in the sense of Guarino (1999)).

Before giving any further consideration to comparison of intensional objects and extensional individuals, however, it is important to clarify some intrinsic properties of

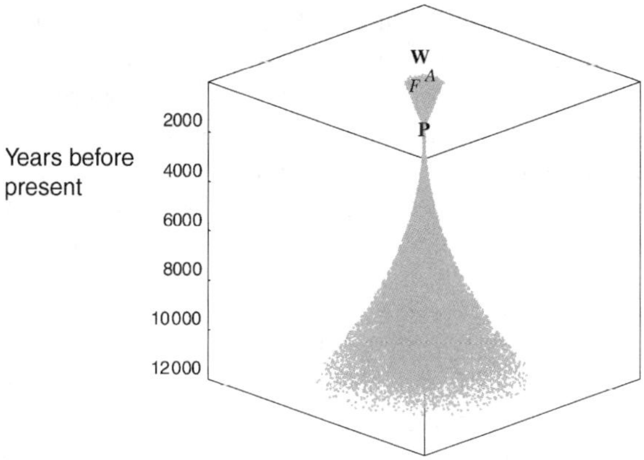

Figure 5.3 *S: a mereological sum (for explanation, see text)*

mereological sums of elements of matter. Consider one such sum, to be called S, the history of which is summarized in Figure 5.3. Although kinds of matter are not being distinguished for current purposes, S was in fact brought together as what might ordinarily be described as sandstone, tile and mortar in the second century AD to form the palaestra or exercise hall of the roman baths in the city of Viroconium (Wroxeter, Salop, UK). The 'Palaestra' (P on the diagram) constitutes an object in the sense of this chapter. It survived for some few hundred years, but the mereological sum S still survives. Some of those particles (in the form of blocks of stone) subsist not only in a large surviving stretch of wall, known as the 'Old Work' (W), but also in the fabric of the nearby church of St Andrew (A) and in farm buildings (F) around the site (all of which are other objects in the sense of this chapter). On the assumption that matter is not destroyed, however, S survives *entire*, even though it is dispersed and of different kinds.

This example makes it clear that being a sum does not require any bringing together or attaching of the atoms and embodies no idea of form or function. It is constancy of parts that define mereological sums: a sum has its parts essentially. Two complementary aspects of mereological sums emerge. The story of S takes its particular form because those parts have been defined as elements of matter. Once an individual is so defined, that individual persists as form and function changes. A GIS user might be content to treat the Palaestra as an object corresponding to a particular sum of matter subject to a time constraint, though the persistence of the sum S will probably generate difficulties as extensional individual and intensional object diverge. The user might possibly want to treat the Palaestra and the Old Work as the same thing, but here extensional definition will not help, as it embraces all manner of intuitively 'extraneous' matter. The individual, S however, appears rather strange and unlikely to be identified as an object.

Just as extensional constancy yields individuals that are not attractive as objects, intuitively attractive objects must correspond, at different times, to different extensional individuals. It is tempting to write that 'as soon as the parts change, there must be a different individual', but this might mislead because, by virtue of the definition of a mereological sum, its parts *cannot* change. This would immediately give rise to problems in identifying an extensional individual corresponding to the Palaestra. If different material were incorporated into the walls of the Palaestra, that intensional object would correspond to a different extensional individual. It is thus immediately obvious that the relationship between objects and individuals is not straightforward. This appears pedantic, because for everyday purposes it is convenient to blur this distinction. Normally, we use two rough criteria to test for constancy of 'things' (continuity in space-time and constancy of parts) and assume that these coincide; they need not. The Palaestra and the Old Work are joined by a continuous curve in space-time. On the other hand, constancy of parts ensures the integrity of the sum S. Moreover, unreflective use of natural language tends to cloak the divergences lurking beneath constancy of name, and may produce curiosa such as Mary Hyde's Little Meadow.

The possibility of divergence of these two rough tests of constancy, evident in the above example, is classically illustrated by the fable of the Ship of Theseus (Jubien, 1997; Varzi, 1998). It has both theoretical and practical implications. The theoretical implications are far-reaching. Jubien (1993) develops an argument from situations of the Palaestra type to reach a conclusion of philosophical consequence – that language does not *refer*. The concept of reference involves words pointing to particular things in the physical world, thereby providing an extra-linguistic foundation for meaning. In the direct reference theory of names (dominant within philosophy), a proper name is considered as a rigid designator, i.e. a term that identifies the same thing in all possible worlds (Kripke, 1980). If, however, chosen objects of discourse do not have a constant relation to extensional individuals, then language cannot refer. The fixity of part-whole relations, Jubien (1993) argues, prevents names serving as rigid designators. Thus he claims names are not designators but simply predicates or (loosely) attributes.

On Jubien's analysis there could not be a physical thing designated the Palaestra, but there could be a sum of matter that has the property of being a wall and also the property of being the Palaestra. Therefore he concludes that 'it may be that our intuitive goal is in some sense to fix a referent but it nevertheless must be true that something else is going on. That something else can only be the defining of a new term' (Jubien, 1993, p. 72).

Reference, it should be concluded is problematic; there is only a rough or vague fit between language and the world. By loosening the relationship between names and matter it becomes easier to solve the riddle of Manchester and Mary Hyde's Little Meadow. More generally, objects emerge when words are used to bundle up matter, that is to bind particular mereological sums. While Jubien's argument problematizes reference, it neither denies the reality of physical matter; nor suggests that stuff is conjured out of language. Sider, reviewing Jubien's (1993) book, summarizes the argument as follows: 'the world consists fundamentally of stuff, which we divide into things in any way that suits our purposes' (Sider, 1999, p. 284).

Given the disjunction between stuff and things, constancy of matter is neither necessary nor sufficient to ensure constancy of the objects of interest to GIS users (Guarino, 1999). The criterion to be used in assessing constancy of objects is not identity, but simply relevant similarity. Starting from a Jubienesque perspective, but focusing more explicitly on social processes, different extensional individuals, identified as parcels of matter, may be said to correspond to the same object (such as a town) at different times by social agreement. In social practice, the mapping of objects onto individuals appears to rest on intensional classes (such as 'town') that are only vaguely defined. ICMs will be incomplete, differ from person to person, and vary according to interests. To the extent that social rather than physical criteria play a significant role in an ICM (for 'neighbourhood' or 'community', for example), the likelihood of vague and contested definitions increases. Critically, the possibility of allowing vague or conflicting definitions to be hidden under a particular linguistic tag allows social action to proceed. Analytic, administrative, and political action continues without requiring either complete intensional or extensional definition. Vague language allows definition to be deferred and other action to proceed.

From such a perspective, definition (clarifying intension) and delimitation (codifying extension) appear as social processes in which representational resources such as GIS are embedded. The referents of terms such as 'town centre' or 'problem estate' may be eventually established through debate. This is a process that might be described using Mallery's (1991) term 'deliberative reference' and increasingly involves GIS use. Given the failure of reference in its strict sense, such forms of 'social reference' bind language to the world through countless tiny acts of operational definition.

5.5 Ways of Worldmaking

The disjunction between language and the world which Jubien's analysis opens up is thus bridged by socially embedded practices – defining new terms. This perspective begs a more critical view of the objects that GIS users such as analysts and administrators 'choose' to define and delimit. Objects defined cease to appear inevitable, and might even start to seem remarkable in the context of all others that might have been recognized. Broadening out the discussion of Section 5.4, the remainder of this chapter treats GIS users as involved (wittingly or unwittingly) in what Goodman (1978) termed 'worldmaking'. Goodman's work prompts the discussion of the structuring and delimitation of geographic objects that follows. Having examined the foundation of constructional systems (of the type touched upon in the last section) in *The Structure of Appearance* (1977, but originally published in 1951), his work moved to a more allusive and discursive exploration of *Ways of Worldmaking* (1978).

In his wide-ranging opening discussion, Goodman characterizes some typical *Ways of Worldmaking*, as *composition and decomposition*, *weighting*, *ordering*, *deletion and supplementation* and *deformation*. He takes pains to explain that these modes are neither exhaustive nor mutually exclusive. The present chapter attempts to explore the relationships forged between objects and individuals through a discussion using Goodman's headings and is intended to be in his spirit. Although Goodman's examples of worldmaking embrace cognitive, social and technical processes, he emphasizes the cognitive. In treating GIS, however, the current discussion is far more concerned with the technical and the social.

5.5.1 Composition and Decomposition

It is Goodman's first mode of worldmaking, 'composition and decomposition', that has been the primary focus of this chapter so far. Worldmaking involves choosing particular mereological sums (more or less consciously). Goodman's suggestion that this is 'normally effected or assisted or consolidated by involving the application of labels: names, predicates, gestures, pictures' (1977, p. 7) has already been exemplified through the discussion of naming in the last section. GIS provides far more resources to assist this process of choice. It may be useful at this point to step back and consider the question 'sums of what?' This involves touching on the primitives of the constructional system itself (considered extensively in Goodman (1977)).

Section 5.4 concerned itself with a constructional system whose basic atoms were infinitesimal elements of matter. Much GIS use might be thought of as a form of worldmaking deploying as atoms not elements of matter, but space-time-reflectance triples (analogous to Goodman's, 1977 'space-time-colour quales') or even elements of area. Remote sensing applications might bear either of these latter interpretations. In this section, therefore, GIS is considered as an operational constructional system. For the sake of exposition, consider a LANDSAT scene held as a raster comprising reflectance measures for $30m^2$ pixels and extending for 300 by 300 pixels. It is just such a foundation that forms the basis for GIS practices adopted by Jelinski and Wu (1996) in exploring aspects of the ecology of the boreal shield of Manitoba. The following paragraphs explore a range of worldmaking practices which are later compared with those adopted by Jelinski and Wu. There is no suggestion that these authors erred; the aim is simply to sketch out a range of constructional possibilities.

By analogy with the discussion of S in Section 5.4, an extensional individual might be defined as a pixel or any combination of pixels within the system. Given that the LANDSAT scene used by Jelinski and Wu extends for 300 by 300 pixels, there are $2^{90000} - 1$ individuals, of which 90 000 are the original pixels and one is the general sum. This provides a universe of extensional individuals (mereological sums) from which particular individuals are 'chosen'. It will always be pertinent to ask how such choices are made. As the pixels are conceived simply as portions of abstract space with entirely arbitrary boundaries, the implied relationship between part and whole is particularly simple. The parts are to be considered as portions of a blank screen; each identical with all others and all identical to the whole, save for their projected attributes. This type of relation between part and whole will be referred to below as the portion:mass model. For the present it will be convenient to restrict consideration of part-whole relations to this strictly mereological form and to introduce Goodman's second mode of Worldmaking – weighting.

5.5.2 Weighting

An obvious approach to choosing particular mereological sums involves deploying a system of classes or kinds such that only atoms with relevant similarity are combined. Restricting sums to atoms of the same kind makes the portion:mass model more intuitively attractive. Goodman's discussion of 'weighting' centres on distinguishing 'relevant and irrelevant kinds' (1978, p. 10). The term 'kind' includes what philosophers term natural kinds. Some natural kinds, denoted by mass nouns like 'clay', 'water' and 'sand', are varieties of matter. Other natural kinds are represented by count nouns or sortals (such as 'tiger' or 'heron'), providing a principle for individuating and counting.

Kinds, however, go much beyond natural kinds to include, for example, roles and classes of artefact. Goodman takes a particularly broad view of kinds. 'Patches of green' might be thought to constitute a relevant kind, or pixels with a particular spectral response. A Humean empiricist admitting the perception of colour, but not the existence of substance (Hume, 1739, Book 1, Part 1, Sect. VI) might find such kinds relevant to the analysis of a LANDSAT scene. Most GI analysts might posit that the pixel values might depend in part upon totally extrinsic conditions (i.e. solar angle), in part upon the observer (i.e. viewing angle) and in part upon various land cover kinds. For the non-Humean confronted with the LANDSAT scene, the first step in identifying individuals might be to attempt to infer land-cover kinds from reflectance values. Operationally this might mean normalizing the reflectance measures associated with each pixel to reduce the effects of solar angle and viewing angle, and hence emphasizing the properties of land-cover kinds (Kimes *et al.*, 1984). (The usual measure would be the Normalized Difference Vegetation Index or NDVI which expresses the difference in the contribution to solar reflectance of the visible and near-infrared part of the spectrum and is calculated as

$$\mathrm{NDVI} = (\mathrm{NIR} - \mathrm{VIS})/(\mathrm{NIR} + \mathrm{VIS})$$

where VIS and NIR are normalized radiance values in visible channels and near-infrared channels respectively.) Strictly, of course, this is only an indexical sign: a continuous measure pointing to land-cover, but without an automatic relation to discrete land-cover kinds.

Inferring land-cover kinds from NDVI can be considered as a step in the process of choosing individuals. It allows identification of mereological sums whose atoms are (likely to be) all of a relevant kind (such as 'water'), Generally, however, there will still be many such sums, i.e. many individuals. In the case of sums of atoms of kinds represented by mass nouns, it is reasonable to treat each as an arbitrarily delimited portion of a mass, to assume complete similarity between each atom and the whole, and also to assume that none has any specific function with respect to the whole. It is far from clear under such circumstances which individuals are appropriate. At one extreme, single pixels might be candidates. At the other extreme, an object such as 'all the world's water' represented by the general sum of pixels of that kind is relatively easy to accept as an individual. Certainly, GIS users may grow accustomed to examining histograms displaying values of a grid or image, indicating, for example, the areas associated with particular kinds such as water, wetland, bog, upland, or plateau. Such descriptions make no demands beyond the mereological level; they indicate the scale of a sum irrespective of its distribution. The very nature of the 'portion:mass' model ensures, however, that any sum of a scale between the individual pixel and 'all the world's water', has as much justification as any other, or in more familiar terminology, that all areal units are modifiable.

Depending on the application, moreover, potentially relevant kinds may proliferate. One reason for this is that kinds are often considered as constituting taxonomies. Thus one kind (such as *picea mariana*, the black spruce of the Manitoban boreal forest) is regarded as an instance of a more general kind (e.g. *picea* – spruce). Individuals can proliferate in turn, because being an atom of a particular kind is also an atom of many more general kinds. The individual defined as the mereological sum of all elements of a particular kind will thus depend upon the taxonomic level at which kinds are specified. Moreover (while this may not be the case when dealing with strictly natural kinds), a range of intensional taxonomies may compete to describe even quasi-natural kinds such as habitats. Thus the EUNIS classification (EUNIS, n.d.) and that due to Olson *et al.* (1983), for example, provide quite different habitat taxonomies. Many relevant kinds, moreover, are defined not by reference to the intrinsic properties of matter, but rather to extrinsic assessments of their value, utility or significance for the observer (such as 'waste' or 'desert'), further increasing the range of possible kinds. This allows the possibility that different cultural or interest groups (such as forestry interests and First Nation communities within the Manitoban boreal shield) may weight the relevance of particular kinds differently and hence identify quite different individuals.

The range of potentially relevant kinds facing the GI analyst is far greater than this suggests, however, as further kinds are created through social action mediated through representational systems. Words can be used to *create* rather than describe kinds; to originate kinds rather than merely assert or negotiate their relevance. Since the publication of Austin's (1975) *How to do Things with Words*, this so-called 'performative' aspect of language has become widely appreciated. Austin himself demonstrated how classes of 'performative' utterances, such as marrying or conferring degrees, create and instantiate kinds. Kinds constituted in this way include the Forest Management Licence Area Agreements (granting commercial rights over the boreal forest) in Manitoba or the right to pick rice by hand (NAFA, 1993). All rights in land represents kinds that may be relevant in the definition of geographic individuals, and which depend upon performative language. It is, therefore, the performative character of language which underlies the 'policy area' curiosum introduced in Section 5.3. The properties created in this way are projections onto matter and extrinsic to it. Patterns of solar reflectance and objects of discourse are thus both partly shaped by analogous projections, rather than simply by the nature of the matter at the earth's surface.

In summary, kinds may be instantiated at different taxonomic levels, different taxonomies may be projected onto matter, embodying classificatory principles based not only on physical structure but social utility, and the identification of relevant kinds will reflect competing interests and values. The recognition of relevant kinds, while providing a principle to guide composition still admits a vast range of mereological sums. The proliferation of potentially relevant kinds implies that, in general, a particular element of space-time will not be occupied by one and only one (conventional) object. Worldmaking practices may strive to achieve such reduction, but only through the power plays in which GIS use is embedded.

5.5.3 Ordering

Ordering represents a further kind of worldmaking practice. Rather than seeking to classify kinds through type – subtype taxonomies such as Linnaean species classifications or hierarchical habitat classifications, ordering is continuous and without branches. Measurement systems provide the paradigm case of orderings, being in Goodman's words, not 'found in the world but built into a world' (1978, p. 14). Besides orderings based on number systems (such as space-time coordinates and measures of spectral response), measurement systems also include the continuous natural language orderings with fuzzy boundaries (e.g. frozen, cold, mild, hot) that have received considerable attention in the GI literature. Recognition of orderings may compete with identification of kinds: continuous measures can displace natural language colour terms and offset any tendency for different languages to partition the visible spectrum in different ways (Crystal, 1987, p. 106).

Worldmaking practices may rest entirely on ordering, to the exclusion of concern with kinds. The procedures *actually* adopted in Jelinski and Wu's (1996) study will serve to exemplify this. That study was not concerned with composing individuals by weighting kinds, but rather identifying the effect of scale and of alternative agglomerations on measures of autocorrelation of NDVI. It could be considered to rest on three levels of ordering. The first is the digital ordering of spectral responses (which in this particular instance the authors were not concerned to map onto vegetational kinds). The second is the spatial ordering of the pixels, upon which the autocorrelation measures depend. The reliance on ordering, however, transcends this. Their research design demands a third (conceptual) ordering that prioritizes cell-size effects and *then* seeks out the effect of varying configuration on autocorrelation measures.

Although the emphasis on ordering is evident in their goal of assessing autocorrelation *at two different scales*, calculation of the autocorrelation measures, of course, demands the recognition of individuals. By defining various individuals composed of the same number of contiguous pixels, Jelinski and Wu (1996) were able to explore configuration effects. The analysis was based on very few individuals: 14 (nine with 100 pixels and five with 16 pixels) of a possible $2^{90000} - 1$ mereological sums. More significantly, their worldmaking choices constrained all these individuals to be: i) conjoint; ii) mutually exclusive; and iii) of equal size. The first two of these conditions are frequently implicit in ICMs for particular geographic objects. Insistence on equality of size is less frequent, but serves to foreground the influence of processes operating at particular scales, potentially extrinsic to the 'objects' that might be considered to occupy the surface of the earth.

From a worldmaking perspective, the object-field dualism so familiar in GIS involves either privileging composition (in combination with weighting), or privileging ordering. On the one hand, an analyst might seek to identify particular mereological sums (extensional individuals) within the scene corresponding to objects of discourse with properties that might explain patterns of reflectance. Alternatively, a sharp focus on ordering, abstracting from the intrinsic properties of the surface, is consistent with the field conception.

In other worldmaking practices, very common within GIS, concern with spatial ordering does not displace concern with kinds, but is combined with it to define extensional individuals. Particularly, it is common to insist that the parts of geographic individuals are topologically connected. Restricting admissible individuals to those whose parts are topologically connected once again constitutes an ordering practice. In analyzing the LANDSAT scene, worldmaking is quite likely to entail seeking out contiguous patches of relevant kinds; the analyst might even supplement such a patch with a boundary.

5.5.4 Deletion and Supplementation

Deletion and supplementation are further worldmaking practices identified by Goodman (1978), and examples of these are particularly commonplace in GIS use. Very straight-forward instances would be the practice of weeding points out of digital representations of line segments (Douglas and Peuker, 1973), or adjusting DEMs for error. Indeed, the area of terrain representation (Hutchinson and Gallant, 1999) is an area where deletion and supplementation practices are much discussed. Map generalization as a whole is another field of activity concerned deeply with deletion and supplementation (see Weibel and Dutton (1999) for an overview, and especially their summary table of generalization operators, p. 136).

The focus of current concern, however, is not processual issues such as line generalization, but the worldmaking practice of adding a boundary to a mereological sum to create a thing from a mass. Inferring a boundary, a familiar GIS operation (e.g. ArcInfo's gridpoly), may provide the key to individuation and counting. The grammar of natural language reflects and guides our assumptions about boundaries. Thus Jackendoff (1992) invokes a general semantic feature he terms '[+ bounded]' to transform a mass noun to a count noun. A lake, might be deemed to be water with a boundary, its status as a count noun being imprinted through the need for an article ('it is water' but 'it is *a* lake').

While kinds represented by mass nouns and kinds represented by count nouns are frequently considered distinct, typically habitat and land cover types bear either grammatical coding. One may thus choose to talk about 'wetland' or 'a wetland' (in the latter case invoking an implicit boundary) in a manner in which one cannot choose to talk about 'clay' or 'a clay'. Thus, for example, the discussion above has (deliberately) referred to water, wetland, bog, upland and plateau – while Jelinski and Wu themselves explain that their study area possesses '*many* wetland*s* and treed bog*s*, and has *an* upland' (1996). As English grammar easily allows either of these worldmaking practices in this context, I will refer to any land cover kind which might be treated grammatically as a mass or a count noun as an ecological kind (respectively unbounded or bounded), and treat the practice of adding (implicit or explicit) boundaries as a matter of choice. At this stage fuzziness is not an issue; insistence that a boundary is fuzzy simply reasserts commitment to the intensional boundary. From the present perspective, a boundary is genuinely a supplement, extrinsic to the substance.

Combination of this form of supplementation with the recognition of relevant kinds generates the inter-relationships between quality, quantity and entity summarized in Figure 5.4. If the boundaries delimiting areas of land or space are arbitrary and the relation between part and whole is simply that between portion and mass, the number of objects increases as their extent decreases. If boundaries are added as a supplement to patches of similar kind, their size and number depends upon the assessment of relevant similarity, which in turn depends upon the constellation of kinds deemed relevant, and of course, granularity (i.e. pixel size). If defining properties of kinds are deemed to include extent (thereby creating a distinction between a wood and a forest, for example), the

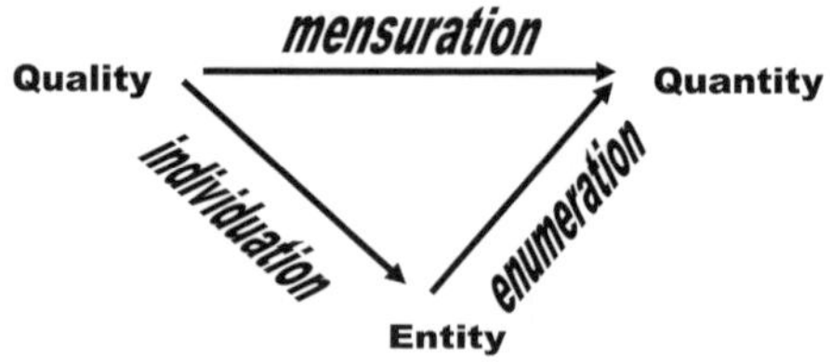

Figure 5.4 Quality, quantity and entity

interdependency becomes more complex. Adding boundaries to kinds denoted by mass terms invokes additional properties – extent, and configuration – both of which might be thought of as properties of the supplement, over and above the basic substance properties. Worldmaking choices may thus be expected to shape the statistical relations between properties of the objects.

In the case of mass kinds, relationships between properties are conceived as applying to qualitatively undifferentiated portions with arbitrary limits. To estimate a statistical relationship between two properties of a mass kind requires ‘packaging’ and so inevitably and properly invokes the modifiable areal unit problem. From this perspective it appears that estimation of such relationships ought to involve sampling from possible mereological sums of atoms of the relevant kind. The solution seems strange because it admits disjoint individuals (but the problem may be stranger as correlating properties – such as wheat yield and rainfall – operationally involves packaging into units that can have no conceptual grounding).

Many other situations arise where objects are similarly conceived. These include field objects such as wheat-yield or rainfall and ‘nominalizations’ such as ‘expenditure’ and ‘income’. All these terms behave grammatically as mass nouns; language in use does not provide them even with implicit boundaries. Relationships are posited between such variables, though the implicit model specifies neither an agent, a geographical scale, nor any principle of individuation. The relationship between wheat-yield and rainfall might be mediated by farm management practice, and that between income and expenditure by economic agents such as firms and households. Analysts can choose to attempt to identify meaningful objects. The practice of critical discourse analysis tends to baulk at nominalizations such as ‘globalization’ or ‘expenditure’ (Fairclough, 1995, 2003; Hodge and Kress, 1993), seeking out an object – an agent responsible for the spending. Nominalization is an instance of what is sometimes termed grammatical metaphor. In this case, a noun is used where a verb might have been thought congruent – arguably an act of deformation disguising agency. GIS practice, on the other hand, tends to treat income or expenditure as attributes of elements of area – arguably a further deformation.

5.6 Ways of Worldmaking: Deformation

Many users have strong views on the capacity of GIS to deform (Openshaw, 1993), and the divergence of practice and opinion is consistent with Goodman’s standpoint that deformation occurs from particular points of view. A typical GI users’ catalogue might include inappropriate surface fitting, extreme generalization, production of stepped surfaces, and misleading map symbology (Monmonier, 1996). This section, however, is limited to considering arguably deformative constructions of part-whole relationships (carrying forward the discussion of Section 5.5). It introduces a broader range of part-whole relations and explores two groups of practices that might be regarded as regarding deformation. The first group, very common in GIS use, involves reduction of human activities to attributes of simple areal objects. This might be held to entail deformation as it displaces both built form (spatially) and human agency (conceptually). The second group of practices posits very different types of intensional object (behaviour settings and actor-networks) which are relatively unfamiliar to GI analysts. These representations might also be held to involve: deformation; behaviour settings, as they are objects whose parts are of radically disparate types; and actor-networks because they undermine the distinction between things and relations.

Deformation always takes place relative to some (implicit) norms, reflecting the (typically implicit) ICMs lying behind the objects that we talk about. These ICMs embody not just mereological and topological properties, but also the physical and chemical properties of kinds, and a whole series of properties at higher ontological levels (on ontological levels, see Guarino (1999). Archetypal physical things might be considered as objects composed of elements of matter bound together by intrinsic forces. More sophisticated objects might be considered to possess internal structures embodying more complex and varied relations between parts and wholes. Individual 'parts' may themselves be highly complex entities (e.g. buildings, people, or households, rather than elements of matter or space-time-attribute triples). As an example, the reader might consider the types of things and relationships that we might typically expect to constitute an object denoted by the term 'school'.

It is deformation in the representation of 'social objects' (such as 'school' or 'community') that form the central thread of this section. Although our common conceptions of such objects may be rich, routinized GIS practices may encourage the adoption of the *mosaic metaphor*, beating the social attributes of individual people to a chromatic blend to be painted over spaces such as administrative or postal areas. In a manner that is intuitively attractive to many users of maps and screens, abstracted space itself is treated as possessing social attributes.

Such mapping of social characteristics is a potentially deformative practice so familiar to GIS users that it appears 'natural' and unremarkable. The spatial fixity of buildings provides the grounding for this practice, but at least two potentially deformative transformations are involved. The first action treats social characteristics as an attribute of buildings, although the human 'parts' are linked only by conventional ties such as that of 'normal residence'. The second action implicitly renders homogenous the distribution of such buildings within each tile of the mosaic, the shading of each marking the transference of attributes to abstracted space. The configuration of dwellings is then effaced and the area is valorized in a manner similar to that in which, in ecological applications, finer grained objects like boulders are subsumed within boulder fields or trees within woods (Jelinski and Wu, 1996).

Application of the mosaic metaphor to socio-economic phenomena involves a series of steps:

- the transfer of attributes from one kind to another (appropriation);
- the erasure of objects and the creation of a mass kind (grinding);
- the creation of ecological objects (considered internally homogenous) by adding boundaries as a supplement (bounding);
- in such a manner as to partition exhaustively a plane.

Each of these steps might be regarded as a deformation. The first two steps together can be thought of as constituting an 'ecological shift', an act of deformation paving the way for the ecological fallacy, and for what Block (n.d.) terms 'the area unit fallacy', under which 'the aggregate characteristics of an overall area are mistakenly applied to each section or neighbourhood within that area'. The imposition of a series of boundaries is from this perspective a further deformative act. Such boundaries limit the applicability of the mass terms, producing a system of differences encapsulated in a mosaic, artificially bolstering the homogeneity of individual tiles, while exaggerating the difference between them.

Most frequently, the definition of the tile boundaries, possibly arbitrary and established by fiat, may precede their 'colouring'. In extreme cases, however, commitment to the mosaic metaphor may be so great as to motivate the construction of a tessellation *de novo*, to receive disembodied attributes. The creation of UK unit postcode 'boundaries' is such a case. Originally developed as an aid to delivering mail, each unit postcode might be considered a 'name', labelling a series of (typically around 16) discrete mail delivery points. As such series do not have boundaries, GIS users wedded to the mosaic metaphor have for many years found them unsatisfactory (Boots, 1999). The ascendancy of the mosaic metaphor has, however, fuelled demand for unit postcode boundary products which has been met by generating Voronoi polygons. A requirement for an exhaustive partitioning of space is thus privileged over concern for the objects which might otherwise be considered to occupy that space.

This example illustrates the way in which the possibilities of deformation spiral thereafter. As a unit postcode relates to a set of disjoint postal delivery points, it has no referent that possesses its boundary essentially. Given this logic, there is no reason for *new* property constructed within a particular synthesized tile to share the associated postcode. On the contrary, postal purposes require that new property on a significant scale will be dubbed with new codes. Only by forcing the worldmaking practices of GIS users to over-ride the logistics of mail delivery will transference of the 'name' to the abstracted space become meaningful. Without this, however, deformation continues to lead to pseudo problems in use. One example affects overlays. In the nature of tessellations that partition a plane around a series of points, the sparser the delivery points, the larger the created tiles become. Hence in the (frequently occurring) extreme, the unit postcode of a single isolated farmhouse will be treated as if it denoted a vast tract of space, allowing the erroneous conclusion that some object (actually a single building) impinges upon a whole series of administrative or other areas. A second instance occurs when, inevitably, eye and brain conspire to lend undue weight to the largest (i.e. emptiest) tiles, prompting concern for the social characteristics of unpopulated areas.

Moreover, far more sophisticated variants of the mosaic model may be developed, with the potential for further deformation, such as that underlying the mosaic of output areas designed for dissemination of the results of the 2001 UK Census. Using a variant of Openshaw's AZP, unit postcode tiles are treated as atoms and combined to form larger tiles whose transferred attributes meet a mandatory population size criterion and which seek to maximize homogeneity of transferred social characteristics. This worldmaking procedure results in a highly constrained set of mereological sums privileging spatial contiguity and seeking to maximize compactness. The mosaic of such individuals corresponds to a particular set of intensional objects; compact socially homogenous 'neighbourhoods' – geometrically defined shells arguably encompassing a 'community' all too easily envisaged as an organism with an internal structure.

Setting the mosaic metaphor aside, however, there are many other ways of conceptualizing relations between parts and wholes, let alone between society and space. Meronomies (that is, idealizations of part:whole relations) may focus on the relations of an individual to its components, to its members, or to the material of which it is constituted, or may specify the nature of the links or relations between the parts. Winston *et al.* (1987) distinguish six types of meronomies which capture functional specialization of parts, spatial cohesion and the degree of dissimilarity between the parts and their whole:

- *portion/mass* (e.g. water/water, forest/forest) with complete similarity between the parts, as discussed above;
- *material/object* (e.g. water/lake) describing the materials comprising an object;
- *member/group* (e.g. tree/forest; house/settlement) in which parts are spatially distinct from one another and do not necessarily have a structural or functional relation with one another;
- *component/integral object* (e.g. door/house) where there is a clear structural and functional relation between the whole and its parts;
- *precise place/area* (e.g. oasis/desert) expressing a fixed spatial relationship, but where there is no clear functional relationship;
- *subactivity/activity* (e.g. pay/buy) describing different sub-activities forming an activity.

Of course, many more could be added. Examples of further classes that appear from some perspectives to involve deformation might include:

- *temporal part/object* (e.g. 'my home') expressing a fixed functional relationship, but with a transitory or unclear spatial relationship;
- *place/object* (e.g. school building/school) identifying the milieu part of a synomorph (see below);
- *activity/object* (e.g. manufacture/factory) identifying the behaviour part of a synomorph (see below);
- *node(actor)/network* (e.g. development site/world city network) identifying a (human or non-human) participant in an actor-network.

The first of these additional classes of relationship has been the subject of protracted philosophical debate. The second and third allow the conceptualization of higher level social objects as 'milieu-behaviour synomorphs' in the fashion of Barker (1968, to be discussed below), while the fourth serves as a reminder that an object may be regarded as part of an actor-network in the spirit of writers such as Callon (1991) or Latour (1993). While the perspectives of Barker and the actor-network theorists differ starkly from each other, they both might be thought of as (productively) deforming familiar ideas of the distinction between things and actions (though in different ways and to different extents), and radically extending the notion of composition. Such objects as these might seem potentially interesting in the social sciences but appear totally divorced from GIS. The rest of this chapter is concerned with the possibility of dialogue.

5.7 Identifying Social Objects

The arguably deformative worldmaking practices of Barker and the actor network theorists potentially provide resources for representing social objects. Whereas archetypal physical things are bound together by intrinsic forces, social entities comprise networks of relations (Giddens, 1984). Social geographic objects may be said to comprise particular mereological sums, but beyond the mereological level might be bound together by specific relationships and activities that are repeatedly *performed*. Although such formulations may seem novel within GIS, just such a conception of a manor is set out in Norden's *Surveior's Dialogue* of 1618:

> *a manor in substance is of lands, wood, meddow, pasture and arable; it is compounded of demeisns and services of long continuance*
>
> (quoted in Kerridge (1967), p. 17)

Roger Barker's ecological psychology represents a systematic attempt at developing such conceptions. In the absence of 'a science of things and occurrences that have both physical and behavioral attributes', Barker introduced classes of intensional objects conceived as having a physical part and a behavioural part (Barker, 1968, p. 19). His work is ontologically radical, or deformative. Motivated by a concern to study human behaviour in its context, he identified a class of geographic objects which he termed 'milieu-behaviour synomorphs' or simply synomorphs. Synomorphy refers to similarity of structure: 'the synomorphy of the boundary of the behaviour and of the boundary of the milieu is striking and fundamental: the boundary of a football field is the boundary of the game' (Barker, 1968, p. 19). They are conceived as micro-social systems composed of people and physical things configured in such a way as to facilitate routine actions within a specific region of space-time. Alongside physical things and human participants, they embody 'setting programs' – bundles of social rules or conventions governing activity. Synomorphs may be grouped into genotypes (Barker, 1968, p. 33) including alongside football games, kinds such as roads, churches, retail stores, restaurants, car-boot sales, bus-stops, and English lessons. The synomorph model thus provides a template for representing a large range of geographic objects.

The focus of ecological psychology has been on a subset of milieu-behaviour synomorphs termed 'behaviour settings'. Like the functional regions defined by geographers, Barker's behaviour settings are bound together by social interaction: a behaviour setting must possess a specified degree of interdependence, and requires a higher degree of interdependence within itself than with other behaviour settings. Interactional interdependence criteria must be met before Barker (1968, p. 21) allows a field to be transformed into an organism. A church may have less than the degree of interdependence required to constitute a behaviour setting. Barker implicitly admits temporal parts and so the 11.00am church service may be allowed as a behaviour setting, and the church as a multiple setting synomorph (regularly repeated events are regarded as the same thing). A street, on the other hand, is likely to have too high a level of interdependence with structurally external synomorphs to be admitted as a behaviour setting (Barker, 1968, p. 25–26). Equally, Barker insists on a physical limit – a 'circumjacent' milieu with unbroken temporal and physical boundaries which might be likened to an exoskeleton. To impose such a boundary would, for Barker, involve deformation. More generally, for Barker, the identification of a behaviour setting was thus very much an empirical issue.

The structural metaphor of the synomorph clearly has potential applicability within GIS. Nevertheless, the possibility of the ideological assertion rather than demonstration of interdependency and synomorphy is never far away. Barker himself uses the term 'community' in an unproblematic way. Taylor, a follower of Barker, moves closer to ideology, holding that such social objects constitute 'freestanding *natural* units of the everyday environment with a recurring pattern of behaviors and a surrounding and supporting physical milieu'. For Taylor, moreover, 'these units organize *community* life' (Taylor, 1998, p. 10 emphasis added). Whether particular worldmaking practices within GIS are considered to involve deformation depends on predispositions with regard to social-spatial synomorphy. Taylor's predilections lead him to favour synomorphy, proposing that streetblocks treated as behaviour settings constitute appropriate objects for geographical analyses of crime (Taylor, 1997, 1998). From his perspective, the popular choice of 'hot-spots' as analytic objects involves deformation (relying on a metaphor in which hot magma rises to the earth's surface). Taylor is thus critical of the use of systems such as STAC (Spatial and Temporal Analysis of Crime) to identify hot spots as circles or ellipses (Taylor, 1998, p. 6). For Block (n.d.), by contrast, it is the deformative nature of arbitrary areal units and the areal unit fallacy which motivates application of STAC thereby abstracting from bounded spaces and instead privileging ordering. The structural metaphor of the synomorph opens up worldmaking possibilities in GIS, but in so doing allows inevitably ideological assertion in much the same manner as the mosaic metaphor that it might potentially displace.

While Barkeresque objects allow a richer representation of social phenomena then the mosaic metaphor, both share an emphasis on boundaries which may justify charges of deformation. It seems almost impossible to treat some social and geographic objects as synomorphs without deformation. Pierce Lewis's galactic megalopolis would be one example. Here the

> *residential subdivision, the shopping centers, the industrial parks seem to float in space; seen together they resemble a galaxy of stars and planets, held together by mutual gravitational attraction, but with large empty spaces between clusters'*
>
> (Lewis, 1983, p. 94).

Another intensional geographic object not easily treated as a synomorph is McGuire and Chan's (2000) NY-LON: a bicontinental megalopolis embracing people in New York and London. The game of international finance is not bounded like a football game. Cities can be represented as networks of people in the spirit of Amin *et al.* (2000), abstracting from the milieu, just as neighbourhoods can be represented as tiles abstracting from behaviour.

The structural metaphor of the 'network' by contrast makes no presumptions about synomorphy. It is a pervasive and currently fashionable worldmaking resource. In human geography it has been revived by Castells (1996), though a myriad of applications in the discipline stretch out from minimum path procedures incorporated within proprietary GIS to analyses of space syntax (Hillier, 1996; Hillier and Hanson, 1984). Batty (2001, and Chapter 11 of this volume) has recently deployed a specific variant of the metaphor, suggesting that cities be considered as 'small worlds', building on Milgram's (1967) work in psychology and Watts' (1999) demonstration of the very widespread applicability of the idea. The network has, moreover, become an increasingly popular organizing metaphor for social objects in the politics and governance literature (Goss, 2001; Heclo, 1978; Rhodes, 1999). In artificial intelligence and cognitive science the same structural metaphor may serve as a powerful tool when it is used in the guise of a semantic network to express relationships between intensional objects (Lehmann, 1992; Sowa, 1991). Not only does the metaphor of a semantic network provide a means of knowledge representation that may be formally equivalent to the predicate calculus, the same network metaphor forms the basis for visualizations ranging from those of Malrieu (1994) to those developed by Chen (1998, 1999) for capturing the relationships between information sources.

These diverse manifestations may, however, fruitfully combine. At a very simple level, the associations encoded in a semantic network might express the subtasks within an activity (such as a shopping trip, or a military operation). As the focus on activities sharpens, each unpacks into a series of more fundamental ones ('parts' in the spirit of Winston *et al.*, 1987). Associating times and resources with these tasks, generates an enhanced network akin to a long-established project-planning tool (PERT). A logically very similar network of times and activities may be used to extend the notion of accessibility from one based on abstract distance to one in which distance is merely a moment of social activity. Such an approach underlies a range of practical applications concerned, for example, with the relation between the movement of dry-bulk material and economic activity (Demilie *et al.*, 1997) or that between public transport services and regional economic performance (Bibby and Capineri, 1998). Similarly, with a sharper focus, nodes themselves may progressively appear more like networks. In the case of a transport network, as scale increases, cities such as Rome or London rather than appearing as highly accessible points become congested networks. In an analogous manner, sharpening the focus on particular social actors such as companies or governments (not to mention social movements) might prompt a shift from a node (agent) metaphor to a network metaphor. Depending on point of view, either the assumption of solidity in the node metaphor, or the assumption of fragility in the network metaphor will involve deformation. When it is recognized that the appearance of solidity may be an *accomplishment* – i.e. that it may be the outcome of a process of 'translation' – (Callon 1986) – concerns may begin to converge with those of actor-network theory.

The notion of an actor-network might be thought of as an extremely creative deployment of the network metaphor, though one which by bringing into question the very idea of 'things' might be held to involve deformation. It is scarcely possible here to provide an account of actor-network theory – the interested reader is referred to Callon (1986) who illustrates key concepts by reference to a potential artefact that never materialized – the French electric vehicle. 'More supple than the notion of system, more historical than the notion of structure, more empirical than the notion of complexity' for a theorist such as Latour, 'the idea of network is the Ariadne's thread of ... interwoven stories'. (Latour, 1993, p. 3). The network of actor-network theory thus has some parallels with a semantic network though it stretches beyond the intensional domain. Crucially, it is concerned with relationships. The network deployed by Callon in his study of the electric vehicle includes government ministries, municipalities, transport companies, and consumers, but also accumulators, fuel-cells, electrodes, electrons, catalysts and electrolytes. A wide variety of published research has deployed the structural metaphor of the actor network (reviewed, for example, by Law and Hassard, 1999), including applications of a geographical slant (Murdoch and Ward, 1997; Smith, 2003).

Both behaviour settings and actor-networks thus provide structural metaphors for representing objects (in the sense of this chapter) whose components include people and inanimate things (Callon, 1991; Wicker, 1987, p. 614). Obviously, the domain of application of actor-network theory is far wider than that of ecological psychology. In contrast to ecological psychology, actor-network theory avoids laying down *a priori* boundaries between context and content (whether geometric or figurative) (Lea *et al.*, 1995, p. 466). Whereas Barker's work seeks out patterns of behaviour with a 'circumjacent' physical boundary, continuous in time and space, synomorphy is not assumed in actor-network theory. By contrast, Law (2000) argues that the network provides an alternative topological system in which elements retain their integrity, not by being a volume within a larger Euclidean volume (i.e. in the manner of the mosaic metaphor), but by virtue of maintaining their position in a set of links or relations. The issue of how 'parts' are enrolled within networks, and the continual processes of definition and redefinition are therefore central to actor-network theory.

The actor-network metaphor thus provides a representational resource which seems at first utterly opposed to those typically embedded within GIS. This antagonism must not be understated. Nevertheless, among the highly heterogeneous elements of which actor-networks are constructed, one finds dateable placeable parts of the physical world that might more conventionally be represented within GIS. Moreover, spatial relations may form part of substantive relations so that Law's 'alternative' topologies may become confounded with physical relations. The actor-world of the French electric vehicle discussed by Callon may seem a very long way from GIS, but it includes a representation of cities. Admittedly, within this actor-world each city is reduced to 'a transport system that must avoid adding to the level of pollution and a town council that seeks to advance towards this goal' (Callon, 1986, p. 29). Even this, however, embodies lower-level geographic objects and relations as

> *the city council must stabilize the elements which hold it together: the middle-class electorate that trusts it, the pedestrian precinct that pushes the flow of traffic to the edge of the town centre, the urban spread and the system of public transport which enables the inhabitants of the suburbs to come and do their shopping in the town centre*
>
> (Callon, 1991, p. 30)

Thus, as Latour stresses, networks are 'not made from some substance different from what they are aggregating. No visible or invisible hand suddenly descends to bring order to dispersed and chaotic individual atoms' (Latour, 1993, p. 122). These atoms are, however, bound into constructions far beyond mereological sums – it is the nature of the relationships that is critical. For Law (2000), actor-network theory shows how regions are constituted by networks. The actor-network perspective continually confronts the representational ploys most readily enacted within GIS, but never abandoning the physical world, continually provides a challenge to develop alternative potentially realizable representations.

5.8 Conclusions

It is hoped that this chapter has succeeded in demonstrating that familiar geographic problems, which GIS use brings into high relief, can be regarded as symptomatic of a broader class of problems deriving from the relations between sign systems and things.

Although analytic focus tends to be on the relationships *between* objects, these are largely shaped by conceptualization and representation of the objects themselves, which are intimately related to the tropes deployed in their representation.

This chapter has implicitly colluded in recognizing the practical value of commonsensical assumptions about a world populated by 'things' (deliberately left unexamined) which participate in 'processes'. These presumptions underlie most of the representational devices discussed, despite their differences. They are evident, for example, in object-orientation, the predicate calculus and the ER model. Also, they are evident explicitly within functional grammar where 'circumstance' is added to 'process' and 'participant', thereby constituting three organizing categories for representing the world. It is neither necessary nor possible to determine whether such frameworks ultimately capture the nature of the world. They certainly capture the way it is described in natural language, and as Niels Bohr remarked 'we are suspended in language in such a way that we cannot say what is up and what is down '(quoted in French and Kennedy (1985), p. 302). Given the importance of natural language supporting the social transformation of the material world, these frameworks will be significant in any event (right or wrong).

A little reflection on the content of this chapter opens up an awareness of the vast range of ways in which the process-participant-circumstance framework may be deployed. Without questioning the reality of matter, or invoking any kind of magic, worlds of qualitatively different overlapping objects may be brought into being to serve as participants (or circumstances). Having demystified names following Jubien, it appears that the world of matter can (using natural language or other representational devices) be decomposed and recomposed into individuals corresponding to objects in any way that suits social purposes (Section 5.4). If the atoms of a constructional system are sufficiently small, the range of possibilities is overwhelming (Section 5.4). Moreover, it might appear that the GIS on the desktop may be treated as an operational constructional system. By ordering, by the selection of kinds, by deletion and supplementation in the manner of Section 5.5, these possibilities are reduced somewhat. By allowing objects to participate in a range of processes (not solely relational ones) and by allowing complex objects with internal structures, the process-participant-circumstance framework permits a vast constellation of possible objects.

Despite this constellation of possibilities, or perhaps overawed by possibility, it may be tempting to dismiss discussion of worldmaking and retreat to the idea of the ordinary thing – the metaphor above all others by which we live. Very frequently, this appears to be convenient. In use, however, practical difficulties soon start to show. The inadequacy of the ordinary thing metaphor underlies the curiosa of Sections 5.1 and 5.2. More generally, reification sets in; that is to say processes and relationships become inadvertently solidified, occluding understanding of objects that are 'performed', and placing them in a domain of 'things' which cannot be remoulded by practical engagement. If regarded as ordinary things, 'neighbourhood', 'community', and social movements inadvertently acquire a solidity that proves quite inappropriate to guiding practical action. Such objects need not be reduced to ordinary things. Neither need they be reduced to delimited portions of the surface of the earth or to the physical structures thereon. This is far from denying their importance in geography: they have a spatial imprint or projection which might be thought of as 'circumstance' in terms of functional grammar.

A retreat to the 'ordinary thing' metaphor would thus be both stultifying and potentially disempowering. It is the related tendency to reduce geography to circumstance, however, that leads to the MAUP. From the perspective of this chapter MAUP arises in situations where *process* is deemed central, where there is little concern with particular *participants* and where there are potentially competing definitions of *circumstance.* Participants may be underspecified or even disappear (in the extreme being considered merely as momentary states arising from the outcome of particular processes, in a manner similar to discussions of nominalizations such as 'expenditure' or 'globalization', for example). It is when our intensional model does not specify participants but allows arbitrary ecological shifts – that is, transmutes participants to circumstances – that the modifiable areal unit problem arises.

Therefore, one response to MAUP may be to avoid ecological shifts by changing analytic scale, specifying participants and their interactions thereby reducing the significance of circumstance. The merit of identifying participants at a lower level depends, however, on purpose and perspective. Although abstraction from participants may reflect an active desire to hide them (the concern of critical discourse analysis), it may reflect the impossibility of identifying them (as when framing regulations to cover future events) or simply a lack of concern about particular participants (with a theoretic or practical justification). (A proposition about the relationship between rainfall and wheat yield might provide a good example.) In changing analytic scale, in GI science as in political science, the process-participant-circumstance model is reapplied, perhaps recasting a participant as a network rather than an ordinary thing, or recasting circumstance as a series of participants.

The nature of the world appears to permit (and the availability of GIS infrastructure to facilitate) the treatment of the same phenomenon variously as participant, process or circumstance. Geographic objects are thus defined in a manner whereby an emphasis on participants competes first with an emphasis on process, and second with an emphasis on circumstance. Worldmaking takes place within this space. Where all but relational *processes* (i.e. processes of being) disappear, GIS catalogues a world of things bound by spatial relations (relational processes), and geography becomes simply a catalogue of circumstances, lending some justification to critiques which suggest the aridity of GIS. Where *circumstance* disappears, geography is eliminated. This will occur when philosophical leanings dictate that absolute space and time – Goodmanian orderings – must give way to what Blaut described as the fusion of relative space and relative time to form a space-time manifold 'or simply process' (Blaut, 1961, p. 43).

In his classic paper, Blaut claims that 'every empirical concept of space must be reducible by a chain of definitions to a concept of process' (Blaut, 1961, p. 43). His suggestion that 'the process involved in a word like area might be the operation of measuring it' clearly suggests a further participant – the analyst who measures, who represents, who superimposes an ordering, who makes a world. This might also be applied to the so-called relational 'processes' of functional grammar (such as 'adjoining' or 'being below'), which seem less like processes than descriptions of states, from the perspective of a particular observer generating them. Observers with representational engines such as GIS thus take centre stage, choosing more or less purposefully from apparent possibilities, engaging in a range of processes – not merely cognitive processes but also material processes such as digitizing and depressing keys to create object texts with a material existence. In the ways illustrated in this paper, and many others they – we – engage in worldmaking. This is not simply a personal and cognitive activity but a social and technical one in which GIS in deeply embedded. Choices within the domain of representation play their part in shaping artefacts, stabilizing social objects (such as 'ICI', the 'Greater Manchester Green Belt' or 'rural England') and creating what Latour (1993) terms 'monsters' or 'hybrids'. Whether the creation and maintenance of such monsters is done in the name of science or of social regulation, its products become part of the world, augmenting it, reducing ambiguity, promoting and pre-empting social actions, and hence reproducing a world which is really constructed. Users of GIS may elect to dismiss such issues from conscious consideration; but the alternative may well be mindless incorporation. It is hoped this chapter will encourage a more critical and creative engagement.

References

Amin, A., Massey, D. and Thrift, N., 2000, *Cities For The Many Not The Few*, Bristol: The Policy Press.

Austin, J.L., 1975, *How to do Things with Words*, second edition, Oxford: Oxford University Press.

Barker, R.G., 1968, *Ecological Psychology: Concepts and Methods for Studying the Environment of Human Behavior*, Stanford, CA: Stanford University Press.

Batty M., 2001, 'Editorial: Cities as small worlds', *Environment and Planning B: Planning and Design*, **28**, 637–638.

Bibby, P. and Capineri, C., 1998, From networks to regional development: Representations of Italian regional disparities, In: *Networks in Transport and Communications: A Policy Approach*, edited by Rietveld, P. and Capineri, C., Aldershot: Ashgate, pp. 193–216.

Blaut, J.M., 1961, Space and process, *The Professional Geographer*, **13**, 1–7.

Block, C.R., n.d, STAC Hot Spot Areas: A statistical tool for law enforcement decisions, available at http://www.icjia.state.il.us/public/pdf/stac/Hotspot.PDF.

Boots, B.N., 1999, 'Spatial tessellations', In: *Geographical Information Systems and Science*, second edition, edited by Longley, P.A., Goodchild, M.F., Maguire, D.J. and Rhind, D.W., New York: Wiley, pp. 503–526.

Callon, M., 1986, The sociology of an actor-network: the case of the electric vehicle, In: *Mapping the Dynamics of Science and Technology: Sociology of Science in the Real World*, edited by Callon, M., Law, J. and Rip, A., London: Macmillan, pp. 19–34.

Callon, M., 1991, Techno-economic networks and irreversibility, In: *A Sociology of Monsters: Essays on Power, Technology and Domination*, edited by Law, J., London: Routledge, pp. 132–161.

Castells, M., 1996, *The Rise of the Network Society*, Oxford, Blackwell.

Chen, C., 1998, 'Bridging the gap: the use of pathfinder networks in visual navigation', *Journal of Visual Languages and Computing*, **9**, 267–286.

Chen, C., 1999, *Information, Visualisation and Virtual Environments*, London, Springer-Verlag.

Crystal, D., 1987, *The Cambridge Encyclopedia of Language*, Cambridge: University Press.

Demilie L., Jourquin, B. and Beuthe, M., 1997, A sensitivity analysis of transportation modes market shares on a multimodal network: the case of dry bulk transports between Benelux/Germany and Spain, In: *Networks in Transport and Communications: A Policy Approach*, edited by Rietveld, P. and Capineri, C. (eds), Aldershot, Ashgate, pp. 235–252.

Douglas, D.H. and Peuker, T.K., 1973, Algorithms for the reduction of points required to represent a digitized line or its caricature, *Canadian Cartographer*, **10**, 112–122.

EUNIS, n.d, European Nature Information System http://eunis.eea.eu.int/habitats.jsp (accessed 2 April 2005).

Fairclough, N., 1989, *Language and Power*, Harlow: Longman.

Fairclough, N., 1995, *Critical Discourse Analysis*, London: Longman.

Fairclough, N., 2003, *Analysing Discourse: Text Analysis for Social Research*, London: Routledge.

French, A.P. and Kennedy, P.J. (eds), 1985, *Niels Bohr: A Centenary Volume*, Cambridge, Mass.: Harvard University Press.

Giddens, A., 1984, *The Constitution of Society: Outline of the Theory of Structuration,* Cambridge: Polity Press.

Goodman, N., 1977, *The Structure of Appearance*, Dordrecht: Reidel (originally published 1951).

Goodman, N., 1978, *Ways of Worldmaking*, Hassocks: Harvester.

Goss, S., 2001, *Making Local Governance Work*, Basingstoke: Palgrave.

Guarino, N., 1999, 'The role of identity conditions in ontology design', In: *Spatial Information Theory: Cognitive and Computational Foundations of Geographic Information Science*, Lecture Notes in Computer Science 1661, edited by Freksa, C. and Mark, D.M., Berlin: Springer.

Heclo, H., 1978, 'Issue networks and the executive establishment', In: *The New American Political System*, edited by King, A., Washington, DC: American Enterprise Institute, pp. 87–124.

Heller, M., 1990, *The Ontology of Physical Objects,* Cambridge: Cambridge University Press.

Hillier, W. and Hansen, J., 1984, *The Social Logic of Space*, Cambridge: Cambridge University Press.

Hillier, W., 1996, *Space is the Machine*, Cambridge: University Press.

Hodge, R. and Kress, G., 1993, *Language as Ideology*, second edition, London: Routledge.

Hoggart, K., 1990, Let's do away with rural, *Journal of Rural Studies*, **6**, 245–257.

Hume, D., 1739, *A Treatise of Human Nature: Being an Attempt to Introduce the Experimental Method of Reasoning into Moral Subjects*, Volume 1, London, John Noon, 1739; available at http://socserv.socsci.mcmaster.ca/~econ/ugcm/3ll3/hume/treatise1.html.

Hutchinson, M.F. and Gallant, J.C., 1999, Representation of terrain, In: *Geographical Information Systems and Science*, second edition edited by Longley, P.A., Goodchild, M.F., Maguire, D.J. and Rhind, D.W., New York: Wiley, pp. 503–526.

Jackendoff, R., 1992, Parts and boundaries, In: *Lexical and Conceptual Semantics*, edited by Levin, B. and Pinker, S., Oxford: Blackwell, pp. 9–45.

Jelinski, D.E. and Wu, J., 1996, 'The modifiable areal unit problem and implications for landscape ecology,' *Landscape Ecology*, **11**, 129–140.

Jubien, M., 1993. *Ontology, Modality and the Fallacy of Reference*, Cambridge: Cambridge University Press.

Jubien, M., 1997, *Contemporary Metaphysics*, Oxford: Blackwell.

Kerridge, E., 1967, *The Agricultural Revolution,* London: Allen and Unwin.

Kimes, D.S., Holben, B.N., Tucker, C.J. and Newcomb, W.W., 1984. Optimal Directional view angles for remote-sensing missions, *International Journal of Remote Sensing*, **5**, 887–908.

Kripke, S., 1980, *Naming and Necessity*, Oxford: Blackwell.

Lakoff, G, 1987, *Women, Fire and Dangerous Things*, Chicago: University of Chicago Press.

Lakoff, G. and Johnson, M., 1980, *Metaphors we Live by*, Chicago: University of Chicago Press.

Latour, B., 1993, *We have never been Modern*, Hemel Hempstead: Harvester Wheatsheaf.

Law, J., 2000, Objects, Spaces, Others, available at http://www.comp.lancs.ac.uk/sociology/papers/law-objects-spaces-others.pdf.

Law, J. and Hassard, J. (eds), 1999, *Actor Network Theory and After*, Oxford: Blackwell.

Lea, M., O'Shea, T. and Fung, P., 1995, 'Constructing the Networked Organization', *Organization Science*, **6**, 462–478.

Lehmann, F. (ed.), 1992, *Semantic Networks in Artificial Intelligence,* Oxford: Pergamon Press.

Lewis, P., 1983. The galactic megalopolis, In: *Beyond the Urban Fringe*, edited by Platt, R.H. and Macinko, G., Minneapolis: University of Minnesota Press, 94–111.

Mallery, J. C., 1991, Semantic content analysis: a new methodology for the RELATUS natural language environment, In: *Artificial Intelligence and International Politics,* edited by Valerie Hudson, M., Boulder: Westview Press, pp. 347–385.

Malrieu. J.P., 1994, Coloured semantic networks for content analysis, *Quality and Quantity*, **28**(1), 55–81.

McGuire, S. and Chan, M., 2000, The NY-LON life, *Newsweek*, 13th November, 40–47.

Milgram, S., 1967, The small world problem, *Psychology Today*, **2**, 60–67.

Monmonier, M., 1996, *How to Lie with Maps*, second edition, Chicago: University of Chicago Press.

Murdoch, J. and Ward, N., 1997, Governmentality and territoriality: the statistical manufacture of Britain's "national farm", *Political Geography*, **16**(4), 307–324.

NAFA, 1993, Strategy: Forest Land and Resources for Aboriginal Peoples; An intervention submitted to the Royal Commission on Aboriginal Peoples Ottawa, Ontario July, 1993, National Aboriginal Forestry Association, http://www.nafaforestry.org/roycom/roycom3.php.

Olson, J.S., Watts, J.A. and Allison, L.J., 1983, *Carbon in Live Vegetation of Major World Ecosystems*, US Department of Energy Publication no. 1997, Environmental Science Division. Oak Ridge National Laboratory, Tennessee, summarized in http://lynx.uio.no/catfolk/mjissues/mjchp_1c.htm.

Openshaw, S., 1984, *The Modifiable Areal Unit Problem*, Concepts and Techniques in Modern Geography 38, Norwich: Geo Books.

Openshaw, S., 1993, GIS crime and criminality, *Environment and Planning A*, **25**, 451–458.

Openshaw, S., 1996, Developing GIS-relevant zone-based spatial analysis methods, In: *Spatial Analysis: Modelling in a GIS Environment*, edited by Longley, P. and Batty, M., Cambridge: GeoInformation International, pp. 55–73.

Parsons, T., 1980, *Nonexistent Objects*, New Haven, CT: Yale University Press.

Quine, W.V.O., 1960, *Word and Object*, Cambridge, Mass. The MIT Press.

Rhodes, R.A.W., 1999, *Understanding Governance: Policy Networks, Governance, Reflexivity and Accountability*, Buckingham: Open University Press.

Searle, J.R., 1995, *The Construction of Social Reality*, Harmondsworth: Penguin.

Sider, T., 1999, Michael Jubien, Ontology, Modality, and the Fallacy of Reference, *Nous*, **38**, 284–294.

Simons, P., 1987, *Parts: a Study in Ontology*, Oxford: Clarendon Press.

Smith, B., 1995, On Drawing Lines on a Map, In: *Spatial Information Theory*, Proceedings of COSIT '95, edited by Frank, A.U., Kuhn, W. and Mark, D.M., Heidelberg: Springer Verlag, pp. 475–484.

Smith, R.G., 2003, World City Actor-Networks, *Progress in Human Geography*, **27**(1), 25–44.

Sowa, J. (ed.), 1991, *Principles of Semantic Networks: Explorations in the Representation of Knowledge*, San Mateo, CA: Kaufmann, M.

Taylor, R.B., 1997, Social Order and Disorder of streetblocks and Neighborhoods: Ecology, microecology and the Systemic Model of Social Disorganization *Journal of Research in Crime and Delinquency*, **33**, 113–155.

Taylor, R.B., 1998, Crime and Small-scale Places: What we Know, What we can prevent and What Else we need to Know, in *Crime and Place: Plenary Papers of the 1997 Conference on Criminal Justice Research and Evaluation*, Washington, D.C.: National Institute of Justice, pp. 1–22.

Varzi, A., 1998, Basic problems of mereotopology, In: *Formal Ontology in Information Systems*, edited by Guarino, N., Amsterdam: IOS Press, pp. 29–38.

Watts, D.J., 1999, *Small Worlds: The Dynamics of Networks between Order and Randomness*, Princeton: Princeton University Press.

Weibel, R. and Dutton, G., 1999, Generalising spatial data and dealing with multiple representations, In: *Geographical Information Systems and Science*, second edition, edited by Longley, P.A., Goodchild, M.F., Maguire, D.J. and Rhind, D.W., New York: Wiley, pp. 125–155.

Wicker, A.W., 1987, Behavior settings reconsidered: Temporal stages, Resources, internal dynamics and context, In: *Handbook of Environmental Psychology*, edited by Stokols, D. and Altman, I., New York: Wiley, Vol. 1, pp. 613–654.

Winston M.E., Chaffin, R. and Hermann, D.J., 1987, A taxonomy of part-whole relations, *Cognitive Science*, **11**, 417–444.

Woolgar, S., 1988, *Science: The Very Idea*, London: Tavistock.

6

Land Use and Land Cover: Contradiction or Complement

Peter Fisher, Alexis J. Comber and Richard Wadsworth

6.1 Introduction

Mapping the land of the earth has been seen as desirable for many years and indeed centuries. That mapping has usually focused on the use to which people put the land. In recent years, increasing confusion has been created between the concepts of land use and land cover, which leads to confusion in analyses and inferences drawn from land classifications. This confusion can be studied by articulating our knowledge of land cover and land use, and in doing so elaborating fundamental ontological and epistemological issues surrounding land cover and land use. In this chapter we explore the nature of, reasons for, and implications of these confusions.

Internationally, there is a growing commitment to record land use and land cover through various land classification programmes and to monitor change. A variety of national, regional and global projects have been initiated. These include the US National Land Cover Dataset (Table 6.1), the Countryside Survey series in the UK, the European CORINE Land Cover mapping (Table 6.2), the International Land Use and Land Cover Change core project run jointly between the International Geosphere-Biosphere Programme and the International Human Dimensions Programme on Global Environmental Change, which includes the USGS Land Characterisation Dataset.

Common to the land classification schemes adopted by these programmes and many others, is a failure to distinguish clearly between land cover and land use. This is problematic when seeking to relate formally different land classifications: land cover and land use are fundamentally distinct. *Land cover* is determined by direct observation while *land use*

Re-presenting GIS Edited by P. Fisher and D. Unwin

Table 6.1 *USGS land use and land cover classification (after Anderson et al., 1976)*

Level 1		Level 2		
1 Urban or built-up land	(Use)	11	Residential	(Use)
		12	Commercial and service	(Use)
		13	Industrial	(Use)
		14	Transportation, communication and utilities	(Use)
		15	Industrial and commercial complexes	(Use)
		16	Mixed urban or built-up land	(Use)
		17	Other urban or built-up land	(Use)
2 Agricultural land	(Use)	21	Cropland and pasture	(Use)
		22	Orchards, groves, vineyards, nurseries, and ornamental horticultural areas	(Use)
		23	Confined feeding areas	(Use)
		24	Other agricultural land	(Use)
3 Rangeland	(Cover)	31	Herbaceous rangeland	(Cover)
		32	Scrub and brush rangeland	(Cover)
		33	Mixed rangeland	(Cover)
4 Forest land	(Cover)	41	Deciduous forest land	(Cover)
		42	Evergreen forest land	(Cover)
		43	Mixed forest land	(Cover)
5 Water	(Cover)	51	Streams and canals	(Cover)
		52	Lakes	(Cover)
		53	Reservoirs	(Use)
		54	Bays and estuaries	(Cover)
6 Wetland	(Cover)	61	Forested wetland	(Cover)
		62	Nonforested wetland	(Cover)
7 Barren land	(Use)	71	Dry salt flats	(Cover)
		72	Beaches	(Cover)
		73	Sandy areas other than beaches	(Cover)
		74	Bare exposed rock	(Cover)
		75	Strip mines, quarries and gravel pits	(Use)
8 Tundra	(Cover)	81	Shrub and brush tundra	(Cover)
		82	Herbaceous tundra	(Cover)
		83	Bare ground tundra	(Cover)
		84	Wet tundra	(Cover)
		85	Mixed tundra	(Cover)
9 Perennial snow or ice	(Cover)	91	Perennial snowfields	(Cover)
		92	Glaciers	(Cover)

requires socio-economic interpretation of the activities that take place on that surface. The mixing of the concepts of land cover and land use has been present for at least the last 25 years (Anderson *et al.*, 1976), and has become so prevalent that classifications of 'pure' land use or land cover are rare even when that is the stated objective (Di Gregorio and Jansen, 2000). Such vague ontologies have had methodological implications for the way that we have sought to integrate information about, for instance, land cover from different classifications: not only do the land cover to land cover relationships have to be established but also those for land cover to *land use* (Fuller *et al.*, 1998). In these circumstances, to be able to relate the information contained within different *land cover* data requires an understanding of the different *land use* ontologies. Because of the

Table 6.2 *The CORINE land cover classification (EEA, 2001)*

Level 1		Level 2		Level 3	
1 Artificial Surfaces	(Cover)	11 Urban fabric	(Cover)	111 Continuous urban fabric	(Cover)
				112 Discontinuous urban fabric	(Cover)
		12 Industrial, commercial and transport units	(Use)	121 Industrial or commercial units	(Use)
				122 Road and rail networks and associated land	(Use)
				123 Port areas	(Use)
				124 Airports	(Use)
		13 Mine, dump and construction sites	(Use)	131 Mineral extraction sites	(Use)
				132 Dump sites	(Use
				133 Construction sites	(Use)
		14 Artificial, non-agricultural vegetated sites	(Cover)	141 Green urban areas	(Cover)
				142 Sport and leisure facilities	(Use)
2 Agricultural areas	(Use)	21 Arable land	(Use)	211 Non-irrigated arable land	(Use)
				212 Permanently irrigated land	(Use)
				213 Rice fields	(Use)
		22 Permanent crops	(Use)	221 Vineyards	(Use)
				222 Fruit trees and berry plantations	(Use)
				223 Olive groves	(Use)
		23 Pastures	(Use)	231 Pastures	(Use)
		24 Heterogeneous agricultural areas	(Use)	241 Annual crops associated with permanent crops	(Use)
				242 Complex cultivation patterns	(Use)
				243 Land principally occupied by agriculture, with significant areas of natural vegetation	(Use)
				244 Agro-forestry areas	(Use)
3 Forest and semi-natural areas	(Cover)	31 Forest	(Cover)	311 Broad-leaved forest	(Cover)
				312 Coniferous forest	(Cover)
				313 Mixed forest	(Cover)

(*Continue*)

Table 6.2 *(Continued)*

Level 1		Level 2		Level 3	
		32 Scrub and/or herbaceous vegetation associations	(Cover)	321 Natural grassland	(Cover)
				322 Moors and heathland	(Cover)
				323 Sclerophyllous vegetation	(Cover)
				324 Transitional woodland-scrub	(Cover)
		33 Open space with little or no vegetation	(Cover)	331 Batches, dunes, and sands	(Cover)
				332 Bare rocks	(Cover)
				333 Sparsely vegetated areas	(Cover)
				334 Burnt areas	(Cover)
				335 Glaciers and perpetual snow	(Cover)
4 Wetlands	(Cover)	41 Inland wetlands	(Cover)	411 Inland marshes	(Cover)
				412 Peat bogs	(Cover)
		42 Maritime wetlands	(Cover)	421 Salt marshes	(Cover)
				422 Salines	(Cover)
				423 Interidal flats	(Cover)
5 Water bodies	(Cover)	51 Inland waters	(Cover)	511 Water courses	(Cover)
				512 Water bodies	(Cover)
		52 Marine waters	(Cover)	521 Coastal lagoons	(Cover)
				522 Estuaries	(Cover)
				523 Sea and ocean	(Cover)

socio-economic aspect of land use, this may necessitate relating linguistic labels that have different cultural, political, linguistic, economic or sociological implications, as well as the botanical or ecological land cover definitions. In general, therefore, land classifications can be seen to suffer from having poorly specified conceptualisations or ontologies.

The aim of this paper is to explore the origins of land cover and land use ontologies, and particularly their confusion and to show that there is a pattern whereby those charged with designing and implementing land classification systems are forced to accommodate the pressures of diverse institutional objectives. It is shown that this has knock-on implications in terms of future analyses and in terms of perpetuating the culture of ontological confusion. The implications of *not* addressing these issues are also described. Finally, a case for clear thinking and the need for scientists to go back and unpick their approaches and the way that they define land classifications is presented.

6.2 Land Use and Land Cover

Land use is a quite distinct concept from land cover, although the two are certainly related. The difference is acknowledged in many documents in this area (Anderson *et al.*, 1976; Campbell, 1981; Di Gregorio and Jansen, 2000), but for completeness definitions are presented here.

Land cover is the physical material at the surface of the earth. It is the material that we see and which directly interacts with electromagnetic radiation and causes the level of reflected energy that we observe as the tone or the digital number at a location in an aerial photograph or satellite image. Land covers include grass, asphalt, trees, bare ground, water, etc. Tone alone (Paine, 1981; Lillesand and Kiefer, 2000) may not be enough to distinguish between the different cover types, but there is a belief, supported by empirical investigation, that with measurement of tone in increasing numbers of discrete wave bands land covers are increasingly becoming separable, although context, pattern and texture may also be used (Paine, 1981; Lillesand and Kiefer, 2000).

Land use, by contrast, is a description of how people *use* the land. Urban and agricultural land uses are two of the most commonly recognised high-level classes of use. Institutional land, sports grounds, residential land, etc. are also all land uses.

Land cover and land use have a complex *many-to-many* relationship (Figure 6.1). Grass, for example, is a land cover type which can occur in any number of land uses: sports grounds, urban parks, residential land, pasture, etc. At the same time, very few areas of homogenous land use have a single land cover; residential land, for example, may contain trees, grass, buildings and asphalt. Indeed a building class is a high level land cover since the material exposed to the electromagnetic radiation may be very variable (slate, tile, asphalt, lead, tin, steel, wood, thatch, etc.).

Some land has a single use, but much is also in multiple uses that cannot be determined from remotely sensed data because the two, or more, uses include the same covers (FAO, 1976). These multiple uses may be simultaneous so that, for example, a single patch of plantation forestry may also be used for several forms of recreation, including hunting and hiking, and even for grazing. Hoeschele (2000) has pointed out exactly this situation in the confusion of land cover and land use and how it may disadvantage subsistence

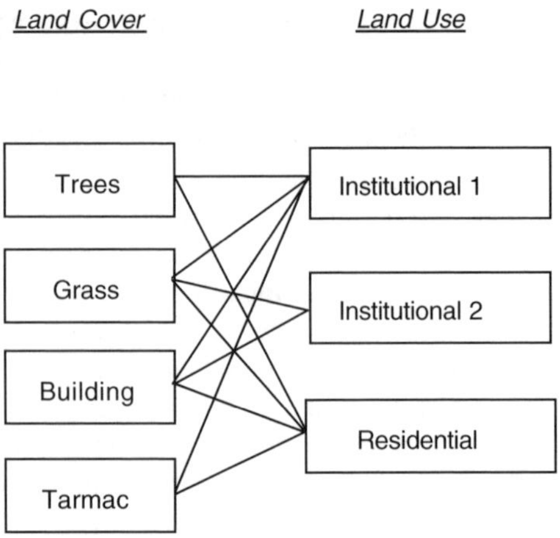

Figure 6.1 *Many-to-many relations of land cover and land use. Many land covers contribute to any one land use and more than one land use can be composed of the same suite of land covers. Furthermore, not all instances of the same land use type will necessarily have the same land covers*

farming using forest land for grazing as opposed to the large land owners who are more interested in the forestry. Multiple land uses may also alternate; the grazing and hunting may not occur contemporaneously for fear of hunters shooting the stock. In summary, a reservoir may provide flood control in the spring and hydro-electric power in the winter, fishing in season and recreational boating all year round.

Land cover is essential for the development of physical models of the environment (climatic and hydrologic). It is, however, not directly useful for most policy and planning purposes (planning of the human or the natural environment), where land use is the relevant phenomenon.

6.3 Origins of Land Cover and Land Use Confusion

In a search through *Geographical Abstracts* (via GEOBASE), it is noteworthy that the earliest mention of 'land cover' was for Waukesha County, Wisconsin, in 1934. In subsequent years little pockets of references to land cover can be found dotted around the world. For instance, the Hunter Valley in New South Wales in the early 1960s, a series of surveys of land cover types adjacent to towns in Massachusetts in 1967, and some land inventory surveys in New Zealand in 1969–1970. There were undoubtedly many further localised surveys recording land cover. However, the overriding trend in land inventory, nationally and locally, was to record land use.

The picture changes at the beginning of the 1970s. We start to see an increase in the rise of land cover applications and surveys being reported, coinciding with the availability of Landsat-1 data and especially the machine processing of that data. This represented a subtle shift in the emphasis of analysis; prior to the availability of medium-resolution satellite imagery, land surveying by interpretation of remotely sensed imagery had used aerial photography. If aircraft sorties were being commissioned, then the resolving power of the imagery (focal length of the lens and altitude) could be specified according to the intended use of the imagery. The interpretation was manual using such phenomena as tone, texture, pattern, context and using local knowledge in many cases (Paine, 1981; Lillesand and Kiefer, 2000). Specification and analysis were *demand* and *application* driven.

The advantages of satellite imagery are its temporal frequency, multi-spectral capability, the vast area covered by each scene, relatively low cost (compared with the alternatives) and digital format. The disadvantages, however, are the widely used, relatively crude methods of by-pixel analysis using only a digital representation of tone, the relatively low spatial resolution of the satellite imagery and, due to these factors, the limited nature of the thematic land use information that can be derived from such imagery. Early imagery from the Landsat series of satellites was interpreted by hand using the same skills as used in aerial photograph interpretation. Machine processing only became common in the late 1980s.

Even with manual interpretation the extent of different land use practices is difficult to infer directly from medium resolution satellite imagery, although it is possible to record the extent of spectrally distinct *land cover* types. Techniques were developed and analyses perfected that would allow different thematic objects to be recognised and identified through by-pixel classification (Campbell, 1996; Jensen, 1996; Mather, 1999; Richards and Jia, 1999). While contextual analysis with ancillary data is increasingly common, methods such as texture analysis and image segmentation, learnt from the skills of manual interpretation, are only now becoming widely available. Such methods are the basis of the UK Land Cover Map 2000 (Smith and Fuller, 2001) although the further potential of such methods to view each segment in its context is still a matter of research (Barr and Barnsley, 1997). Through the use of some training data, relatively reliable thematic attribute allocation is possible. Computer techniques have become standardised, although the thresholds derived from training datasets for class allocation are still very application- or scene-specific. The techniques employed by those involved in recording land cover from remotely sensed imagery have become increasingly dependent on the imagery available to them. Rather than being commissioners of data, the data specifications are determined by those who commission the platforms and sensors. Without any control over the specification process, by the end of the 1970s land surveying using remotely sensed data and associated techniques could be seen as *data* driven.

We suggest that the change from land use to land cover mapping is also a consequence of what is possible with automated processing of satellite imagery. Thus we now have a period when the product delivered from this mapping is land cover rather than land use, and is data driven, as opposed to demand or application driven.

6.4 Accommodating Diverse Institutional Objectives

The availability of medium resolution satellite imagery from the SPOT and Landsat platforms and the associated development of data driven techniques for the analysis of remotely sensed imagery, occurred in parallel with a desire of governments to be able to perform wide ranging national inventories of their land resource. Governments wanted to be able to better manage their land resource for a host of reasons: planning, development, environmental management, etc. The problem they faced was that diverse land mapping initiatives were driven by multiple government agencies, each with their own data specifications, data collection methodologies and classification schemes for recording land stock and its spatial location. There was, however, a perceived need for consistent information that could be compared across time, space and at different levels of aggregation. The problem of accommodating these needs raises further data and methodological issues, and, whilst it has been repeated many times subsequently, was first exemplified by what has been perhaps the most influential work in the area, the widely quoted work of Anderson *et al.* (1976), outlining the USGS Land Use and Land Cover Classification.

The remit of Anderson *et al.* (1976) was to develop a national land classification scheme that could be derived from image processing (manual and machine interpretation) of remotely sensed data, could accommodate data at a variety of resolutions, and could integrate the existing classifications used by various Federal agencies and departments. They proposed a 4 tier land use/cover classification system, the top 2 levels of which were explicitly specified (Table 6.1), with the lower 2 levels to be developed by the users. This was to allow detailed local information to be aggregated into consistent Level 1 and Level 2 land information in order to 'meet the principal objective of providing a Land use and Land cover classification system for use in land use planning and management activities' (Anderson *et al.*, 1976, p. 8).

The recommendations and method of classification development of the USGS system as described in Anderson *et al.* (1976) have been copied and adopted by many other national and international land monitoring initiatives. Perhaps the single best known of these is the desired figure of 85% accuracy for land use and land cover mapping, below which, it was suggested, the mapping would not be useful (quoted earlier by Anderson, 1971). This has received some critical comment in the literature (Congalton and Green, 1999), not least because it has no real basis for being of significance in either statistical theory or experience. The origin of the 85% figure is somewhat of a mystery, and the authors would very much like to hear from anyone who can provide any information. Preliminary enquiries on email lists suggest that the figure was based entirely on Anderson's professional judgement, with a view to estimated costs, and no objective documented evidence (Duane Marble, pers. comm.; USGS Standards Team, pers. comm.; Bob Chen, pers. comm.).

Unfortunately another lasting memorial of the work of Anderson *et al.* (1976) is the continuing confusion of land cover and land use in countless subsequent classifications schemes; most classification schemes incorporate both land use *and* land cover elements. Classifications are also still data driven in that they identify land use and land cover features that can be determined spectrally. Furthermore, the idea that certain land use types can be identified with 'minimal reliance on (external) supplemental information' (Anderson *et al.*, 1976, p. 3) is prevalent despite the subjective and context-dependent nature of land use. Similarly, the ideas of compatibility with existing land classifications and satisfying diverse user interests have been adopted by subsequent land monitoring initiatives. Anderson *et al.* (1976) state that 'special attention has been paid to the definitions of land use categories used by other agencies, to the extent that they are useful in categorising data obtained from remote sensor sources' (Anderson *et al.*, 1976, p. 7).

These include, especially, the Standard Land Use Coding Manual published by the US Urban Renewal Administration and the Bureau of Public Roads (1965), the inventory of Major Uses of Land made by the Economic Research Service of the US Department of Agriculture (Frey, 1973), and the national inventory of soil and water conservation needs, initiated in 1956 by agencies of the US Departments of Agriculture and Interior (all cited by Anderson *et al.*, 1976).

However, Anderson *et al.* (1976) identify the problems of adopting such approaches to the design and implementation of land classifications:

- 'There is no one ideal classification of land use and land cover, and it is unlikely that one could ever be developed' (Anderson *et al.*, 1976, p. 4).
- 'In almost any classification process it is rare to find the clearly defined classes that one would like' (Anderson *et al.*, 1976, p. 4).
- 'Each classification is made to suit the needs of the user and few users will be satisfied with an inventory that does not meet most of their needs' (Anderson *et al.*, 1976, p. 4).

They note the fundamental ontological differences between land use (activities) and land cover (surface) when they point out that these concepts are closely related and have been used interchangeably (although not in a strict sense). The principal points of departure between the USGS system they describe and others that have gone before is the emphasis placed on the remotely sensed data as the primary data source. Whilst some land uses can be inferred directly from land cover others require contextual information: 'Certain land uses such as pasture cannot be separated consistently and accurately using remotely sensed data sources appropriate to the more general levels of the classification' (Anderson *et al.*, 1976, p. 7). The logical conclusion is that land use 'must be interpreted using land cover as the principal surrogate, in addition to the image interpreter's customary references to pattern, geographic location, and so forth' (Anderson *et al.*, 1976, p. 9). A further significant point made is that different agencies have their own specific use of various definitions and that the terminology used for some classes will not be compatible with the same terminology used in other classifications such as the Standard Land Use Coding Manual published by the US Urban Renewal Administration and the Bureau of Public Roads (1965).

Anderson *et al.* (1976) identify the problems of having to accommodate diverse mapping interests. They state that the approach to land use and land cover classification embodied in the system is 'resource-oriented' in contrast to the people-oriented classification of the Standard Land Use Coding Manual (US Urban Renewal Administration and the Bureau of Public Roads, 1965), which assigns 7 out of 9 classes to traditional land use classes (urban, transport, recreation, etc.), but these 7 only cover 5% of the national area. In this context, the USGS classification can be termed 'resource-oriented' in that it addresses the remaining 95% of the land. Implicit in this position is:

- the need for a planning tool;
- the urban bias of the Standard Land Use Coding Manual;
- the need for a resource-oriented classification, whose primary emphasis is on the remaining 95%.

These points are addressed by the Land Use and Land Cover Classification with 8 of the 9 classes concerning non-urban land. In general, with large amounts of the USA being open agricultural and range lands which have homogenous land use, and the relatively small amount of heterogeneous urban land, broadly speaking the range and agricultural land are largely classified by land cover while the urban areas are classified by land use.

In summary, perhaps the real cause of the inclusion of Land Use and Land Cover as a combined concept in the same classification system, when they are incompatible, is that there are actually *cartographic* objectives to the USGS classification. Specifically, there is a desire for a spatially even density of information across the nation so that a map will have approximately the same level of detail in all areas. There is a reluctance to include relatively small areas of urban development adjacent to vast areas of homogenously classified rangeland, so the rangeland needs to be broken up by land cover.

6.5 Implications of the USGS Classification

The failure to address the different ontologies of land use and land cover is manifest in much of the work that followed the pioneering work of Anderson *et al.* (1976). These show three common themes:

1. *Confusing land use and land cover.* This has severe implications when seeking to relate one classification system to another or when seeking to aggregate different levels in the classification. In the hierarchy of land use and land cover definitions (Table 6.1) proposed by Anderson *et al.* (1976), there are many-to-many linkages between Level 1 and Level 2. However, as stated previously, land use has a socio-economic aspect incorporating definitions that have cultural, political or economic significance and meanings. Different land monitoring initiatives and analyses have sought repeatedly to combine different classification schemes that have land use and land cover confusions embodied in them (e.g. MLURI, 1993; EEA, 2001; Fuller *et al.*, 1994; Mackey, *et al.*, 1998; Di Gregorio and Jansen, 2000) whilst repeating the explicit warnings in Anderson *et al.* (1976) about confusing land cover and land use (Fuller *et al.*, 1994; Wyatt *et al.*, 1994, 1998; MLURI, 1993; Di Gregorio and Jansen, 2000).

   Examination of the class hierarchy in Table 6.1, shows that some Level 2 classes which can be viewed as land use are located squarely in Level 1 cover classes (Strip mines, etc., and reservoirs which are parts of water and barren land). Apart from these two examples, however, the Level 1 use classes are composed of Level 2 use classes, and similarly for cover classes. Similar confusion is visible in the CORINE classification (Table 6.2).
2. *Developing of classification to meet user needs.* Typically, classification systems cannot be imposed on users. They are all developed with particular applications in mind, for particular purposes. They are established through a process of consultation and committees with the intention of meeting the needs of participating agencies. Examples include LCSS (Di Gregorio and Jansen, 2000), Land Cover of Scotland 1988 (MLURI, 1993), the National Countryside Monitoring Scheme (Mackey *et al.*, 1998).
3. *Changing techniques.* As methods advance and develop, the techniques change for analysis and for our understanding of the processing that may be necessary to

extract the defined land use or cover classes. Methods are not consistent through time, and particularly there have been significant changes in methods of change detection and monitoring (e.g. EEA, 2001; Wyatt *et al.*, 1998; Fuller *et al.*, 1994, Di Gregorio and Jansen, 2000). Methods of automated classification have also changed dramatically (compare the methods documented by Jensen, 1996, and Tso and Mather, 2001).

These themes have common origins: the desire for a single all-encompassing classification system for the area of concern; an analysis that is data driven; and project management by steering committees of interested parties, whose specific concerns have to be negotiated, reconciled and compromised.

Unfortunately, the adoption of the land classification paradigm embodied by the USGS approach described in Anderson *et al.* (1976) marrying as it does the capabilities of the satellite imagery, the diverse ambitions of the institutions represented on the steering committees of projects and the development of a hybrid land use/land cover classification scheme – results in the user needs *not* being satisfied despite the consultation of users and their influence via steering committees. This is because of the inevitable compromises by all parties involved that are necessary if the classification is going to serve them equally. The effect can be seen at both the national, regional and international level where policy objectives are shaped by the land classification paradigm embodied by the USGS approach (Anderson *et al.*, 1976).

The LCSS/Africover project (Di Gregorio and Jansen, 2000) was developed in response to the need for standardisation of data collection, to 'develop a common integrated approach and a methodology that is applicable at any scale' as described in UNCED's Agenda 21 (LCSS, 2001). The system incorporates a series of classifiers, one of which is 'artificiality of cover' applied to all possible combinations of classifiers. The implication is that *every* combination of classifiers incorporates a land use component.

The rationale behind the European Environment Agency's CORINE land cover inventory is expressed in terms of coordinated information gathering, environmental monitoring, and a consistent pan-European framework (EEA, 2001). Descriptions and specifications of the inventory treat land cover as the elemental building block to achieve these objectives. When these land cover monitoring concepts are elaborated, it is with the language of the environment: 'state of natural areas', abundances of 'wild fauna and flora' and 'water resources', 'gradual desertification' and 'drying-up of wetlands' (EEA, 2001) After this rationale for the CORINE land use inventory comes the following disclaimer:

> *'Most nomenclatures used for mapping or statistics relating to space are land use nomenclatures produced for the purpose of compiling an inventory of human activities. The terminology available (to designate the items in a land cover nomenclature) therefore relates to land use, so specific terminology for land cover inventories has yet to be developed'.*

The classification system defines 1 out of 5 Level 1 classes 6 out of 15 Level 2 classes and 19 out of 44 classes at Level 3 as land use (Table 6.2). Furthermore, more than in the USGS classification, there is confusion within the classes as they move through the hierarchy. Thus the Artificial surfaces in the Level 1 class, splits into two cover classes and two use classes at Level 2, one of which splits into a cover and a use class at Level 3; the confusion is complete.

6.6 Conclusions – A Case for Clear Thinking

The definitions of land use and land cover as stated in this article are not controversial and are not original. Indeed, the distinctions are clearly stated by Anderson *et al.* (1976) and many others since. In revisiting that publication and presenting a critical reading, we do claim, however, to have arrived at an understanding of just why the two terms are intertwined as if they are synonyms in spite of the clear and stated understanding that they are not. Why others should have perpetuated this confusion is less clear.

The remit of those charged with determining the USGS land classification in 1976 was to develop a national land classification system to be applied to data at a variety of resolutions, that could be integrated with the existing classifications. Combined with pressure for consistency and coordination of effort were the rapid developments of computing and digital technology in land mapping initiatives. This resulted in the wholesale adoption of the recommendations made by Anderson *et al.* (1976) by a discipline in its infancy, with the concomitant mapping implications described above: land use/cover confusion, 85% as a benchmark for accuracy and classifications designed to meet a multitude of institutional objectives. These are reflected in the way that even land mapping projects that acknowledge the definitions of land use and land cover go on the present classification schemes that confuse those definitions even when projects are called 'cover' only (e.g. LCM, LCS, LCSS).

The adoption of such approaches for recording land stocks at single instances in time in order to achieve a particular set of policy objectives is uncontroversial. However, there is an increasing demand for combinable or interoperable spatial data which means clear semantic definitions of terms (Bishr, 1998). For land cover, this means relating policy and ecological definitions to reflectance thresholds, neither of which is unproblematic. Such problems are small when compared to those encountered when seeking to relate different land use classifications. In these instances, the relationships amongst the socio-cultural aspects of the nomenclatures have to be established, and the alternate and multiple land uses accommodated before data can be integrated.

We would like to emphasise that Anderson *et al.* (1976) are not the perpetrators of the problem described here, rather they responded to the demands of their steering committee of US Federal agencies that had their own agendas. They must have been under considerable pressure to come up with such a hybrid system because of the substantial caveats they included: land use and land cover are not the same, and yet here is a system that treats them as if they are. Anderson *et al.* (1976) went as far as they could to highlight the problem. The criticism is of those who have adopted the land use/land cover approach articulated by Anderson *et al.* (1976) without fully thinking through the implications of doing so. Indeed, the land cover/land use couplet has been adopted by much subsequent work and has become the *modus operandi* for many land surveys where the differences between land cover and land use are frequently noted, but not accommodated.

The broader issue raised by this analysis reflects a general trend in modern science. We have seen a shift away from approaches and methodologies tailored rigorously to particular analytical approaches and specific applications, towards ones of convenience, national inventory and reuse of secondary information. These are stimulated by the availability of increasing computational power and digital data, which has resulted in reduced conceptual thinking about the nature of the analyses we perform on our data.

If computational speed and power were not so readily available it is not difficult to imagine that the emergence of data driven science would have been harder and as a consequence the definitions, methods, etc., reflected upon earlier.

Classification systems that confuse land cover and land use cannot be universally applied across different resolutions of data because of the epistemological differences in how we obtain land classification from data at different scales. There is a need to formalise the separation of land cover and land use ontologies to allow better comparison of land mapping initiatives in the future.

We conclude with a plea for clearer thinking and a request that we re-evaluate our approaches. Land use and land cover is just one theme in many spatial analyses executed within Geographical Information Systems. Many other themes are used and they too may be shrouded by the similarly complex issues of poor definition.

Acknowledgements

This paper describes work done within the REVIGIS project funded by the European Commission, Project Number IST-1999-14189. We wish to thank our partners in the project, especially Andrew Frank, Robert Jeansoulin, Arko Lucieer, Alfred Stein, Nic Wilson, Mike Worboys and Barry Wyatt.

References

Anderson, J.R., 1971, Land use classification schemes used in selected recent geographic applications of remote sensing, *Photogrammetric Engineering*, **37**(4), 379–387.

Anderson, J.R., Hardy, E.E., Roach, J.T. and Witmer, R.E., 1976, *A Land Use and Land Cover Classification System for Use with Remote Sensor Data*, U.S. Geological Survey, Professional Paper 964, p. 28, Reston, VA.

Barr, S. and Barnsley, M., 1997, A region-based, graph-theoretic data model for the inference of second-order thematic information from remotely-sensed images, *International Journal of Geographical Information Science*, **11**(6), 555–576.

Bishr, Y., 1998, Overcoming the semantic and other barriers to GIS interoperability, *International Journal of Geographical Information Science*, **12**(4), 299–314.

Campbell J.B., 1981, Spatial correlation-effects upon accuracy of supervised classification of land cover, *Photogrammetric Engineering And Remote Sensing*, **47**(3), 355–363.

Campbell, J.B., 1996, *Introduction to Remote Sensing*, 2nd edition. Taylor & Francis, London.

Congalton, R.G. and Green, K., 1999, *Assessing the Accuracy of Remotely Sensed Data: Principles and Practices*, Lewis, Boca Raton.

Di Gregorio, A. and Jansen, L.J.M., 2000, *Land Cover Classification System (LCCS): Classification concepts and user manual*, Environment and Natural Resources Service (SDRN), FAO, Rome.

EEA (2001) CORINE Land Cover – Technical guide, European Environment Agency http://etc.satellus.se/the_data/Technical_Guide/ [available 27/08/01].

FAO (1976) *A Framework for Land Evaluation*, Soils Bulletin 32. Food and Agriculture Organisation of the United Nations, Rome.

Frey, H.T., 1973, Major uses of land in the United States – Summary for 1969: Agricultural Economics Report 247, Economic Research Service, US Department of Agriculture.

Fuller, R.M., Groom, G.B. and Jones, A.R., 1994, The land cover map of Great Britain – an automated classification of Landsat thematic mapper data, *Photogrammetric Engineering And Remote Sensing*, **60**(5), 553–562.

Fuller, R.M., Wyatt, B.K. and Barr, C.J., 1998, Countryside survey from ground and space: different perspectives, complementary results, *Journal Of Environmental Management*, **54**(2), 101–126.
Hoeschele, W., 2000, Geographic information engineering and social ground truth in Attappadi, Kerala State, India, *Annals of the Association of American Geographers*, **90**(2), 293–321.
Jensen, J.R., 1996, *Introductory Digital Image Processing: A remote sensing perspective*, 2nd edition, Prentice Hall, Englewood Cliffs, NJ.
LCSS, 2001, http://www.lccs-info.org/objectives.htm [Available 22/08/01].
Lillesand, T.M. and Kiefer, R.W., 2000, *Remote Sensing and Image Interpretation*, 4th Edition, John Wiley and Sons, Ltd, Chichester.
Mackey, E.C., Shewry, M.C. and Tudor, G.J., 1998, *Land Cover Change: Scotland from the 1940s to the 1980s*, The Stationery Office, Edinburgh.
Mather, P.M., 1999, *Computer Processing of Remotely Sensed Images: An Introduction*, John Wiley and Sons, Ltd, Chichester.
MLURI, 1993, *The Land cover of Scotland 1988 Final Report*, MLURI, Aberdeen.
Paine, D.P., 1981, *Aerial Photography and Image Interpretation for Resource Management*, John Wiley and Sons, Inc., New York.
Richards, J.A. and Jia, X., 1999, *Remote Sensing Digital Image Analysis: An Introduction*, Springer Verlag, Berlin.
Smith, G.M. and Fuller, R.M., 2001, An integrated approach to land cover mapping: the land cover of Jersey, *International Journal of Remote Sensing* **22**, 3123–3142.
Tso, B. and Mather, P.M., 2001, *Classification Methods for Remotely Sensed Data*, Taylor & Francis, London.
US Urban Renewal Administration, Housing, and Home Finance Agency, and the Bureau of Public Roads (1965) *Standard Land Use Coding Manual: A Standard System for Identifying and Coding Land Use Activities*, Washington, D.C.
Wyatt, B.K., Greatorex-Davies, N.G., Hill, M.O., Parr, T.W., Bunce, R.G.H. and Fuller, R.M., 1994, *Comparison of Land Cover Definitions*, Countryside 1990 Series: Volume 3. Department of the Environment, London.
Wyatt, B., Gerard, F. and Fuller, R., 1998, *Correspondence to other themes as a basis for integrated approaches*, Final Report July 1998, Unpublished contract report to the European Topic Centre on Land cover, European Environment Agency.

7

Transformation of Geographic Information using Crisp, Fuzzy and Rough Semantics

Ola Ahlqvist

7.1 Introduction

The continuing development of computers and computer software to assist in decision-making has brought additional possibilities but also new challenges to the discipline of geographic information science. Some of these possibilities and challenges involve the representation of uncertainty and imprecision in geographic knowledge.

One important problem with answering geographic questions is that we almost never have exactly the information we seek at hand. In order to reach an answer to our question we try to make use of existing geographic knowledge, often in the form of digital databases.

Existing geographic knowledge can be seen as summarized and organized data, but changing knowledge into data and back again is no straightforward task. A prerequisite in order to use models or analytical tools upon any piece of geographic information is to have a full understanding of the origin and context of each dataset used.

We also need to bear in mind the fact that a lot of knowledge used in geographic decision-making comes with some degree of uncertainty. Randomness, vagueness and imprecision in geographic information means that analyses made will be of a non-deterministic nature.

Thus, most attempts that use GIS to do geographic analyses take data that use several different conceptualizations and come from different contexts, and this carries with it

Re-presenting GIS Edited by P. Fisher and D. Unwin

resulting uncertainties that need to be considered. This analysis process depends upon a transformation of all information into a desired output conceptualization and context with a declared level of uncertainty and confidence.

Trying to find automated methods for that kind of transformation has been the focus of much work. Recently the question of GI-integration has been put into a comprehensive framework of interoperability of geographic information systems (Vckovski, 1998; Goodchild *et al.*, 1999). Interoperability has earlier been understood as the capability to transfer data from one computer system to another. It is only recently that the discussion about interoperability has been at a more general level, concerned with the establishment of a smooth interface between multiple information sources (Harvey, 1999). Problems of interoperability can be created by, for example, different geometric syntactic representations, difference in class hierarchies, or different semantics (Bishr, 1998). Since different applications, here considered as a user context, have different worldviews and semantics, interoperability at the application level is very much a semantic problem (Bishr *et al.*, 1999).

From a geographical and more general aspect of representing deeper meaning, geographic models of the real world have a tradition of being space-time centered where descriptions of space have dominated. Is the world made up by discrete objects (atomic) or is it a continuum of named attributes (plenum)? The 'atomic' and 'plenum' spatial views lead either to a worldview focused on objects or on fields (Couclelis, 1992). These separate space ontologies are probably at work in parallel in our own 'external models' of reality. The traditional map, for example, uses symbols/objects to communicate intangible geographical features such as an atmospheric low pressure on a weather forecast map. From a semiotic perspective, Sowa (2000) holds symbolic and image-like reasoning as two necessary components of a complete system of reasoning. Theoretically, geographic information systems have the ability to incorporate both plenum/image like/field and atomic/symbolic/object views, usually represented by raster and vectors respectively.

This theoretical basis is reflected in several proposed conceptual modeling frameworks for geographical databases (Bishr, 1998; Livingstone and Raper, 1994; Mark and Frank, 1996; Mennis *et al.*, 2000; Nyerges, 1991b; Peuquet, 1994; Sinton, 1978; Usery, 1996). Most of these works suggest that it is necessary to represent simultaneously field-based, object-based and time-based views to be able to provide a full description of a geographic phenomenon. Although some of the proposed frameworks have never been extended towards the actual database representation, those that were have proposed set theoretic approaches for its implementation (Bishr, 1998; Livingstone and Raper, 1994; Mennis *et al.*, 2000; Usery, 1996).

Related frameworks have also been proposed, such as mereotopology (Smith, 1996) that uses mereology as an alternative to set theory to describe topological relations between parts and wholes of things. One reason to search for alternatives to set theory has been its limited capability to express semantic ambiguity of categories (Mark and Frank, 1996). Kuhn (1999) also points out that existing approaches to semantic modeling such as semantic networks and first order logic are too limited for a rich and deep description of semantic meaning. That motivated him to suggest a connection between semantic nets and algebra that combines the better of these two worlds, a direction proposed as early as 1984 by Andrew Frank (Kuhn, 1999).

On the other hand, fuzzy (Zadeh, 1965) and rough (Pawlak, 1991) extensions of traditional set theory seem to be viable techniques capable of handling the types of category uncertainty or imprecision that previously were problematic from a representational viewpoint. For example, Usery (1996) demonstrates how to represent vague features as fuzzy sets, and Ahlqvist *et al.* (2000) describe the use of rough set theory to represent classification ambiguity.

This paper suggests an implementation structure for conceptual modeling that combines object and plenum representations where concept vagueness and ambiguity can be considered using a combination of crisp, fuzzy and rough representation techniques.

7.2 A Worked Example on Semantic Level Modeling and Uncertainty

The map data used in this illustration was taken from two separate mappings covering the same area. These two vegetation maps had been produced for nature preservation tasks. However, they used different vegetation classification systems (Figure 7.1). The left-hand map from 1971 uses 3 classes: wet, mesic; and dry vegetation. This map will be called VEG3. The map in Figure 7.1 shown on the right is from 1986 and uses 35 classes; the classification focus is on types of tree cover and understory species composition. This will be called VEG35.

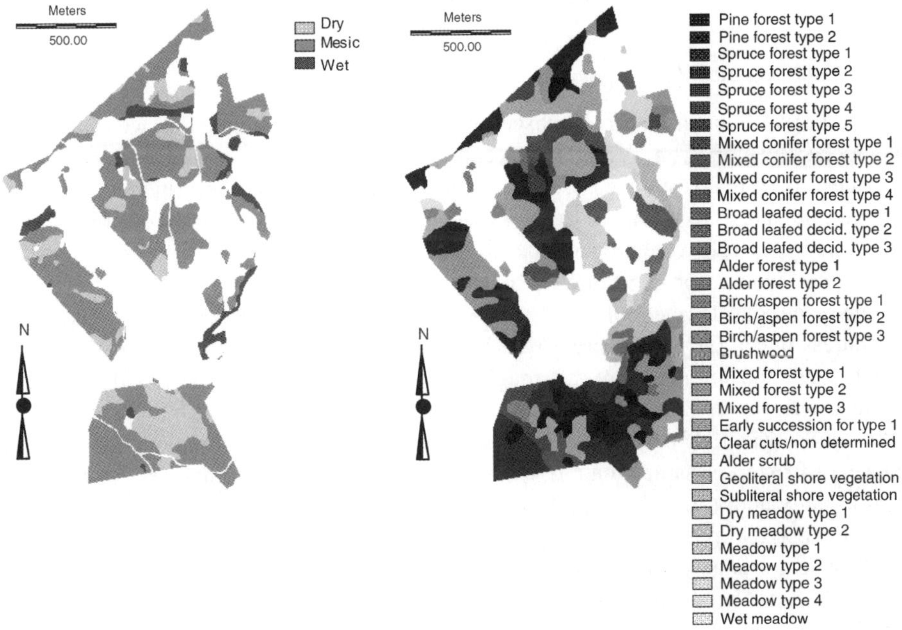

Figure 7.1 *Original maps used in this study. Left, the VEG3 map and right, the VEG35 map (Reproduced from Ahlqvist, O., Keukelaar, J. and Oukbir, K., 2003: Rough and fuzzy geographical data integration, International Journal of Geographical Information Science,* **17***(3), pp. 223–234. Figure 1, page 228)*

If we want to compare these two maps in order to see any changes over time, the standard procedure would be to translate one of the classification systems into the other. This illustrates a general situation where we want to transform information from one context to another. Thus, a reclassification from 35 classes to 3 would be a straightforward task if only one-to-one or many-to-one relations exist. But the correspondence between two concepts from two different contexts usually does not have this simple relation (see Chapter 6). It is probably not an overstatement to propose that, generally, most such relations are not of a simple crisp nature. In all these cases there will be some degree of uncertainty, and the major challenge for a general conceptual modeling framework is to be able to handle this uncertainty.

Although uncertainty may have various sources, randomness and imprecision are two major types of uncertainty that are of importance in spatial knowledge representation and inference (Leung, 1997; Fisher, 1999). Randomness can be expressed as a probability, which is a well studied, understood and also well documented mathematical and statistical idea (Fisher, 1999). Its implementation within the field of geographic information science is also fairly well developed and particularly suited to cases with semantically well-defined objects.

Among the suite of methods to represent uncertainty in knowledge due to poorly defined objects or imprecision, some useful extensions to standard sets have been put forward, for example, fuzzy sets (Zadeh, 1965) and rough sets (Pawlak, 1982). These extensions seem to be able to handle cases of vagueness and indiscernibility (Duckham *et al.*, 2000, Worboys and Clementini, 2001).

7.2.1 Indiscernibility

Let us start with the problem of indiscernibility. If we look at Figure 7.2, we can see that the reclassification from VEG35 concepts onto the ordered concepts of VEG3 will include a lot of overlapping cases. For example, from the original information in VEG35 we have no clue as to whether a certain vegetation unit classified as Ash forest is of a wet or mesic type. The resolution of the original information, with respect to wetness, is simply not enough for us to make a unanimous decision.

The recent development of rough set theory (Pawlak, 1991) has provided a viable tool to address uncertainty that arises from inexact, noisy or incomplete information. Also, rough sets and the concept of rough classification have demonstrated promising applications for geographic information handling (Ahlqvist *et al.*, 2000; Schneider, 1995; Worboys, 1998). A viable solution to our reclassification problem is to use the concept of rough classification (Ahlqvist *et al.*, 2000), based on rough set theory, where we basically admit a 'maybe' part of each class.

Following Ahlqvist *et al.* (2003), rough sets are based on approximation spaces $(\mathcal{U}, \theta)$ where θ is an equivalence relation imposing a granularity on a finite universe $\mathcal{U}$. $\mathcal{U}/\theta = \{E_i\}$ is the set of equivalence classes; the equivalence class containing x may also be denoted $[x]_\theta$ or $[x]$. A definable set in this universe is a union of equivalence classes. A rough set then, is a pair of definable sets, (L, U), such that $L \subseteq U$. Such a rough set is an approximation of any set A such that $L \subseteq A \subseteq U$. $\mathcal{R}_\theta(\mathcal{U})$ denotes the set of all rough sets of $(\mathcal{U}, \theta)$. Given $X = (L, U) \in \mathcal{R}_\theta(\mathcal{U})$, then $\underset{\sim}{X} = L$ and $\tilde{X} = U$. In

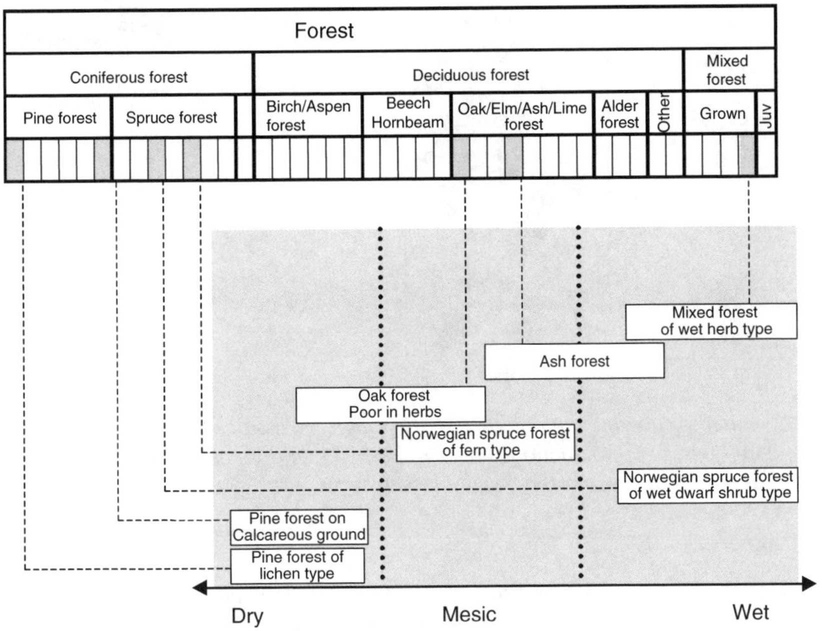

Figure 7.2 *The taxonomy of forest vegetation classes VEG35 is mapped onto a classification of water availability. Some of the vegetation-types are highlighted as enlarged boxes to exemplify the mapping*

this definition, L is called the lower approximation, and U is called the upper approximation.

The meaning of the upper and lower approximations is that, for a certain point $x \in u$, if $x \notin U$, then $x \notin X$, if $x \in L$, then $x \in X$, and if $x \in U - L$ (the maybe part), then we are not sure if $x \in X$ or not (Ahlqvist *et al.*, 2003).

Following Ahlqvist *et al.* (2000), this can be extended to something called a rough classification. A rough classification is a set of rough sets, $\{X_i | X_i \in \mathcal{R}(u), i \in I\}$, with the additional restriction that $\forall_i \forall_j \neq i : \underset{\sim}{X}_i \cup \tilde{X}_j = \emptyset$.

The result of a reclassification from VEG35 to VEG3 using rough classification would look like Figure 7.3. Here, lower approximations, i.e. certain areas, are colored black and areas of uncertainty, the 'maybe' part, are shaded for each of the three classes in the rough classification. What happens here is that, for example, the Ash forest class mentioned above, is reclassified into 'maybe wet' and 'maybe mesic'. This representation can make it a bit hard to visualize the outcome of a rough classification but since this example only has three resulting classes, we may use one image per class in the visualization. Obviously, the information about wetness in the VEG35 classification has not enough detail to discern between the different classes in VEG3. But still, the rough classification is in many cases at least able to exclude one of the classes as an impossible alternative, leaving only two alternatives to choose from.

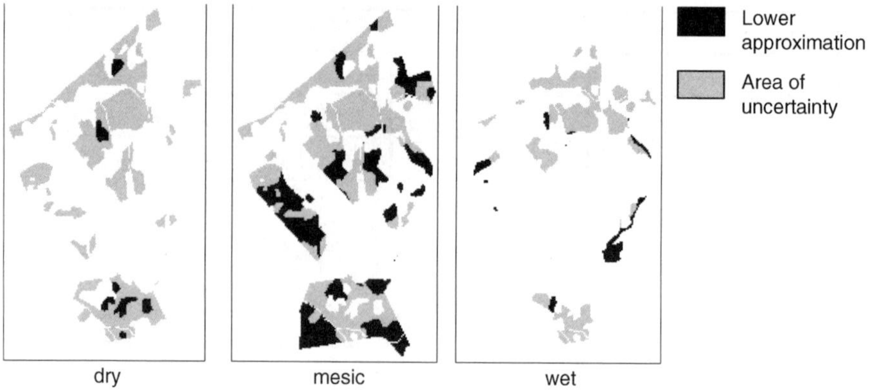

Figure 7.3 Images showing the outcome of the rough reclassification from the VEG35 to the VEG3 system. Lower approximations are colored black and areas of uncertainty are shaded for each of the three classes in the rough classification

7.2.2 Vagueness

We can also imagine this reclassification being a problem primarily due to vagueness. The VEG3 classification system is based upon vegetation moisture. Thus, we could use information about soil water content to produce a classification of the area based only on wetness. A topographically based wetness index can be derived using a digital elevation model over an area (Moore *et al.*, 1991). Normally, the result will be a raster-based

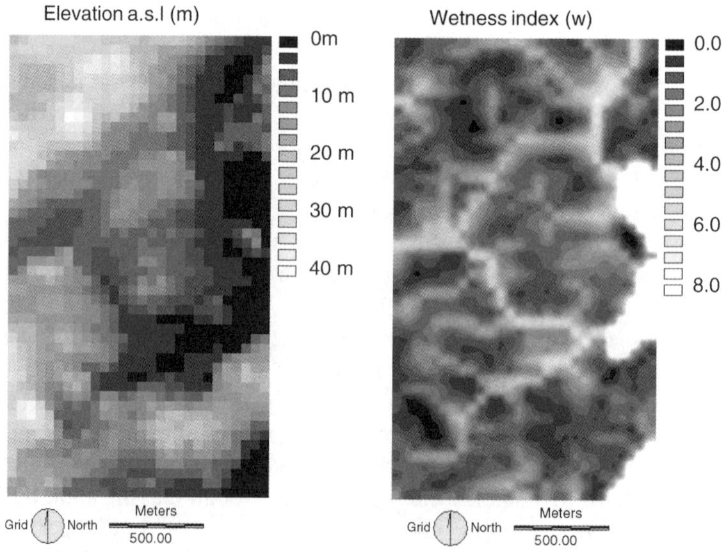

Figure 7.4 Original digital elevation model over the study area together with the wetness index image produced (Reproduced from Ahlqvist, O., Keukelaar, J. and Oukbir, K., 2003: Rough and fuzzy geographical data integration, International Journal of Geographical Information Science, **17**(3), pp. 223–234, Figure 2, page 230)

representation of the continuous attribute, wetness index, over space (Figure 7.4). Dry areas show in black and wet conditions in increasingly whiter shades.

A classification into dry, mesic and wet classes can be done, but the question of whether a certain wetness index value is 'wet' is not unproblematic. It raises some uncertainty that has to do with the degree to which the concept 'wet' matches the actual spatial unit with a wetness value of, say, 3.5. This type of inexact knowledge led to the development of fuzzy sets and fuzzy logic (Zadeh, 1965). Fuzzy sets are a suitable representational tool to accommodate graded and subjective statements such as 'wet' to express partial belonging to a specific concept.

Formally, a fuzzy set F on a universe $\mathcal{U}$ is defined by a membership function $\mu_F : \mathcal{U} \rightarrow [0,1]$, which indicates the degree of membership of a data point in F. This is also an extension of the standard mathematical set; a standard set could be defined by a membership function $\mu_C : \mathcal{U} \rightarrow \{0,1\}$. A fuzzy classification is a set of fuzzy sets $F = \{D_i | i \in I\}$. We will assume the additional restriction on these fuzzy classes $\forall i\ \forall j \neq i \forall x \in \mathcal{U} : \mu_{D_i}(x) = 1 \Rightarrow \mu_{D_j}(x) = 0$.

In implementations of fuzzy systems we may translate the grade of belongingness of an object (here a pixel) to the concept (wet) into membership values, which can be used in propositions and inference. This approach enables a translation from linguistic terms of interpretation judgments into fuzzy memberships for further use in analyses. As such, fuzzy sets and fuzzy logic are suitable for use in representation and inference with vague types of imprecise human knowledge.

The membership functions displayed in Figure 7.5 were used to produce a fuzzy classification. This consists of three fuzzy classes: wet; mesic; and dry, each being a fuzzy set and represented here by the three images displayed in Figure 7.5. These three images show the membership values at every pixel location in each of the three fuzzy classes. In a sense, they convey a location-based view of the target classification.

Recent developments have shown that it is also possible to negotiate these two classification types, fuzzy and rough, using a representation that is also capable of incorporating crisp classifications (Ahlqvist *et al.*, 2003). This joint representation, called a rough fuzzy (RF) set, is based on approximation spaces $(\mathcal{U}, \theta)$. The universe is $\mathcal{U}$, with an equivalence relation θ; equivalence classes are denoted $[x]_\theta$ or $[x]$. A rough fuzzy definable set, or RF-definable set, is a fuzzy set defined by $\mu : \mathcal{U} \rightarrow [0,1]$ such that $\forall x \forall y \in [x] : \mu(y) - \mu(x)$. A rough fuzzy set (RF-set) is a pair of RF-definable sets (L, U), such that $L \subseteq U$. We say that (L, U) approximates the fuzzy set F if $L \subseteq F \subset U$. We denote by $\mathcal{J}_\theta(U)$, the set of all rough fuzzy sets of the approximation space $(\mathcal{U}, \theta)$. Given $X = (L, U) \in \mathcal{J}_\theta(U)$, then $\underset{\sim}{X} = L$ and $\tilde{X} = U$. RF-sets can be expressed as $(\mu_{\underset{\sim}{X}}\ \mu_{\tilde{X}})$, where $\mu_{\underset{\sim}{X}}$ indicates the degree of necessary membership, while $\mu_{\tilde{X}}$ indicates the degree of possible membership. In this way we are able to express both uncertainty due to indiscernibility as well as uncertainty due to vagueness. In the same way as with rough sets and rough classifications we can define a rough fuzzy classification. A RF-classification is a set of RF-sets $\{X_i | X_i \in \mathcal{J}_\theta(U), i \in I\}$, with the additional restriction that $\forall i\ \forall j \neq i \forall x \in \mathcal{U} : \mu_{\underset{\sim}{X}_i}(x) = 1 \rightarrow \mu_{\tilde{X}_j}(x) = 0$ (Ahlqvist *et al.*, 2003).

In the case of the vegetation classification example, the two sources of information, the rough classification derived from the VEG35 classification and the fuzzy classification derived from the wetness information, can be integrated. This can be achieved by, first, converting both sources into rough fuzzy classifications and, second, joining the two by

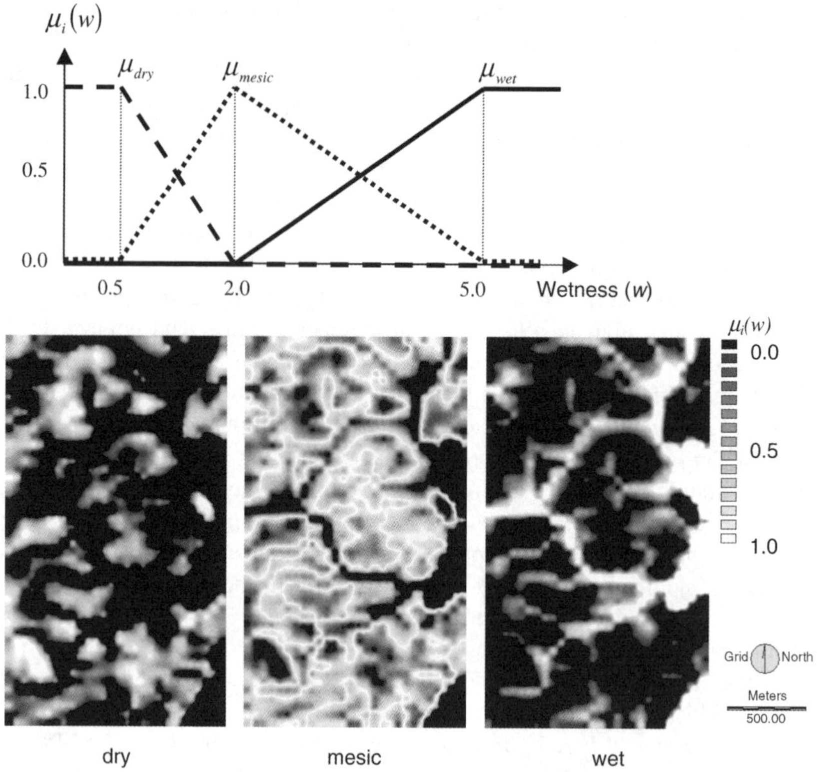

Figure 7.5 *Membership functions for the fuzzy reclassification (top) and the three images showing the three fuzzy classes* $\{dry, mesic, wet\}$. *In the images, black areas represent no membership, and higher degree of membership is increasingly brighter reaching white at full membership (Reproduced from Ahlqvist, O., Keukelaar, J. and Oukbir, K., 2003: Rough and fuzzy geographical data integration, International Journal of Geographical Information Science,* **17***(3), pp. 223–234, Figure 3, page 231)*

an intersection operation (Ahlqvist *et al.*, 2003). There is still the problem of visualizing the spatial outcome of these types of semantic uncertainty. In this case, the result can be presented as a set of six images (Figure 7.6) where grades of membership in these images can be interpreted as the degree of possibility and necessity for the target classes.

7.3 Discussion and Further Possibilities

The given examples demonstrate spatial analysis of graded and indiscernible semantic relationships due to different types of semantic uncertainty. They also point at the interrelation between the thematic and the spatial dimension of geographic information. The identified semantic uncertainty defined in the thematic dimension as rough and fuzzy classifications had rough and fuzzy spatial implications (Figure 7.6). Goodchild (1995) noted that for poorly defined features it is not always possible to separate attribute

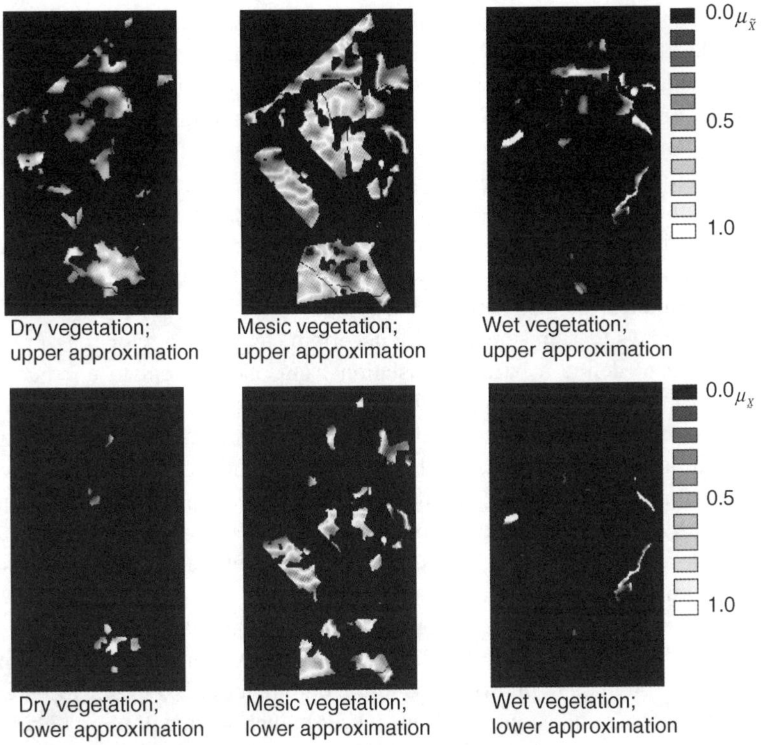

Figure 7.6 *The resulting rough fuzzy classification showing (upper row) the degree of possible membership and (lower row) necessary membership in each of the target classes: dry, mesic and wet vegetation, respectively (Reproduced from Ahlqvist, O., Keukelaar, J. and Oukbir, K., 2003: Rough and fuzzy geographical data integration, International Journal of Geographical Information Science,* ***17****(3), pp. 223–234, Figure 5, page 232)*

accuracy from spatial accuracy. For example, in the case of vegetation maps it is subject to discussion whether the location of a boundary between two vegetation types is uncertain due to the problem of measuring the exact location of the vegetation types or if it is due to the problem of discerning between the two vegetation types at the correct location (Goodchild, 1995; Painho, 1995).

We can also see a clear parallel with the work of Mark (1993) on conceptual boundaries between similar categories of water bodies in the English (lake, pond, lagoon) and French (lac, étang, lagune) languages. Mark showed in his experiment that many categories for water entities may be discriminated using size, spatial relation to the ocean, salinity, presence of marshes at the edges, and whether it is man-made or not. An important observation in these examples is that a mapping between similar concepts, but from different contexts, often cannot be achieved through a crisp relation. Components of uncertainty related to graded concepts and indiscernibility between competing concepts cannot be fully described by a crisp binary relation.

My worked example uses one thematic dimension – moisture. The example by Mark (1993) explicitly illustrated the use of three thematic dimensions, or controlling factors: size; edge marshiness; and man-madeness. Theoretically, the thematic dimensionality could be increased infinitely creating a multidimensional network of concept meanings provided by mappings between similar concepts and between concepts and controlling factors. Nyerges (1991a) outlined a similar framework that he called a heterarchy of concepts, based on multiple and interconnected conceptual hierarchies, forming a multidimensional knowledge framework of concept meanings. Thus, a heterarchy would be something similar to what linguists would call connected word maps. It has been suggested that such a heterarchy of concepts could be explored by, for example, generalization operations (Nyerges, 1991a) or be used to provide input to predictive geographical models (Nyerges, 1991b).

My example shows that, in semantically uncertain cases, a fuzzy or rough representation can be used to define a partial translation from one concept to another, possibly maintaining a truthful representation of the semantic uncertainty in this representation. It also illustrates how multiple sources with different types of semantic uncertainty can be used to explain a specific concept, similar to the ideas put forward by Nyerges (1991a, 1991b). Therefore, I would suggest that a joint collection of multiple sources of knowledge with interconnected concepts, using a richer collection of set theoretic constructs such as fuzzy and rough sets, creates a general framework for transforming information from one user context to another.

In order to develop this idea it is important to address the related problem of achieving GIS interoperability at a semantic level. Several suggested interoperability approaches cited above are based upon definition of common ontologies (Bishr *et al.*, 1999; Gahegan, 1999) or metaclasses (Livingstone and Raper, 1994). Such work will ultimately become a matter of getting groups of people together to negotiate their disagreements and, consequently, the issue of integrating different individual worldviews turns into what has been formalized as part of the sociology of science theory as Group or Organizational Decision Support Systems (King and Star, 1990). Bishr *et al.* (1999) use the term 'geospatial information community' to mean a group of spatial data producers and users who share an ontology of real-world phenomena. However, King and Star (1990) take a broader stance, using a social metaphor rather than a psychological one, as in, for example, Smith and Mark (1998), and address the entire decision-making process in which 'due process' and the construction of 'boundary objects' is of particular importance (Star, 1989).

Due process can be explained as the constant struggle of groups and organizations to recognize, gather and weigh evidence from heterogeneous conflicting sources (King and Star, 1990). Boundary objects is a structure for coordinating distributed work that not only involves heterogeneous actors, elements, and goals but also incorporates different research methods, values, and languages. A boundary object both supplies common points of reference as well as differences to enhance participant understanding of what world views other participants hold, and why they hold them.

This theory has recently been brought into geographic information science by Harvey (1997) and further discussed by Harvey and Chrisman (1998), Chrisman (1999a) and Harvey (1999). It seems from their examples of wetlands mappings in the United States and the ATKIS standard database model in Germany that a definition of common ontologies and schema integration can at best reach some kind of associations and partial matching. Again this can hardly be represented by approaches based on binary relations but it can be constructively moved further if viewed from the ideas of due process and boundary objects.

Vague, inconsistent, ambiguous and illogical information open the domain for concept negotiation, and there is enough proof that these situations are successfully handled within, for example, organizational decision processes (King and Star, 1990). Four types of boundary objects have been identified: repositories, ideal types; coincident boundaries; and standardized forms. Repositories are 'piles' of objects that are indexed in a standardized form such as a library or a museum. The ideal type or platonic object may be fairly vague but a good enough abstraction from all included domains of participants such as an atlas or a diagram. Coincident boundaries are terrain objects that have the same boundaries but different internal contents such as the delineation of counties. The last type of boundary objects, standardized forms or labels, are methods of common communication such as the standardized form used by a national census.

It is argued that boundary objects may serve as a mediator in negotiations around which similarities and differences can be articulated (King and Star, 1990; Harvey and Chrisman, 1998). If it turns out to be possible to formalize the idea of boundary objects into something that can explicitly represent commonalities as well as differences, it is hoped this would serve as a better means to represent deeper meaning in geographic databases.

I would suggest thinking about the spatial tessellation used in a geographic information analysis as 'boundary objects' (Star, 1989) for the negotiation of different aspects of uncertainty. Essentially, the spatial unit acts as a boundary object 'terrain with coincident boundary' within which similarities and differences are articulated and negotiated by the overlay or multi-band operators.

For example, a multi-spectral image classification process uses the boundary object idea. Each location pixel acts as a boundary object for the evaluation of the information from the different spectral bands in image data. Also a 'standard' multi-criteria evaluation performed as an overlay operation in a geographic information system makes use of the boundary objects. The overlay operation evaluates the joint outcome for each spatial unit, be it a polygon or a pixel. The situation of a polygon overlay, following the transformational view of GIS operations described by Chrisman (1999b), is actually a two-step process where the first step consists of identifying the lines that define the spatial limits of the boundary objects, and the second step is the actual negotiation process.

In the given example of translation of vegetation classes and wetness information, it seems reasonable to think that the demonstrated technique may be used in the kind of negotiation that can be expected from divergent viewpoints held by different people in organizations. Such negotiations may often reach agreements around vague or imprecise terms as suggested by King and Star (1990). In the case of this experiment, space served as a boundary object around which the different aspects of uncertainty where integrated. The vague concepts 'wet', 'mesic' and 'dry' could be assigned graded boundaries using fuzzy classification and imprecise information was declared using rough classification.

Translated into an applied situation, different organizations may, according to King and Star (1990), exchange their information only through the process of continuous identification, gathering and weighing of heterogeneous information. This is implemented here through rough-fuzzy classifications and would be a simplified articulation of a 'due process' (Star, 1989; King and Star, 1990), which also has been described in a similar geographic setting by Harvey (1999). Within the due process, boundary objects sit in the middle of a group of participants trying to negotiate their divergent viewpoints. I would argue that the use of space as a boundary object will make it possible to apply the multi-criteria evaluation framework to performing concept mediation and thus perform a transformation between contexts both within and between organizations. This remains to be fully tested, as does whether other types of boundary objects such as 'repositories', 'ideal types' or 'standardized forms' prove to be suitable for geographic applications.

To conclude, I would suggest that the whole idea of boundary objects and due process can be used for coordinating a specific objective, for example, in a data transformation process from one context to another. Use of a richer collection of set theoretic approaches that can be integrated into joint representations, such as the rough-fuzzy data integration approach presented here, will contribute to a successful implementation of this proposition.

References

Ahlqvist, O., Keukelaar, J. and Oukbir, K., 2000, Rough classification and accuracy assessment, *International Journal of Geographical Information Science*, **14**(5), pp. 475–496.

Ahlqvist, O., Keukelaar, J. and Oukbir, K., 2003, Rough and fuzzy geographical data integration, *International Journal of Geographical Information Science*, **17**(3), pp. 223–234.

Bishr, Y., 1998, Overcoming the semantic and other barriers to GIS interoperability, *International Journal of Geographical Information Science*, **12**(4), pp. 299–314.

Bishr, Y.A., Pundt, H., Kuhn, W. and Radwan, M., 1999, Probing the concept of information communities – a first step toward semantic interoperability, In: *Interoperating Geographic Information Systems*, edited by Goodchild, M., Egenhofer, M., Fegeas, R. and Kottman, C., Kluwer Academic Publishers, Boston/Dordrecht/London, pp. 55–69.

Chrisman, N.S., 1999a, *Trading Zones versus Boundary Objects: incomplete translations of technical expertise*, Presentation at Society for the Social Studies of Science, San Diego CA October 1999, URL: http://faculty.washington.edu/chrisman/Present/4S99.pdf.

Chrisman, N., 1999b, A transformational approach to GIS operations, *International Journal of Geographical Information Science*, **13**(7), pp. 617–637.

Couclelis H., 1992, People manipulate objects (but cultivate fields): Beyond the raster-vector debate in GIS, In: Lecture notes in computer science 639, *Theories and Methods of Spatio-temporal Reasoning in Geographic Space*, edited by Frank, A.U., Campari, I. and Formentini, U., Springer, Berlin, pp. 65–77.

Duckham, M., Mason, K., Stell, J.G. and Worboys, M.F., 2000, *A Formal Ontological Approach to Imperfection in Geographic Information*, paper presentation at GISRUK 2000, York, April 2000.

Fisher, P.F., 1999, Models of uncertainty in spatial data, In: *Geographical Information Systems and Science – Principles and Technical Issues*, second edition, edited by Longley, P.A., John Wiley and Sons, Inc., pp. 191–205.

Gahegan, M.N., 1999, Characterizing the semantic content of geographic data, models and systems, In: *Interoperating Geographic Information Systems*, edited by Goodchild, M., Egenhofer, M., Fegeas, R. and Kottman, C., Kluwer Academic Publishers, Boston/Dordrecht/London, pp. 71–83.

Goodchild, M.F., 1995, Attribute accuracy, In: *Elements of Spatial Data Quality*, edited by Guptill, S.C. and Morrison, J.L., Elsevier Science Ltd, Oxford, pp. 59–79.

Goodchild, M.F., Egenhofer, M., Fegeas, R. and Kottman, C., 1999, *Interoperation Geographic Information Systems*, Kluwer Academic Publishers, Boston/Dordrecht/London.

Harvey, F., 1997, Improving multi-purpose GIS design: participative design, In: *Spatial Information Theory: A Theoretical Basis for GIS*, edited by Hirtle, S.C. and Frank, A.U., Springer Verlag, Berlin, pp. 313–328.

Harvey, F., 1999, Designing for interoperability: overcoming semantic differences, In: *Interoperating Geographic Information Systems*, edited by Goodchild, M.F., Egenhofer, M., Fegeas, R. and Kottman, C., Kluwer Academic Publishers, Boston/Dordrecht/London, pp. 85–97.

Harvey, F. and Chrisman, N., 1998, Boundary objects and the social construction of GIS technology, *Environment and Planning A*, **39**(9), pp. 1683–1694.

King, J.L. and Star, S.L., 1990, Conceptual foundations for the development of organizational decision support systems, *Proceedings of the Twenty-Third Annual Hawaii International Conference on System Sciences*, Kailua-Kona, HI, USA, 2–5 Jan. 1990, IEEE Comput. Soc. Press, Los Alamitos, CA, USA 1990, pp. 143–151.

Kuhn, W., 1999, An algebraic interpretation of semantic networks, In: *Spatial Information Theory. Cognitive and Computational Foundations of Geographic Information Science*, Proceedings, International conference COSIT'99, edited by Freksa, C. and Mark, D.M., Stade, Germany, August 25–29, 1999, pp. 331–347.

Leung, Y., 1997, *Intelligent Spatial Decision Support Systems*, Springer-Verlag, Berlin.

Livingstone D. and Raper J., 1994, Modelling environmental systems with GIS: theoretical barriers to progress, In: *Innovations in GIS 1*, edited by Worboys, M.F., Taylor and Francis, London, pp. 229–240.

Mark, D.M., 1993, Toward a theoretical framework for geographic entity types, In: *Spatial Information Theory: A Theoretical Basis for GIS*, Proceedings European conference, COSIT'93, edited by Frank, A.U. and Campari, I., Marciana Marina, Elba Island, Italy, September 19–22, Berlin, Springer-Verlag, pp. 270–283.

Mark D.M. and Frank A.U., 1996, Experiential and formal models of geographic space, *Environment and Planning B: Planning and design*, **23**, pp. 3–24.

Mennis, J.L., Peuquet, D.J. and Qian, L., 2000, A conceptual framework for incorporating cognitive principles into geographical database representation, *International Journal of Geographical Information Science*, **14**(6), pp. 501–520.

Moore, I.D., Grayson, R.B. and Ladson, A.R., 1991, Digital terrain modeling: A review of hydrological, geomorphological and biological applications. *Hydrologic Processes*, **5**(1), pp. 3–30.

Nyerges T.L., 1991a, Representing geographical meaning, In: *Map Generalization; Making Rules for Knowledge Representation*, edited by Buttenfield, B.P. and McMaster, R.B., Longman Scientific and Technical, Essex, England, pp. 59–85.

Nyerges, T.L., 1991b, Geographic information abstractions: conceptual clarity for geographical modeling, *Environment and Planning A*, **23**, pp. 1483–1499.

Painho M., 1995, The effects of generalisation on attribute accuracy on natural resource maps, In: *GIS and Generalisation; Methodology and Practice*, edited by Müller, J-C., Lagrange, J-P. and Weibel, R., Taylor and Francis, London, pp. 194–206.

Pawlak, Z., 1982, Rough sets, *International Journal of Computer and Information Sciences*, **11**(5), pp. 341–356.

Pawlak, Z., 1991, *Rough Sets: Theoretical Aspects of Reasoning about Data*, Kluwer Academic Publishers, Dordrecht/Boston.

Peuquet D.J., 1994, It's about time: A conceptual framework for the representation of temporal dynamics in geographic information systems, *Annals of the Association of American Geographers*, **84**(3), pp. 441–461.

Schneider, M., 1995, *Spatial Data Types for Database Systems*, PhD Thesis, Fernuniversität, Hagen, Germany.

Sinton, D.F., 1978, The inherent structure of information as a constraint to analysis: Mapped thematic data as a case study, In: *Harvard Papers on GIS*, Volume 7, edited by Dutton, G., Addison-Wesley, Reading, MA, pp. 1–17.

Smith, B., 1996, Mereotopology: A theory of parts and boundaries, *Data and Knowledge Engineering*, **20**, pp. 287–303.

Smith, B. and Mark, D.M., 1998, Ontology and geographic kinds, In: *Proceedings. 8th International Symposium on Spatial Data Handling (SDH'98)*, edited by Poiker, T.K. and Chrisman, N., 12–15 July 1998, Vancouver, International Geographical Union pp. 308–320.

Sowa, J.F., 2000, *Knowledge Representation: Logical Philosophical and Computational Foundations*, Pacific Grove, CA, USA.

Star, S.L., 1989. The structure of ill-structured solutions: Boundary objects and heterogeneous distributed problem solving, In: *Distributed Artificial Intelligence*, edited by Gasser, L. and Huhns, M.N., Research notes in artificial intelligence, Pitman, London, England, pp. 37–54.

Usery, E.L., 1996, A conceptual framework and fuzzy set implementation for geographic features, In: *Geographic Objects with Indeterminate Boundaries*, edited by Burrough, P.A. and Frank, A.U., Taylor and Francis, London, pp. 71–85.

Vckovski, A., 1998, Guest editorial, Special issue: Interoperability in GIS, *International Journal of Geographic Information Science*, **12**(4), pp. 297–298.

Worboys, M.F., 1998, Computation with imprecise geospatial data. *Computers, Environment and Urban Systems*, **22**(2), pp. 85–106.

Worboys, M.F. and Clementini, E., 2001, Integration of imperfect spatial information, *Journal of Visual Languages and Computing*, **12**, pp. 61–80.

Zadeh, L.A., 1965, Fuzzy sets, *Information and Control*, **8**, pp. 338–353.

8

Uncertainty and Geographic Information: Computational and Critical Convergence

Matt Duckham and Joanne Sharp

8.1 Introduction

The relationship between critical geography and geoinformatics has recently entered a new phase of cordiality and constructive discourse, exemplified by the NCGIA Initiative 19 (Harris and Weiner, 1996), the Varenius project (Goodchild *et al.*, 1999) and a variety of journal special issues on topics such as GIS and society (Harvey, 2000) and feminist geography and GIS (Kwan, 2002). This new-found *entente* has opened the way to the re-evaluation of existing GIS-based research and to much closer cooperation between the two disciplines for the exploration of new areas of common interest. Within this context, this chapter represents an attempt to fuse certain aspects of critical thinking on GIS with the development of new computational approaches to uncertainty in geographic information. Underlying this fusion is the contention that both subjects stand to gain from closer cooperation (Sheppard, 1995). Such a fusion may provide a route to a more complete style of GIS-based research, acknowledging some of the important societal issues connected with GIS at the same time as accommodating some of the practical restrictions connected with computational systems.

Amongst the variety of different critiques of GIS surveyed by Sheppard *et al.* (1999), is the tendency of GIS to promote certain types of representations, or 'ways of knowing' (Pickles, 1999), over others, for example in the association of GIS with positivist modes of thought (Lake, 1993). In an attempt to combat such tendencies, Pickles (1999) calls for

Re-presenting GIS Edited by P. Fisher and D. Unwin

a more collaborative 'pluralistic GIS', representing multiple perspectives and containing conflicting information. Bondi and Domosh (1992) highlight the contradictions that arise when GIS ignores the pluralistic and treats geographic information as objective, uncontroversially 'true', and free of any context. Similarly, Haraway (1988) insists upon the recognition of the partiality of perspectives inherent in any representation or conceptualisation of a situation. Many critical geographers and geoinformatics researchers have taken these critiques as being indicative of a fundamental disjuncture between the two types of knowledge. While considered elegant and ideologically correct by fellow critical geographers, such attacks have made it all too easy for geoinformatics researchers to evade responsibility for the important underlying conceptual issues. Geoinformatics researchers may claim that political and ethical questions should be dealt with by 'those better versed in philosophy and the social/human side of geography' (Mike Goodchild, quoted in Schuurman, 1999), or simply ignore critical human geographers as ill-informed on the practices of GIS. On the other hand, to critical theorists, for whom all knowledge should be contextualised, it can appear that geoinformatics researchers are intent on ignoring theory, social context and effect, and instead are content to work on the technical challenges offered by developing technology and algorithms.

However, this chapter argues that there is some consonance between critical geography research and current geoinformatics research into uncertainty in geographic information. Following from the critiques of Haraway, it seems clear that GIS knowledge is not somehow 'wrong' but that it is situated and limited, as are all knowledges. Examination of the ways in which uncertainty and imperfection can be represented in GIS provides one way of addressing some of these critical concerns.

8.2 Uncertainty in Geographic Information

Some of the more hyperbolic celebrations of GIS offer glimpses of a utopian future built using perfect, global geographic information (Abler, 1993; Gore, 1998). In actuality, such visions are not particularly credible for one simple reason: the inherent uncertainty surrounding any information about our geographic environment. The impression of certainty usually conveyed by GIS is at odds with the uncertain nature of geographic information, a contradiction that has been acknowledged as an important research topic for nearly two decades (Chrisman, 1983; Goodchild and Gopal, 1989). An established body of research in the geoinformatics literature deals with many different aspects of uncertainty in geographic information (Fisher, 1999; Longley *et al.*, 2001). While the misperception that improvements in technology are inexorably moving towards the elimination of uncertainty in GIS is still occasionally evident in some geoinformatics and critical geography literature, most literature recognises the fundamental nature of uncertainty in geographic information (Goodchild, 1995).

8.2.1 Imperfection in Geographic Information

Uncertainty arises because information about our geographic environments is always imperfect. Specifically, geographic information is always to some extent *imprecise*, *inaccurate* or *vague*. Imprecision results from incompleteness or lack of detail in information about a geographic environment. As humans we can never observe all the rich detail in our environment. Despite the assistance of measuring instruments, the information produced using those instruments can never completely describe even the physical characteristics of a geographic environment (Veregin, 1999). Furthermore, there is every reason to believe that these unobserved details matter. Work on dynamical systems indicates that even with relatively simple systems, such details can have a profound effect upon the eventual state of a system (Pietgen *et al*., 1992).

Inaccuracy results from incorrectness or error in observations. Traditionally, accuracy has been defined in the statistical sense of deviation of an observation from the 'true' value, or some observation 'taken to be true'. In practice, since such references to truth are highly problematic, accuracy is commonly assessed by comparing observations with an independent data source of higher accuracy. While this weaker definition introduces an 'awkward aspect of circularity' (Goodchild and Jeansoulin, 1998), it is usually possible to make a subjective assessment of the relative accuracies of different datasets. A more realistic non-circular definition of accuracy is as 'the deviation of an observed value from that indicated by an independently derived dataset believed to be more reliable'. Our belief in the reliability of a dataset is usually founded on the use of more advanced instrumentation or because it is known that greater care and attention was taken during data collection. Whichever definition of accuracy you subscribe to, humans are certainly prone to make mistakes and while we can improve our performance using technology and instrumentation, instruments are also not infallible. When they do fail, the resulting information is inaccurate in the sense that it does not accord well with other, more careful observations of our geographic environment. Another important and often ignored source of inaccuracy is the simplifying assumption that our geographic environment is static. Most geographic information is not explicitly temporal, and as a consequence becomes inaccurate as the dynamic environment changes over time. The most careful observations of a dynamic environment are likely to appear inaccurate if the change over time has not been recorded or accounted for.

Vagueness concerns concepts that exhibit borderline cases, having no clearly defined boundary (Keefe and Smith, 1996). Taking 'tallness' as an example of a vague concept, it is usually possible for someone in a room full of people to identify those individuals who are 'tall' and those who are 'not tall'. However, we would also expect some borderline cases, where it is simply not possible to determine whether borderline individuals are 'tall' or 'not tall'. Many common geographical concepts are vague (Fisher, 2000), such as 'near', 'north', 'mountain' and 'developing world'. Vague concepts offer particular problems for classical logic, which is ill-equipped to deal with vagueness. It is entirely possible to invent precise definitions of vague concepts. This is commonly attempted in land cover surveys, for example, where vague concepts such as 'broad-leaved forest' may be precisely defined according to tree species mix, height of trees, percentage crown cover, total area, etc. (Bossard *et al*., 2000). However, such *precisifications* do not fundamentally address the vague nature of concepts like 'broad-leaved forest'. Arbitrarily designating the boundaries of a vague concept is never entirely satisfactory as it quickly leads to absurdity. To illustrate, if our definition of 'broad-leaved forest' is dependent on precise thresholds for attributes, such as density of trees, it is entirely possible that minute changes in the geographical environment, for example the germination of one new tree seedling, can cause the classification of an area to dramatically change, in this case from 'not broad-leaved forest' to 'broad-leaved forest'. This type of behaviour runs strongly counter to most people's intuition about categories like 'forest' and 'tall'. This informal illustration is related to the *sorites paradox*, which has been puzzled over since ancient Greek times (Cargile, 1969).

8.3 Formal Models of Imperfection

There exists a widening variety of different formal models that are being used to manage and describe uncertainty in geographic information. This section informally introduces a selection of different models currently used in geoinformatics, and indicates the different interpretations of imprecision, inaccuracy and vagueness that each model can support.

8.3.1 Stochastic Models

By far the most commonly used model of imperfection in geographic information is the stochastic model. Adopting a stochastic model assumes imperfection is the result of random variation. In the stochastic model of imperfection, observations are drawn from a population of possible observations with predictable characteristics under central limit theorem. The mean value of the population is an estimate of the 'true' value. Accuracy can be represented by deviation from this 'true' value, for example using the root mean squared error (RMSE). Precision can be represented by the spread of observations, for example using standard deviation. There exists no similar common stochastic interpretation for vagueness since the stochastic model assumes phenomena are essentially crisp and knowable. Adopting a stochastic model of imperfection in geographic information carries the advantage of a well-understood theory of natural variation dating back more than two centuries. Modern statistical techniques offer a powerful and sophisticated arsenal of analysis tools that can be applied to great effect.

There is now a significant body of research devoted to the development and understanding of stochastic models of imperfect geographic information. Bivariate statistics are commonly used to model the accuracy of planar spatial objects (Ehlers and Shi, 1996; Leung and Yan, 1998; Shi, 1998). Geostatistical techniques can be used to provide a detailed stochastic model of the inaccuracy and imprecision of field-based geographic information (Heuvelink, 1998; Journel and Huijbregts, 1978). Nevertheless, while the stochastic model offers a highly sophisticated and detailed model of certain types of imperfection, it depends on highly sophisticated and detailed assumptions that often require considerable effort to maintain. For example, the assumption of statistical independence in the stochastic model is commonly violated by spatial autocorrelation in geographic information. As a result, more flexible models of imperfection are often needed.

8.3.2 Fuzzy Set Theory

Interest in the application of fuzzy set theory to imperfection in geographic information has steadily grown over the past two decades. In classical logic, elements are classified as either *in* or *out* (Boolean truth values *true* or *false*) of a particular set X. In a fuzzy set, each element is identified with a real number from the interval [0,1] that describes the degree of membership of that element to the set X.

Unlike stochastic models, fuzzy set theory naturally lends itself to a vague interpretation as well as inaccurate and imprecise interpretations. Fuzzy set theory has been successfully used, for example, to describe the accuracy of land cover classifications (Gopal and Woodcock, 1994; Woodcock and Gopal, 2000). Similarly, imprecision in geographical boundaries (Altman, 1994; Leung, 1987) and vagueness in geographic objects (Usery, 1996) and map classifications (Oberthür, *et al.*, 2000) have all been tackled using fuzzy set-based interpretations.

While fuzzy set theory can be used to model vagueness, imprecision and inaccuracy, difficulties remain. Most importantly, the assignment of fuzzy membership values is not clearly understood (Keefe and Smith, 1996). Attempts have been made to map observations to fuzzy membership values using linguistic values (Gopal and Woodcock, 1994; Leung, 1987) or cluster analysis, a technique called fuzzy k-means (Burrough and McDonnell, 1998). However, the assignment of fuzzy membership values is always to some extent subjective, a feature often regarded as a weakness in the application of fuzzy set theory. At the same time, this interface between the formal and subjective makes fuzzy set theory particularly interesting to some critical theorists.

8.3.3 Three-valued Logic

In contrast to sets in classical set theory (often termed crisp sets), three-valued logic classifies elements as either *in*, *out* or *indeterminate* members of a set X. One of the most common three-valued logical systems is rough set theory (Pawlak, 1982). A *rough set* comprises of a *lower* and *upper* approximation. Informally, the lower approximation $\underline{S}(Z)$ contains all those elements that are definitely in the set Z. The upper approximation $\overline{S}(Z)$ contains all those elements that are possibly in the set Z.

Rough sets can assume a variety of interpretations, including inaccuracy, imprecision and vagueness (Duckham *et al.*, 2001). Rough sets have been used to model imprecision in the spatial and semantic components of geographic information (Worboys, 1998) and to address accuracy issues in thematic spatial information (Ahlqvist *et al.*, 2000). Rough sets are commonly used in a vague interpretation, for example when applied to perception of vague concepts such as nearness (Duckham and Worboys, 2001; Worboys, 2001). However, rough sets are not the only possible three-valued logic. Another three-valued logic might be constructed from a fuzzy set, where membership values were restricted to three values {0,0.5,1.0}, for example. Worboys and Clementini (2001) describe a variety of different three-valued logics tailored to deal with the integration of information exhibiting particular imprecision, inaccuracy and vagueness characteristics. Three-valued logic has the advantage of simplicity, but may not be sophisticated enough to provide an adequate model of imperfection in many cases, especially when compared with the stochastic model.

8.3.4 Other Logical Models

While the stochastic model, fuzzy set theory and three-valued logics are the most common formal models of imperfect geographic information, there exists a relatively rich literature of alternative logical models that might equally be applied to problems of reasoning with imperfect spatial information. For example, three-valued logics are themselves examples of a more general class of *multi-valued* logics. Roy and Stell (2001) explore the application of three-, four-, and six-valued logics to uncertainty in spatial regions. The truth values of these higher-valued logics form a rich variety of lattice structures that support a range of expressive interpretations.

A different approach is the application of *supervaluation theory* (Fine, 1975). Supervaluation is an attempt to retain some of the inferential power of classical logic, at the same time addressing some of the difficulties presented to classical logic by vague concepts. For example, in classical logic for any proposition P, $P \vee \neg P$ is a tautology, commonly known as the principle of excluded middle. In many non-classical logics, such as rough set theory described above, $P \vee \neg P$ is not tautologous and as a consequence we lose some of classical logic's powerful mechanisms for consistency checking. Supervaluation provides a framework with which to discuss the classical truth or falsity of statements across sets of specifications. Each specification can be thought of as a different (classical) universe. A statement that is true in all possible universes is said to be *super-true*, while one that is false in all possible universes is *super-false*. Thus, within a particular specification, classical logic holds and $P \vee \neg P$ is a tautology, even though across many specifications a statement may be neither super-true nor super-false. Bennett (2001) describes how supervaluation semantics can be applied to resolve some of the difficulties posed by vague geographic phenomena such as 'forest', while other interpretations of supervaluation theory are entirely feasible.

The same problem is tackled from the opposite direction by *paraconsistent* logics (Besnard *et al.*, 1997). Classical logic is *explosive* in that anything can be inferred from a contradiction, i.e. for any propositions P and Q, $P \wedge \neg P \rightarrow Q$. So using classical logic, given the information that someone is both 'tall' and 'not tall' (or that someone is neither 'tall' nor 'not tall') we might infer, say, that 'the Earth is flat'. In short, classical logic breaks down in the presence of inconsistency. Paraconsistent logics are weakened forms of classical logic that limit the range of inferences that can be made from a statement. As a result, paraconsistent logics are not explosive.

Multi-valued logics, supervaluation and paraconsistent logics represent just three non-classical logical models that are beginning to be explored in the context of geoinformatics, but many more candidate logical models exist. GIS are just beginning to incorporate some of the more established non-classical modes, such as fuzzy and rough set theories. The following section suggests an architecture for achieving computation using a wide range of non-classical logics within a GIS.

8.4 Imperfection and Computation

The discussion above provides a flavour of the rich diversity of different formal models that can potentially be applied to imperfection in geographic information. It should now be evident that no single formalism can offer a complete view of imperfection, to the exclusion of all others. Furthermore, it seems likely that no single model can be discarded out of hand: each model has its strengths and weaknesses. Consequently, a forward step in the development of GIS would be to specify a computational framework able to allow the integration of these various models. Such integration is possible by recognising that the underlying reason for applying the different formal models is the ability to resolve inconsistencies in observations.

8.4.1 Inconsistency in Observations

The common theme underlying the diverse formal models of imperfection in geographic information discussed above is that they are used as mechanisms for resolving inconsistency in observations. For example, Figure 8.1 illustrates the relationship between inconsistent information and the stochastic model of imperfection. In attempting to determine the location of the feature, four different observations are made. These observations are related in that they are all observations of the location of the same feature. They are inconsistent in that not all the observations are in agreement. The stochastic model can be used to resolve this inconsistency by assuming that any inconsistency is as a result of natural variation. In a GIS, this would usually translate into storing not the underlying inconsistent observations, but instead the derived characteristics of the population from which the inconsistent observations were drawn, i.e. the mean ('true') location and some measure of spread, such as the standard deviation or the size of the 95% confidence interval.

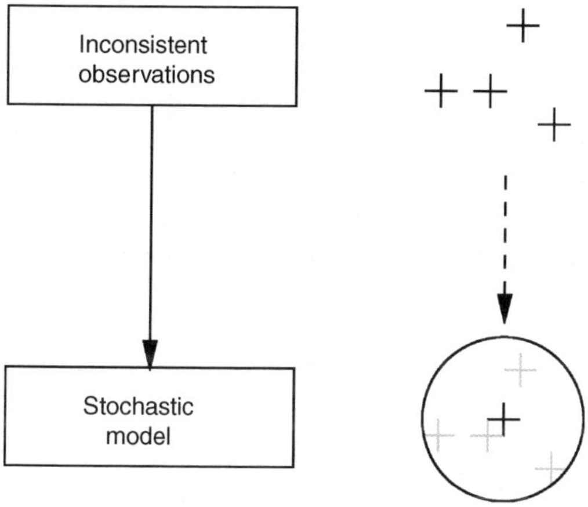

Figure 8.1 *Stochastic model of inconsistency*

In a similar way, other formal models are used to provide a single consistent representation of inconsistent observations. For example, Figure 8.2 (after Worboys and Clementini, 2001) shows two different inconsistent observations of a spatial region, such as a forest. For clarity, these two observations have been drawn as disconnected, but they are intended to be partially overlapping. Combining these two observations using three-valued logic semantics results in a region with a broad boundary. The dark coloured core of the resulting region contains those locations which were forest in both the original observations and so is definitely part of the forest (the lower approximation of the forest $\underline{S}(Forest)$). The lighter shaded penumbra contains those locations that were in one or other but not both of the observations. The penumbra plus the core forms the region that

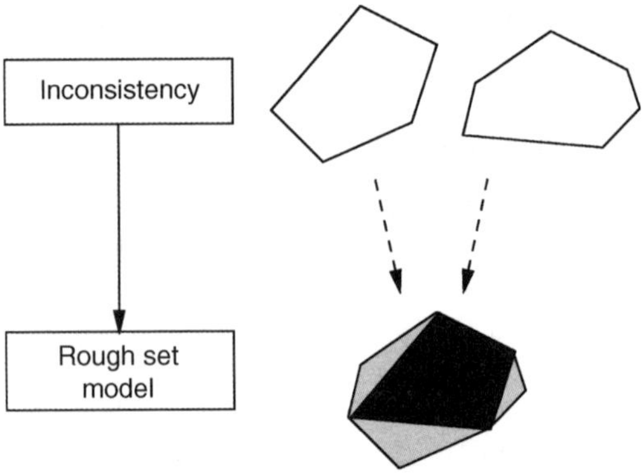

Figure 8.2 *Rough set model of inconsistency*

is possibly part of the forest (the upper approximation of the forest $\overline{S}(Forest)$). Everything outside the upper approximation is definitely not part of the forest.

Within both GIS and databases more generally, inconsistency is usually resolved or eradicated at the earliest opportunity, and the underlying observations are discarded. The development of relational database systems set particular store by consistency issues (Date, 1990). Over recent years, the growth in usage of object-oriented (OO) databases, modelling and software development techniques has in part been due to the improved support for consistency afforded by OO systems. Unfortunately, the information transformations used to resolve such inconsistency are commonly unidirectional. It will not usually be possible to deduce the original inconsistent observations after the inconsistency has been resolved (i.e., we cannot travel against the direction of the arrows in Figures 8.1 and 8.2).

In light of this, it makes sense to consider developing database systems able to support inconsistent information, alongside the functionality to derive multiple realisations of that information using a range of different models of imperfection. A number of authors have already highlighted the importance of the retention of original surveyed observations in data quality management (Heuvelink, 1998). The concept of a *measurement-based GIS* has been proposed as a system capable of storing these possibly inconsistent surveyed observations and resolving any inconsistency on-the-fly using automated survey adjustment techniques (Goodchild, 1999). Along the same lines, the more general concept of a *consistency-based GIS* should be able to store inconsistent geographic information as well as provide various mechanisms based on the formal models described above in order to derive multiple consistent realisations of that information. Such a consistency-based GIS would have several advantages from a computational perspective:

- Parsimony: no information is discarded by the system, so information loss is minimised.
- Multiple-realisations: multiple different realisations of the stored information can be generated based on different formal models applied to the inconsistent information in the database.

- Context sensitivity: by retaining all available information it is possible to produce different on-the-fly realisations of the inconsistency suited to a particular user's context.
- Updates: new information can be more easily added to the database, and revised realisations generated in the light of this new information.

Arguably the most important of these from a computational perspective is the information loss associated with premature resolution of inconsistency (Cholvy and Hunter, 1997).

8.5 Computational and Critical Convergence

The discussion so far has attempted to provide an overview of current work on uncertainty, and outline some of the reasons why access to a range of formal methods for dealing with inconsistency is increasingly important. While this discussion has adopted a primarily computational perspective, there exist clear areas of commonality between the computational aspects of uncertainty and the issues of concern within critical geography, such as context and partial perspectives (Section 8.1).

By representing and reporting imperfections in geographic information as opposed to simply trying to remove imperfections, geoinformatics researchers are deliberately trying to provide a form of context for geographic information – to situate the knowledge produced in the context of its production. Using any of the formal models of imperfection described in Section 8.3 it is possible to communicate both knowledge about a geographic environment and meta-level contextual information about the status of that knowledge. While such formal models are still limited to the representation of quite specific types of context, they can at the very least indicate to users that the information to which it refers is not meant to be entirely 'truthful' or 'objective'. In turn, users have more opportunity to form a personal opinion on their certainty in information. This is very much in the spirit of Haraway's influential critique of the scientific discourse that claims total knowledge, in that it presents its results as being situated and partial.

The discussion in Section 8.4.1 argues that there are clear computational reasons for wanting to take these elements of context a step further. By adapting GIS architectures to allow the storage and manipulation of inconsistent information, it should be possible to increase the flexibility of the system and minimise loss of valuable information. While motivated by computational concerns, the idea of using rather than attempting to resolve away inconsistency in GIS is an implicit acceptance of the conflict and difference characteristic of real world situations. Developing such a GIS able to operate using inconsistent information would seem to be a first step on the road to pluralism and the inclusion of diverse partial perspectives into the way we use geographic information.

While inconsistency can be related to diversity, the production of a single, consistent, consensual view of geographic information remains central to many computational, logical and decision making processes. A consistency-based GIS aims to provide the functionality to resolve inconsistency and produce consistent realisations of information when necessary. Currently this task is performed on behalf of the GIS user, usually by the data producer or surveying organisation that captured the data. The uni-directional nature of the information transformations represented in Figures 8.1 and 8.2 effectively 'locks in' one model of uncertainty with the data. Providing users with the mechanisms necessary to perform these transformations for themselves enables users to construct multiple consistent realisations of information, using different formal models. In turn, this may help users achieve a greater understanding of the characteristics and limitations of a dataset.

At the same time, particular users or applications may require particular models of uncertainty. Some models, such as fuzzy set theory, offer much greater room for the inclusion of subjectivity. Others, like the stochastic model, are much more prescriptive. With a consistency-based GIS, there is no need for the generic 'one-size-fits-all' approach to uncertainty common in contemporary geographic information provision (for example, simply providing a global RMSE value with digital elevation model). Users can select which model(s) of uncertainty are appropriate to their particular expertise, application or preferences.

The ideas put forward do not require a revolution in geoinformatics, nor do they suggest that critical geographic thought is without relevance to GIS. Indeed, it is highly unlikely that the reinterpretation of existing technology sketched in this paper would be radical enough to satisfy many of the more strident critics of GIS. Rather, this paper is an attempt to draw out some of the links between the two disciplines. Sieber (2000) argues that the relationship between GIS and grass roots organisations needs to be re-examined to account for the sometimes beneficial (as well as detrimental or conforming) effects of GIS. Sieber (2000) points to examples where grass roots organisations have effected their own changes upon the GIS technology, rather than the converse, in accordance with their own objectives. In a similar vein, this paper has argued that some of the levers needed by critical geographers to influence the future development of GIS already exist, and are waiting to be pulled. Working with existing research trends in computation and geoinformatics may enable critical geography to be much more effective in directing that development in the future.

Acknowledgements

This chapter was written while Matt Duckham was working at the Department of Computer Science, Keele University, UK, on the EPSRC funded project 'Managing Vagueness, Uncertainty and Granularity in Spatial Information', GR/M 56685. He would like to acknowledge valuable comments from Antony Galton, Nadine Schuurman, John Stell, David O'Sullivan, Mike Worboys and members of the original ESRC meeting at Loughborough organised by David Unwin in the preparation of this chapter.

References

Abler, R., 1993, Everything in its place: GPS, GIS, and geography in the 1990s, *Professional Geographer*, **45**(2), 131–139.

Ahlqvist, O., Keukelaar, J. and Oukbir, K., 2000, Rough classification and accuracy assessment. *International Journal of Geographical Information Science*, **14**(5), 475–496.

Altman, D., 1994, Fuzzy set theoretic approaches for handling imprecision in spatial analysis, *International Journal of Geographical Information Systems*, **8**(3), 271–289.

Bennett, B., 2001, What is a forest? On the vagueness of certain geographic concepts, *Topoi*, **20**, 189–201.

Besnard, P., del Cerro, L., Gabbay, D. and Hunter, A., 1997, Logical handling of inconsistent and default information. In: *Uncertainty Management in Information Systems*, edited by Smets, P. and Motro, A., Boston, Kluwer, ch. 11, pp. 325–341.

Bondi, L. and Domosh, M., 1992, Other figures in other places: on feminism, postmodernism and geography, *Environment and Planning D: Society and Space*, **10**(2), 199–213.

Bossard, M., Feranec, J. and Otahel, J., 2000, *CORINE land cover technical guide – Addendum 2000*, Tech. rep., European Environment Agency, Technical report No. 40.

Burrough, P. and McDonnell, R., 1998, *Principles of Geographical Information Systems*, second edition. Clarendon, Oxford.

Cargile, J., 1969, The sorites paradox. *British Journal for the Philosophy of Science*, **20**, 193–202.

Cholvy, L. and Hunter, A., 1997, Information fusion in logic: A brief overview. In: *Qualitative and Quantitative Practical Reasoning* (ECSQARU'97/FAPR'97) Lecture Notes in Computer Science, 1244, Springer-Verlag, Berlin, pp. 86–95.

Chrisman, N., 1983, The role of quality information in the long term functioning of a geographic information system, *Cartographica*, **21**(2/3), 79–87.

Date, C., 1990, *An Introduction to Database Systems*, fifth edition, vol. 1. Addison-Wesley, Massachusetts.

Duckham, M., Mason, K., Stell, J. and Worboys, M., 2001, A formal approach to imperfection in geographic information, *Computers, Environment and Urban Systems*, **25**(1), 89–103.

Duckham, M. and Worboys, M. Computational structure in three-valued nearness relations. In: *Spatial Information Theory: COSIT '01*, edited by Montello, D., 2001, vol. 2205 of Lecture Notes in Computer Science, Springer-Verlag, Berlin, pp 76–91.

Ehlers, M. and Shi, W., 1975, Error modelling for integrated GIS, *Cartographica*, **33**, 1 (1996), 11–21.

Fine, Vagueness, K., truth and logic, *Synthese*, **30**, 265–300.

Fisher, P., 1999, Models of uncertainty in spatial data. In: *Geographical Information Systems*, edited by Longley, P., Goodchild, M., Maguire, D. and Rhind, D., second edition, vol. 1, Longman, Essex, ch. 13, pp. 191–205.

Fisher, P., 2000, Sorites paradox and vague geographies, *Fuzzy Sets and Systems*, **113**(1), 7–18.

Goodchild, M., 1995, Sharing imperfect data. In: *Sharing Geographic Information*, edited by Onsrud, H. and Rushton, G., Rutgers, New Jersey, pp. 413–425.

Goodchild, M. 1999, Measurement-based GIS. In: *Proceedings of the International Symposium on Spatial Data Quality*, edited by Shi, W., Goodchild, M. and Fisher, P. London: Taylor and Francis, pp. 1–9.

Goodchild, M., Egenhofer, M., Kemp, K., Mark, D. and Sheppard, E., 1999, Introduction to the Varenius project, *International Journal of Geographical Information Science*, **13**(9), 731–745.

Goodchild, M. and Gopal, S. Eds., 1989, *Accuracy of Spatial Databases*, London: Taylor and Francis.

Goodchild, M. and Jeansoulin, R., 1998, Editorial. *GeoInformatica*, **2**(3), 211–214.

Gopal, S. and Woodcock, C., 1994, Theory and methods for accuracy assessment of thematic maps using fuzzy sets, *Photogrammetric Engineering and Remote Sensing*, **60**(2), 181–188.

Gore, A., 1998, *The Digital Earth: Understanding our planet in the 21st century*, http://www.digitaearth.gov/VP19980131.html. Accessed 15th August 2001.

Haraway, D., 1988, Situated knowledges: The science question in feminism and the privilege of the partial perspective, *Feminist Studies*, **14**, 575–599.

Harris, T. and Weiner, D., 1996, *GIS and society: The social implications of how people, space and environment are represented in GIS*, Tech. rep., National Center for Geographic Information and Analysis, NCGIA report 96–7.

Harvey, F., 2000, The social construction of geographical information systems, *International Journal of Geographical Information Science*, **14**(8), 711–713.

Heuvelink, G., 1998, *Error Propagation in Environmental Modelling with GIS*, London: Taylor and Francis.

Journel, A. and Huijbregts, C. J., 1978, *Mining Geostatistics*, London: Academic Press.

Keefe, R. and Smith, P., 1996, Introduction: Theories of vagueness. In: *Vagueness: A reader*, edited by Keefe, R. and Smith, P. Eds. MIT Press, Massachusetts, ch. 1, pp. 1–57.

Kwan, M-P., 2002, Introduction: Feminist geography and GIS, *Gender, Place and Culture*, **9**(3), 261–262.

Lake, R., 1993, Planning and applied geography: positivism, ethics and geographic information systems, *Progress in Human Geography*, **17**(3), 404–413.

Leung, Y., 1987, On the imprecision of boundaries, *Geographical Analysis*, **19**(2), 125–151.

Leung, Y. and Yan, J., 1998, A locational error model for spatial features, *International Journal of Geographical Information Science*, **12**(6), 607–620.

Longley, P., Goodchild, M., Maguire, D. and Rhind, D. Eds., 2001, *Geographic Information Systems and Science*, John Wiley & Sons, Inc., New York.

Oberthür, T., Dobernmann, A. and Aylward, M., 2000, Using auxiliary information to adjust fuzzy membership values for improved mapping of soil qualities, *International Journal of Geographical Information Science*, **14**(5), 431–454.

Pawlak, Z., 1982, Rough sets, *International Journal of Computer and Information Science*, **11**, 341–356.

Peitgen, H., Jurgens, H. and Saupe, D., 1992, *Chaos and Fractals: New Frontiers of Science*, Springer-Verlag, New York.

Pickles, J., 1999, Arguments, debates, and dialogues: The GIS-social theory debate and the concern for alternatives. In: *Geographical Information Systems*, edited by Longley, P., Goodchild, M., Maguire, D. and Rhind, D., second edition, vol. 1, John Wiley and Sons, Inc., New York, ch. 4, pp. 49–60.

Roy, A. and Stell, J., 2001, Spatial relations between indeterminate regions, *International Journal of Approximate Reasoning*, **27**(3), 205–234.

Schuurman, H., 1999, Speaking with the enemy? An interview with Michael Goodchild. Environmental and Planning D: Society and Space, **17**(1), 1–15.

Sheppard, E., 1995, GIS and society: towards a research agenda. *Cartography and Geographic Information Systems*, **22**(1), 5–16.

Sheppard, E., Couclelis, H., Graham, S., Harrington, J. and Onsrud, H., 1999, Geographies of the information society, *International Journal of Geographical Information Science*, **13**(8), 797–823.

Shi, W., 1998, A generic statistical approach for modelling error of geometric features in GIS, *International Journal of Geographical Information Systems*, **12**(2), 131–143.

Sieber, R., 2000, Conforming (to) the opposition: the social construction of geographic information systems in social movements, *International Journal of Geographical Information Science*, **14**(8), 775–793.

Usery, L., 1996, A conceptual framework and fuzzy set implementation for geographic features. In: *Geographic Objects with Indeterminate Boundaries*, edited by Burrough, P. and Frank, A., London: Taylor and Francis, ch. 5, pp. 71–85.

Veregin, H., 1999, Data quality parameters. In: *Geographical Information Systems*, edited by Longley, P., Goodchild, M., Maguire, D. and Rhind, D. Eds, second edition, vol. 1, London, Wiley pp. 177–189.

Woodcock, C. and Gopal, S., 2000, Fuzzy set theory and thematic maps: Accuracy assessment and area estimation, *International Journal of Geographical Information Science*, **14**(2), 153–172.

Worboys, M., 1998, Computation with imprecise geospatial data, *Computers, Environment and Urban Systems*, **22**(2), 85–106.

Worboys, M., 2001, Nearness relations in environmental space, *International Journal of Geographical Information Science*, **15**(7), 633–651.

Worboys, M. and Clementini, E., 2001, Integration of imperfect spatial information, *Journal of Visual Languages and Computing*, **12**(1), 61–80.

Part II

Not Just Space

9

Not Just Space: An Introduction

Michael Batty, Antony Galton and Marcos Llobera

9.1 Geographical Space Is...

9.1.1 ...Not Just a Mathematical Continuum

One of the great achievements of the western intellectual tradition has been to pin down an abstract notion of *bare space* as an infinitely extended and infinitely subdivisible continuum in which each point can be specified by means of a tuple of numerical coordinates. This conception of space has proved to be enormously fruitful in providing a formal framework within which many aspects of the physical world can be modelled. Not the least of its achievements has been to provide a vehicle for supporting geographical knowledge in the form of map projections precisely specified in mathematical terms.

There is much more to geographical knowledge, however, than the specification of positions by means of coordinates, and this has been recognised, at least implicitly, throughout the development of geography. Nonetheless, this traditional mathematical conception of space is widely perceived as constituting the dominant influence in providing a theoretical basis for GI Science, and it is the purpose of Part II to draw attention to a number of ways in which it falls dramatically short of the kind of rich and highly structured conceptions of space that are required to do justice to all the concerns of geography.

9.1.2 ... But Structured by Qualitative Relations Amongst Regions and Objects

Geographical space is divided into regions and populated by many kinds of objects – indeed, regions themselves may be conceived as abstract objects, and their existence is

Re-presenting GIS Edited by P. Fisher and D. Unwin

entirely dependent on the existence of more concrete objects. And just as the notion of an object is problematic, so too is the notion of a region.

Assuming, however, that we have somehow arrived at a clear conception of objects, regions, and their properties and relations, we have in place the ingredients for a conception of space that is altogether richer than the bare mathematical continuum: it is a space that is at least in part *constituted* by the objects and regions it contains. To the extent that objects and regions are social constructs (and there can be no denying that at least *some* are), so too will be this richer conception of space. This is all too obvious in the case of the partition of land into political and administrative regions, which are created entirely by human fiat, with different social groups sometimes disagreeing, with tragic effects, on the right way to effect the division.

Leaving aside such issues for the time being, consider what might be an appropriate vocabulary in which to carry out a discourse about regions and objects in space. Going beyond the purely quantitative language of coordinates and their associated mathematical techniques, everyday language provides a rich vocabulary of terms for spatial properties and relations at the level of objects and regions, terms such as 'near' and 'far', 'left' and 'right', 'north', 'south', 'east' and 'west', 'inside' and 'outside', 'visible' and 'hidden', 'whole' and 'part', 'separated', 'in contact', and 'overlapping'. These terms are all qualitative in nature, and the forms of reasoning and representation that employ them may likewise be described as qualitative. Although in everyday language the rules governing their use are unstated and perhaps not precisely statable, it is evident that our usually fluent mastery of these terms forms an important component of our everyday 'naive' spatial knowledge. It has therefore been recognised that if GI Science is to engage with this body of knowledge, it is necessary to formulate adequate theories of these qualitative spatial terms and the relationships between them.

Some examples of such work are the Region Connection Calculus of Randell *et al.*, (1992) and the 4-intersection and 9-intersection frameworks of Egenhofer and Franzosa (1995), both of which handle qualitative connection relations between spatial regions. Also, the work on cardinal directions (Freksa, 1992), qualitative distance (Hernández *et al.*, 1995), ordering (Schlieder, 1995; Isli and Cohn, 1998; Balbiani and Osmani, 2000), qualitative coordinates (Kulik and Klippel, 1999), connectivity (Galton, 2000), lines of sight (Galton, 1994; Randell *et al.*, 2001), and various aspects of shape (Schlieder, 1996; Galton, 2000). Much of this work has been pursued in the context of Artificial Intelligence and Cognitive Science, both of these fields having recognised, virtually since their inception, that there is a need for systematic approaches to the formalisation of qualitative representation and reasoning. Galton's paper (Chapter 10) subsumes work of this kind under a general Qualitative State Space (QSS) paradigm; the chapter notes that although this approach has attracted a wide following and has produced some interesting results, it is not yet clear how far it will be applicable to the concerns of GI Science. There is certainly a need for further research into this issue but our theories of space need to recognise other paradigms and means of representation.

9.1.3 ...With Complex Interactions

As soon as space is partitioned, the mathematical continuum loses its purity but acquires a degree of richness that is represented by sets of relations where space itself is composed of discrete and identifiable objects. Simple partitions based on adjacency determine rudimentary networks which, if regular, provide a basis for classifying tessellations which tile the plane. The emphasis in representing spatial objects as tessellations, regular or otherwise, is not however on their relations *per se*; it is more on the nature of their boundaries and on the efficiency of space packing. In fact, relations that imply more than simple adjacency, which is a local property, such as those essential to diffusion and interaction, are more likely to be based on action-at-a-distance where space is once again treated as a continuum. Yet this is the image of a theoretical geography that was crafted a generation or more ago, and the advantage of treating space discretely, so that attributes of uniqueness can be associated with such objects, has raised our awareness that more complex sets of relations and interactions need to be defined.

Our focus is now much more on studying the dual of the spatial problem – the network of connections that ties spatial objects together. In fact if attention is transferred away from spatial objects *per se* to the networks that bind them together, then a whole new world of interactions opens up. Indeed it might be argued that the intrinsic property of space is not the point but the line that links points together – the network that provides space with its relative location and differentiates it from similar elements that form its composition. Networks also embody processes in that specific interactions take place across them and within them, while the network can also be seen as the basis for change in the spatial composition and its associated attributes. The need for this kind of refocusing strikes at the very heart of GI Science. In the past, GI Science has been peculiarly deficient at dealing with network systems. Systems of relations and processes of interaction defined between locations are hard, if not impossible, to represent in commercial or proprietary GIS. This makes transportation analysis, for example, difficult to enable, despite network and routing algorithms being featured in conventional systems. Just as GIS has never broached the time dimension, it has rarely dealt with spatial interaction, instead concentrating on location, location, location.

Much of this is changing, if only for the reason that the substantive systems to which GI Science is applicable are increasingly ones in which networks of relations predominate. The world is rapidly becoming global in that from every local position, global action can be initiated. Hierarchies and networks dominate our lives. In the last 30 years, the world has switched from a concern for central action and control to decentralised inter-action. The network has quickly become the icon of our times, just as systems with local action have become the baseline models for our analysis. Kevin Kelly (1994) in his book *Out of Control* sums this up admirably when he says: 'The symbol of science for the next century is the dynamical Net'. Explorations and insights into the structure of space in formal terms have until quite recently been largely the prerogative of those working with the continuum where various physical and geometrical principles based on scaling and interaction have been derived. For example, densities and diffusion of all kinds seem to follow scaling rules, and in the human domain these have been articulated in various theories of social physics, which form the cornerstone of location theory and regional science.

Network structures, on the other hand, have been largely restricted to descriptive science. Geographic researchers working with the theory of graphs, for example, using them to find order in social and transportation networks, have largely confined themselves to descriptive analysis, where the emphasis has been on developing measures of structure based on counting elements within graphs which are ordered with respect to proximity. Most of the measures that we have are based on clustering, path lengths, density of nodes and so on. Only recently has this begun to change as researchers define different types of network structures which pertain to different types of problems. Two important characterisations of network structure are guiding these developments. The first mirrors the search for structure in social physics and depends on ideas in scaling while the second defines a particular class of networks in which local and global proximity is maximised; these have been called *small worlds* (Watts, 1999). Much of this analysis is being driven by an intense interest in the fast evolving, organically growing network of networks – the Internet – which seems to scale in ways that are very similar to scaling in other distributions such as city sizes, word frequencies, and incomes. In short, the nodes in the Internet scale in terms of power laws with very few really large nodes and very many small ones. Nodes, of course, depend on links and thus the analysis is of in-degrees and out-degrees measured by such elements as hyperlinks in web pages and physical links between hardware. The Internet also seems to be a small world where any member can reach any other relatively quickly but at the same time remain part of a dense local clustered group. This can be formally measured for networks in terms of the average path length and the clustering index which need to be compared to other baseline network structures such as random graphs for the small world quality to be appreciated.

Small worlds clearly characterise many networks of interest to GI Science. Classic examples are social networks where weak ties bind distant peoples together, while the very structure of cities and regions and the hierarchy of settlement from metropolis to village can be articulated in network terms as small worlds. Within cities, the ability to reach distant people while at the same time remaining part of a localised clustered group – the village in the city – depends on transportation links, and as cities grow, new technologies of movement ensure that they remain coherent and that proximity is maximised. The emergence of global cities depends on maintaining this proximity, and air travel and the Internet are among recent transportation technologies that keep the world together. Much spatial analysis is being informed by this network paradigm. One of the best developed areas is spatial epidemiology, which is now being extended to deal with network structures. The traditional ideas about the diffusion of epidemics locally are being supplanted by ideas – borne out in reality – which suggest that epidemics often spread on networks that are small worlds. The ways in which peoples and animal populations interact is not simply local but in accordance with networks that have global proximate qualities.

These new ideas are reviewed by Batty (Chapter 11), but our appreciation of spatial structure within GI Science has as yet barely begun to embrace them. What is clearly required in this science is a synthesis with existing ideas of spatial representation: first, in the domain of the mathematical continuum, where classical interaction based on diffusion has not been present in the representational side of this science; and second, ideas about how the continuum is structured in terms of networks need to be taken on board. It is here that ideas about network scaling and small worlds will find their place, while at the same time extending the science to deal not only with the real and the immediately observable but with the less visible – with cyberspace and the many related spaces that are now identifiable as our theory begins to define them. This, of course, relates to how we perceive space, and the way we interpret it through various theories of cognition. It takes us back to qualitative relations and the ideas of Galton (Chapter 10) but it also takes us on to space, spatial objects, relations, and interactions which are ...

9.1.4 ...Perceived by Humans

Why, as I follow a certain mountain trail, does a 12th century polychrome church seem to rise out of nowhere? How come this happens following the path in one direction and not in the other? Where, along the path, does this feeling emerge? Generally: why and how do people encounter the world as they do? These are all questions that have been the focus of extensive narratives in social and cultural geography, and in other disciplines such as anthropology, archaeology and sociology, but that have yet to be fully addressed by GI Science.

Instead, the majority of GI Science initiatives (e.g. NCGIA'S *Varenius* project) have been concerned with *retrieving* and *formalising* spatial concepts that humans employ when they refer to, and/or navigate in, the world. This has been done in the hope of providing a theoretical cognitive foundation for GIS, eventually leading towards the design of new, more intuitive, interfaces that would enable a wider range of people to use and manipulate spatial information. These studies are based on and follow the approach predicated by cognitive science, an approach grounded mostly on the assumption that people learn and operate in the world by means of mental representations (*schemata* or models) generated from information of the 'outside' world obtained through the senses. Thus, cognitive theory presupposes that all information is representational, in some way or another, and that it may be readily transmitted. According to this view, people share a common view of the world, i.e. perceive the same environment, because they manage to share the same mental *schemata*. This way of thinking, ultimately originating with Descartes, maintains a strict separation between the mind and the body, favouring the latter.

In the past few years, this view has been challenged on several fronts: in anthropology, geography, artificial intelligence, philosophy (see Bourdieu, 1990; Ingold, 2000; Thrift, 1996), particularly when it relates to how we operate in the world in the context of everyday life. This alternative view is based on the notion that people engage with the world on the basis of their previous relationships with the environment. People share similar, not identical, perceptions because they share the context (i.e., time-space and motivation) in which these relationships are generated, not necessarily the content itself. Thus, for the most part, they operate in the world using a kind of practical knowledge which is best understood as the knowledge we associate with a skill, knowledge that has been forged through the repeated performance of routinised daily activities and learnt through trial-and-error. This is the type of knowledge that we employ on our daily tasks, which by and large is not represented in our heads, or communicated through any sort of formal instruction. Like any skilful practitioner, we conduct our daily life not so much guided by rules, propositions or any other mental representations but by consulting directly with the world. This is made possible by a sense one has built from previous experiences of a particular situation. Whenever a situation deviates from the optimal body–environment relationship established previously, we tend to 'guide' our activity closer towards our acquired optimum in order to alleviate this tension.

Based on this latter view, it is possible to propose an alternative approach to the one currently found among most GI Science initiatives and studies focused on human spatial experience. In this new approach the human body is not peripheral, as claimed by Mark *et al.* (1999, p. 757), but central to our exploration of space. The incorporation or the simulation of the body, in any possible measure, is an essential component to understand how we perceive and experience the world, for it is precisely our physical make-up that conditions the way we encounter the world. It is the reason why, in a gathering, we usually maintain social communication by being at the edge of a circle rather than in the centre, and why neural networks deprived of a human-like body, without up/down and front/back distinctions, will never be capable of building an understanding of the world similar to ours (Dreyfus, 1995).

Emphasis is on using and developing GI Science concepts and techniques that will help us understand how perception is constructed out of the interaction between the body and the environment, rather than on finding an overarching set of universal mental spatial concepts. Thus, the stress is on developing and employing GI Science methods in their heuristic, exploratory mode rather than in their prescriptive one, and exploring the structure of the world (in our case a simplified version of it) as it is encountered by an individual *within* it, moving through it. At a very basic level, as a starting point, this entails the adoption of a *situated* and *mobile* frame of reference: the development of a *body-centred* geography. Such a perspective is implicit, or rather essential, in the praxis of architects, landscape designers and urbanists, and forms the foundation of subjects like ecological psychology, but with a few exceptions (Thiel, 1997) it has never been the basis for the development of concepts and methods aimed at describing or analysing space. To this day, most spatial analytical techniques depend on a single fixed frame of reference and on two-dimensional spatial representations. The possibility, and indeed the desire, to adopt non-traditional spatial representations have already been acknowledged by Raper (2001) as a vital element in the development of GI Science.

The adoption of this perspective, and subsequent development, will become an endeavour that can only be successful if tackled by an interdisciplinary effort. This process, however, may well help to bridge over some of the existing divide currently present between GI Science and contemporary spatial theory, as well as precipitate many new insights and challenges (both theoretical and technical). Llobera (Chapter 12) illustrates this thesis with examples, focusing on the study of visual space, which are used to illustrate some of these possibilities.

References

Balbiani, P. and Osmani, A., 2000, A model for reasoning about topological relationships between cyclic intervals, In: *Proceedings of the Seventh International Conference on Principles of Knowledge Representation and Reasoning (KR2000)*, edited by Cohn A.G., Giunchiglia F. and Selman B. eds, San Francisco, CA: Morgan Kaufmann, pp. 378–385.

Bourdieu, P., 1990, *The Logic of Practice*, Cambridge: Polity Press.

Brooks, R., 1991, 'Intelligence without representation', *Artificial Intelligence*, **47**, pp. 139–159.

Clark, A., 1997, *Being There: Putting Brain, Body, and World Together Again*, Cambridge, MA: MIT Press.

Dreyfus, H.L., 1992, *Being-in-the-World: A Commentary on Heidegger's Being and Time, Division I*, Cambridge, MA: MIT Press.

Dreyfus, H.L., 1995, The current relevance of Merleau-Ponty's phenomenology of embodiment, *Filozofska Istrazivanja*, **15**(3), pp. 385–399.

Egenhofer, M.J. and Franzosa, R.D., 1995, On the equivalence of topological relations, *International Journal of Geographical Information Science*, **9**, 133–152.

Freksa, C., 1992, Using orientation information for qualitative spatial reasoning, In: *Theories and Methods of Spatio-Temporal Reasoning in Geographic Space*, Vol. 639 of Lecture Notes in Computer Science, edited by Frank, A.U., Campari, I. and Formentini, U. Berlin: Springer-Verlag, pp. 162–178.

Galton, A.P., 1994, Lines of sight, In: *AI and Cognitive Science'94*, edited by Keane, M., Cunningham, P., Brady, M. and Byrne, R., Dublin University Press, pp. 103–113.

Galton, A.P., 2000, *Qualitative Spatial Change*, Oxford: Oxford University Press.

Hernández, D., Clementini, E. and Di Felice, P., 1995, Qualitative distances, in Frank, A.U. and Kuhn, W. eds, *Spatial Information Theory: A Theoretical Basis for GIS*, Vol. 988 of Lecture Notes in Computer Science, New York: Springer-Verlag, pp. 45–57.

Ingold, T., 2000, *The Perception of the Environment: Essays in Livelihood, Dwelling and Skill*, London: Routledge.

Isli, A. and Cohn, A.G., 1998, Qualitative spatial reasoning à la Allen: An algebra for cyclic ordering of 2D orientations, In: *Proceedings of the 12th International Workshop on Qualitative Reasoning*, edited by Zhao, F. and Yip, K., AAAI Press, pp. 65–76.

Kelly, K., 1994, *Out of Control: The New Biology of Machines*, London: Fourth Estate.

Kulik, L. and Klippel, A., 1999, Reasoning about cardinal directions using grids as qualitative coordinates, In: *Spatial Information Theory: Cognitive and Computational Foundations of Geographic Science*, edited by Freksa, C. and Mark, D.M., New York: Springer-Verlag, pp. 205–220.

Mark, D.M., Freska, C., Hirtle, S.C., Lloyd, R. and Tversky, B., 1999, Cognitive models of geographical space, *International Journal of Information Science*, **13**(8), pp. 747–774.

Randell, D.A., Cui, Z. and Cohn, A.G., 1992, A spatial logic based on regions and connection, In: *Principles of Knowledge Representation and Reasoning: Proceedings of the Third International Conference* (KR'92), edited by Nebel, B., Rich, C. and Swartout, W., San Mateo, CA: Morgan Kaufmann, pp. 165–76.

Randell, D., Witkowski, M. and Shanahan, M., 2001, From images to bodies: Modelling and exploiting spatial occlusion and motion parallax, In: *Proceedings of the Seventeenth International Joint Conference on Artificial Intelligence* (IJCAI'01), edited by Nebel, B., San Francisco, CA: Morgan Kaufmann, pp. 57–63.

Raper, J., 2000, *Multidimensional Geographic Information Science*, London: Taylor and Francis.

Schlieder, C., 1995, Reasoning about ordering, In: *Spatial Information Theory: A Theoretical Basis for GIS*, Vol. 988 of Lecture Notes in Computer Science, edited by Frank, A.U. and Kuhn, W., New York: Springer-Verlag, pp. 341–349.

Schlieder, C., 1996, Qualitative shape representations, In: *Geographic Objects with Indeterminate Boundaries*, edited by Burrough, P.A. and Frank, A.U., London: Taylor and Francis, pp. 123–140.

Thiel, P., 1997, *People, Paths and Purposes. Notations for a Participatory Envirotecture*, Seattle and London: University of Washington Press.

Thrift, N., 1996, *Spatial Formations*, Theory, Culture and Society Series. London: Sage.

Watts, D.J., 1999, *Small Worlds: The Dynamics of Networks Between Order and Randomness*, Princeton, NJ: Princeton University Press.

10

The QSS Framework for Modelling Qualitative Change: Prospects and Problems

Antony Galton

10.1 What is Qualitative Change?

It is not easy to pin down exactly what is meant by 'qualitative' (Galton, 2000b). If we focus on change, we might say that a qualitative change happens when a different verbal, non-numerical description of a state of affairs applies at one time but not at another; for example, at one time Great Britain was physically connected to mainland Europe, and now it is not. The emphasis on verbal descriptions ties in with the issue of salience: we give different descriptions to situations whose difference is salient for us. If the only way we have of characterising the difference between two situations is by means of measurement then that is an indication that the difference is not salient. Salience can to some extent be equated with importance: salient differences are important for us (we speak of 'differences in degree' and 'differences in kind' and are particularly interested when the former turns into the latter). This is not to say that purely quantitative distinctions are unimportant but their importance belongs to a more technical setting.

One approach to this is to consider the following relation on some set of 'situations': *there is no salient difference between situation X and situation Y.* I shall write this 'X *qsame* Y' ('X is qualitatively the same as Y'). This will be context-relative, of course, because salience is a function of contexts, among other things (context includes: who is *using* the information – e.g., some colour differences are not salient to the colour-blind; also, what is the information being used for?). The relation *qsame* is obviously reflexive

Re-presenting GIS Edited by P. Fisher and D. Unwin

and symmetric. An important question is whether or not it is transitive. The answer is that sometimes it is and sometimes it is not.

An example of non-transitive *qsame* is the distinction between 'upland' and 'lowland': the one grades into the other. As we pass from the plain to the hills, any two points sufficiently close are not saliently different as regards elevation (they are *qsame*), yet points far enough apart are. Non-transitive *qsame* relations correspond to gradations in reality. Changes in such relations are typically gradual, with no sharp transitions.

An example of a transitive *qsame* comes from the distinction between 'connected' and 'disconnected'. With respect to this distinction, any situation in which region A is connected to region B is *qsame* to any other situation in which these regions are connected; and situations in which A is disconnected from B are *qsame* to each other but not to any of the former class of situations. At least, this is the case if we accept the idealisation that connectedness is an all-or-nothing affair, with any two regions being determinately either connected or disconnected. In this case, changes in transitive *qsame* relations will be sudden, giving the appearance of discontinuity even though all the underlying physical changes are continuous.

Since *qsame* is always reflexive and symmetric, a transitive *qsame* will be an equivalence relation, partitioning its domain into a set of jointly exhaustive and pairwise disjoint classes (the equivalence classes). Within each class, all situations are qualitatively the same, but between classes they are qualitatively different. When a situation changes, there is no qualitative change so long as the changes are confined to one equivalence class, but as soon as a class boundary is crossed, there is a sudden discontinuous change. The discontinuity comes from the discreteness of the set of descriptions we apply to the phenomena (corresponding to the equivalence classes); it does not contradict the continuity of the underlying phenomena, which emerges when we go beyond qualitative description and describe the situation in terms of precise numerical measurements.

Non-transitive *qsame* relations can be made transitive by drawing sharp boundaries, which may be more or less arbitrary. An example is elevation bands, e.g. 0–99 m, 100–199 m, 200–299 m, and so on. For some purposes we might regard all points within a given band as having the same 'qualitative elevation'; on a map, different colours may be assigned to different bands; or the boundaries between the bands may be indicated by contour lines. If the elevation at a point changes, then even if in reality the exact numerical value of the elevation changes continuously, the resulting changes in qualitative elevation will be sudden, occurring at isolated instants as the boundaries between the bands are crossed. These bands do not correspond to any salient features in reality: it is a kind of artificial salience imposed on the continuous reality. This can be helpful but of course it can also be misleading.

10.2 The Qualitative State Space Framework

In recent years, the investigation of these transitive *qsame* relations – though not under that description – has been a major endeavour within the Qualitative Spatial Reasoning community (which has affiliations with Artificial Intelligence, Computer Vision, Cognitive Science, Linguistics, and Geographical Information Science). The general programme may be described as follows: for each spatial attribute of interest to one or other of these communities, first define a qualitative version of the attribute, then investigate its 'conceptual neighbourhood' structure; that is, determine which values of the attribute are neighbours in the sense that one can be transformed directly to the other under continuous change in the underlying quantitative attribute. To illustrate this using our artificial example of 'qualitative elevation', the value 0–99 m is a conceptual neighbour of the value 100–199 m because the former can change directly into the latter under conditions of continuous increase in elevation; but it is not a conceptual neighbour of the value 200–299 m.

Many of the attributes that have been investigated have been relational in character, that is the attribute applies to a situation in which there are two (or possibly more) objects of interest and we want to describe how they are related. The example of connectedness is of this kind: a situation for which the applicability of the attribute 'connected' is in question must contain two objects, such that we are interested in whether one is connected to the other. For such relational attributes, another important area of investigation has been into how they behave under composition. For example, if A is connected to B, and B is connected to C, what can we say about the connectedness or otherwise of A and C? In this case, the answer is 'nothing', but in other cases significant inferences may be drawn: for example, if A contains B as a part, and B is connected to C, then A must be connected to C as well. In general, investigating questions of this kind can provide further insights into the nature of the qualitative relations under consideration and, it is hoped, form a worthwhile basis for automating the process of reasoning about them. Note, however, that for the purpose of describing qualitative spatial change these composition tables are less important than the conceptual neighbourhood diagrams.

The qualitative descriptors that occupy the nodes of conceptual neighbourhood diagrams may be regarded as representing possible *states* of the system under consideration; the arcs of the diagram represent possible qualitative transitions between these states. Thus, the system described by a conceptual neighbourhood diagram may be described as a Qualitative State-Space (QSS). Since qualitative states are sometimes referred to as modes, the term Mode Space may also be used. General properties of such systems have been investigated from a topological point of view by Ligozat (1994, 1999) and Galton (1997, 2000a, 2000b, 2001a).

In the following subsections, I survey briefly a range of work falling under this general description. For a more complete overview, see Cohn and Hazarika (2001).

10.2.1 Qualitative Physics

Qualitative physics (Weld and de Kleer, 1990) is a branch of Artificial Intelligence devoted to qualitative representation and reasoning about physical systems. It includes the important idea of an *envisionment* of a system's qualitative behaviour. System states that are qualitatively identical (from some point of view) are collected together into a single *qualitative state* or *mode* (Davis, 1990). As the system evolves continuously, it undergoes abrupt transitions from mode to mode, but not all mode transitions are physically possible. An *envisionment diagram* for the system contains a node for each mode, with directed arcs joining modes between which a direct transition is possible. A simple example is the envisionment diagram for a simple pendulum, described in terms of its qualitative position and velocity at each moment (Figure 10.1). Each of these

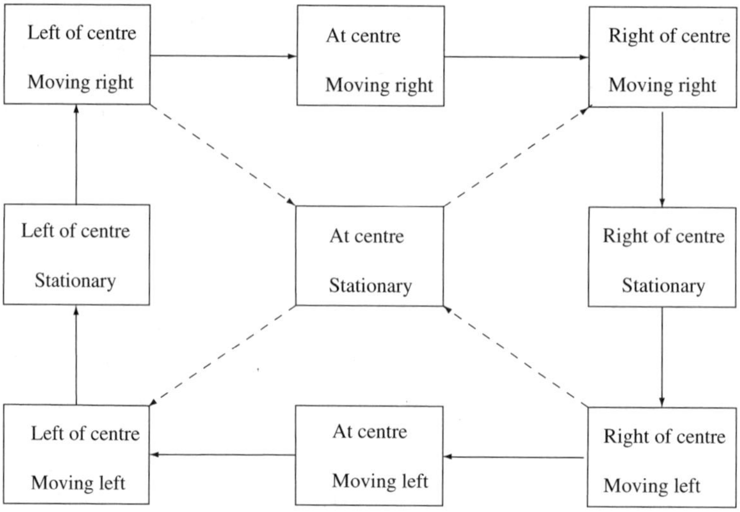

Figure 10.1 *Envisionment diagram for a simple pendulum*

attributes can in principle take any real value, but qualitatively what matters is whether they are positive (measured to the right from the equilibrium position, say), negative or zero, giving nine distinct modes altogether. In the diagram, the solid arrows show the trajectory followed through the state space when the only force acting is gravity; the broken arrows show additional possibilities in the presence of additional imposed forces (e.g., damping or externally imposed impulses).

10.2.2 Mereotopological Relations on Regions

'Mereotopology' refers to the abstract study of parthood and connection. A considerable amount of work has been done on this in the context of philosophy (ontology), cognitive science, and artificial intelligence (Smith, 1993, 1994; Asher and Vieu, 1995; Pratt and Lemon, 1997; Cohn and Varzi, 1998). Some of this work falls explicitly within the QSS framework.

The Region Connection Calculus (RCC). Randell *et al.* (1992) identified a set of eight jointly exhaustive and pairwise disjoint relations on spatial regions. The relation of connection is taken as the primitive in terms of which the other relations are defined. A key auxiliary relation is 'part of', which is defined by the rule: *A is part of B if every region connected to A is also connected to B*. Six of the eight relations are now defined as follows:

- A and B are *disconnected* (DC) if they are not connected.
- A and B are *externally connected* (EC) if they are connected but have no common part.
- A and B are *partially overlapping* (PO) if they have a common part but neither is a part of the other.
- A and B are *equal* (EQ) if each is a part of the other.
- A is a *tangential proper part* of B (TPP) if A is part of, but not equal to, B, and there is a region that is externally connected to both A and B.
- A is a *non-tangential proper part* of A (NTPP) if A is a part of, but not equal to B, and no region is externally connected to both A and B.

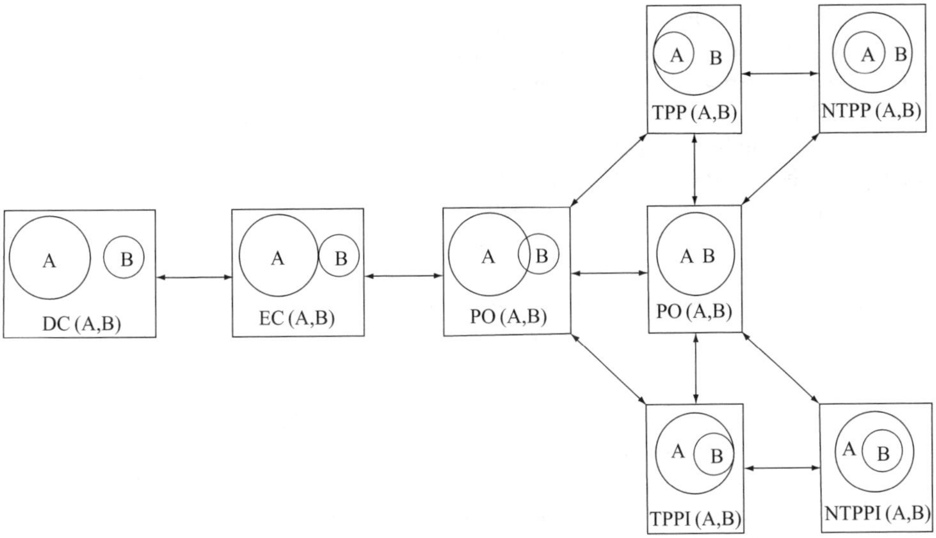

Figure 10.2 *The conceptual neighbourhood diagram for the Region Connection Calculus (RCC)*

The remaining two relations are the inverses of TPP and NTPP, denoted TPPI and NTPPI. This system of eight relations is called the Region Connection Calculus (RCC), and comes with a conceptual neighbourhood diagram (Figure 10.2) and a composition table. It has been used to provide qualitative descriptions of scenarios involving spatial change such as the absorption of food by an amoeba (Cui *et al.*, 1992) and the operation of a force pump (Randell and Cohn, 1992). This work has been extended to handle convexity (Cohn *et al.*, 1997) and vague regions (the 'egg-yolk' theory – see p. 140).

The 9-intersection Framework. Egenhofer and Franzosa (1991) developed a similar calculus from a different starting point: the qualitative mereotopological relations between regions in the plane are classified by considering the intersections between the interior, boundary and intersection of one region with the interior, boundary, and intersection of the other. This gives a matrix of nine entries, each of which is either 'empty' or 'non-empty'. The resulting set of relations is called the 9-intersection framework. An example is the relation in which the exterior of A only intersects the exterior of B, the boundary of A intersects both the exterior and boundary of B, while the interior of A intersects the interior, boundary and exterior of B: this relation is called 'covers' in Egenhofer's terminology, but it is identical to the RCC relation TPPI ('has as a tangential proper part'). In Egenhofer's matrix notation, this is represented as

$$\begin{pmatrix} 1 & 0 & 1 \\ 1 & 1 & 1 \\ 0 & 0 & 1 \end{pmatrix}$$

where the three rows correspond to the boundary, interior and exterior of A and the columns to the boundary, interior and exterior of B; the entries 0 and 1 mean 'empty intersection' and 'non-empty intersection' respectively.

The 9-intersection framework was developed from the simpler, and correspondingly less expressive, 4-intersection framework (Egenhofer, 1989), in which only the intersections between interiors and boundaries are considered. As with RCC, these relations are structured by conceptual neighbourhood, and this structure is used for reasoning about changes in the qualitative relations between regions arising from continuous changes in the regions themselves (Egenhofer and Al-Taha, 1992).

The 9-intersection framework has been used to characterise relationships between regions in discrete space (Egenhofer and Sharma, 1993), and Egenhofer has investigated the 9-intersection relation on the sphere, in which more relations are possible (e.g., with two disjoint hemispheres, the interior of each intersects only the exterior of the other, and the two boundaries intersect only each other – a relation not possible for bounded regions in the plane).

Refinements to Handle Multiple Overlap. Egenhofer and Franzosa (1995) consider a more expressive formalism which is powerful enough to characterise relations between planar regions up to full topological equivalence. Whereas the 9-intersection framework does not, for example, distinguish regions that overlap in a single connected component from regions that overlap in more than one component, the more expressive formalism of Egenhofer and Franzosa (1995) does. Of course, in order to characterise the topological character of a relation completely in this way, the formalism is of necessity rather cumbersome, and it is questionable whether it would be useful in practical situations. A compromise notation, closely modelled on the 4-intersection framework, but much more expressive, was presented by Galton (1998). Instead of considering interior and boundary, this system uses interior and exterior; and instead of simply saying whether or not two components overlap, we say in how many connected components they do so. An example is shown in Figure 10.3, where in the left-hand diagram region A overlaps B in two distinct components in such a way that the part of A outside B and the part of B outside A have one component each, while the region exterior to both A and B has two components; by contrast, in the right-hand diagram, all four areas have two components each. Galton (1998) emphasised strongly the topic of change, investigating in detail the nature of the possible paths through the conceptual neighbourhood diagram.

Regions with Indeterminate Boundaries. Further complications arise from considering relations between regions that are not themselves determinately specified. Cohn and Gotts (1996) considered the case of regions characterised in terms of 'inner' and 'outer'

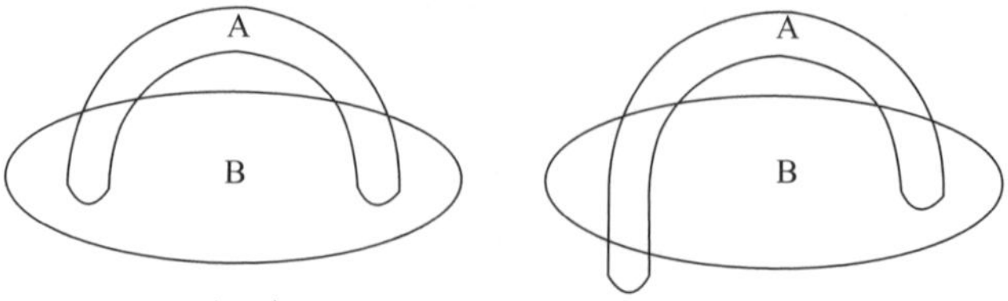

$$\begin{pmatrix} 2 & 1 \\ 1 & 2 \end{pmatrix} \qquad \begin{pmatrix} 2 & 2 \\ 2 & 2 \end{pmatrix}$$

Figure 10.3 *Counting connected components to determine modes of overlap* boundaries, such that everything within the inner boundary is definitely in the region, and everything outside the outer boundary is definitely outside it, the status of the area between the two boundaries being regarded as in some way indeterminate (e.g., because it is genuinely intermediate in character, or because there is lack of information about where the true boundary lies, or lack of clear criteria for determining this). The region is thought of as an egg, with the inner boundary separating the yolk from the white. This is similar to the point of view of Rough Set Theory (Pawlak, 1982, 1986), but the basis is provided by mereotopology rather than set theory. The relationship between two regions arises from the relationship between the yolk and white of one and the yolk and white of the other. There are various dependencies among these, leading to a total of 46 relations in all, duly pinned down in the framework of a conceptual neighbourhood diagram and a composition table.

A similar, but subtly different, scheme was devised by Clementini and Di Felice (1996). They applied Egenhofer's 9-intersection framework, but on the understanding that the boundary of each region may be 'broad', that is, ribbon-like rather than strictly of zero breadth. This leads to a set of 44 relations rather than the 46 of Cohn *et al.* (1997). But in both schemes a conceptual neighbourhood diagram is produced, and not surprisingly the two diagrams differ only in minor points of detail.

10.2.3 Miscellaneous Other Work

Other areas in which the Qualitative State Space or similar frameworks have been used in spatial reasoning include:

- cardinal directions (Frank, 1992, 1996; Freksa, 1992);
- qualitative distance (Hernández *et al.*, 1995);
- ordering (Schlieder, 1995; Isli and Cohn, 1998; Balbiani and Osmani, 2000);
- qualitative coordinates (Kulik and Klippel, 1999);
- connectivity (Galton, 2000a);
- lines of sight (Galton, 1994; Randell *et al.*, 2001);
- various aspects of shape (Schlieder, 1996, Galton and Meathrel, 1999, Galton, 2000b, Gottfried, 2003).

10.3 Can the QSS Framework be Applied to GIS?

Work of the kind described in the previous section could, in principle, be used to support high-level spatial reasoning tasks such as:

1. inferring new spatial relations from a given set of qualitative spatial data (using the composition tables);
2. interpolating possible/likely qualitative change histories between given end-points (using the conceptual neighbourhood diagrams);
3. predicting possible future qualitative changes from the present situation (using conceptual neighbourhood diagrams);
4. checking consistency in a qualitative spatio-temporal database (using both composition tables and conceptual neighbourhood diagrams).

For many of these domains of qualitative spatial description, these ideas remain at best programmatic at the current stage of research. In particular, the Qualitative State Space framework has not, to the best of my knowledge, so far been incorporated in any commercially viable system.

In the remainder of this section, I shall examine some fundamental questions concerning how the work described in the previous section can contribute to Geographic Information Systems development.

10.3.1 Why Should We Work with Qualitative Information?

Several justifications for seeking to work with purely qualitative information are commonly given and I discuss three of these below. It is not clear how far these proposed justifications stand up to detailed examination, especially in the context of developing practical applications.

Unavailability of Quantitative Data. Data for input into a geographical information system might take the form of natural language descriptions without any precise numerical coordinates or other values. As currently constituted, GISs are unable to handle data of this kind; what is needed is a way of representing qualitative data and of making valid (or plausible) inferences from such data.

Cognitive Salience of Qualitative Data. Even if quantitative data are available in abundance, the human user of an information system wants to be presented with results that are intelligible without the necessity of detailed further analysis. Visualisation in the form of maps, charts and diagrams will often help here, but in many cases something more closely akin to a natural language description is appropriate, and for this we need to be able to present the salient features of the underlying quantitative data in qualitative terms. This will not be possible unless the system itself has some ability to compute with qualitative, as opposed to numerical, data.

Computational Complexity of Working with Quantitative Data. The idea here is that working at the quantitative level is often unnecessary on the grounds that the desired results can be obtained much more simply, and with smaller data storage requirements, by means of purely qualitative reasoning. This is perhaps belied by the known complexity results for qualitative reasoning in the QSS framework (e.g., consistency checking in RCC is NP-complete (Renz and Nebel, 1999)), although a certain amount of work has been done in identifying tractable sub-problems of computationally hard spatial reasoning problems (Wolter and Zakharayaschev, 2000).

10.3.2 Relationship between Low-level and High-level Data

A good deal of geographical data are in the form of arrays of continuous field-values (raster data), whereas from an informatic perspective it is often more natural to think in terms of discrete objects and their attributes (vector data) (Peuquet, 1984, Couclelis, 1992, Galton, 2001a, 2001b). Raster data are usually more quantitative in nature: not only are the field values typically numerical (and in most cases free to vary continuously), but the spatial locations to which they are assigned are also specified numerically (e.g., by means of grid coordinates). Vector data *may* contain numerical elements (e.g., exact specification of location as one of an object's attributes), but it is not intrinsically a quantitative format and indeed lends itself much more readily to qualitative analysis.

How does the QSS framework fit into this dichotomy? Suppose we wish to reason about regions using the RCC relations. First, a region is an object and is naturally specified using vector data. It may be derived from some underlying raster data, but the RCC theory itself has nothing to say about this: the assumption is that the regions are already given as objects before the RCC theory gains purchase. Similar remarks apply to all the systems within the QSS framework. The question therefore arises as to how such systems can operate in the context of 'real' data, that is, data *collected* from the world rather than fabricated at a computer terminal.

Imagine a geographical information system that is able both to handle low-level raster data and to perform high-level reasoning along the lines suggested by the systems developed within the QSS framework – one might for example be interested at looking at the changing relationships between various regions defined in terms of vegetation cover, land use, human population, etc. Such a system requires the capability of extracting suitable high-level qualitative information from the raster data. This is not simply a matter of raster-to-vector conversion – a process that has been the subject of considerable research effort and, as I understand it, is still not performable in a clean and reliable way. There are also the subtler questions as to how to identify which regions are of interest, how they should be characterised, and how they are related in terms of the available higher-level descriptors. None of this is addressed by the QSS framework as such; rather it is taken for granted that the high-level descriptions are available.

Even if the process of converting low-level data to high-level information is successfully (or even partially) automated, it is far from clear whether high-level reasoning on the outputs from such conversions will deliver information that could not just as efficiently be extracted directly from the (assumed available) low-level data. To make the point as sharply as possible, I give a highly simplistic example: if we know that region A is a proper part of region B, which is disjoint from region C, then we can infer that region A is disjoint from region C. If we are able to extract these relations from low-level data, then having extracted the two premises of the inference we can reason qualitatively to draw the conclusion; but in this situation we could surely have extracted the conclusion directly from the low-level data! Although this is a very simple-minded example, the point it illustrates would probably apply more widely; it would appear that although a good deal of research effort has gone into developing high-level reasoning systems based on the QSS framework (including work by the present author!), their potential usefulness or otherwise for operating in the context of real geographical information systems has yet to be adequately evaluated.

An alternative scenario is one in which information is entered initially into the system in a suitably high-level form for QSS-based systems to work on it. This is what is done in practice in most actual implementations of QSS-based systems. Any connection with lower-level data involving numerical values assigned to actual locations has to be mediated through the human users of the system, and once again the benefit of automating the high-level reasoning is not clear. The supposed justifications (ability to handle uncertain or incomplete information, efficiency of working at the higher-level, etc.) have already been discussed.

10.3.3 Handling the Time Dimension

As discussed above, reasoning with conceptual neighbourhoods enables one, in principle, to handle qualitative change. Once again, however, there are important questions as to how the high-level reasoning engages with real data.

By analogy with the raster and vector data models for spatial information, there appear to be two main ways in which the time dimension can be handled in an information system (Galton, 2001b). In 'raster time', temporal information is presented in the form of a temporally indexed succession of 'snapshots': the world at t_1, the world at t_2, etc. In 'vector time', on the other hand, temporal information is presented in the form of named states and/or events, each including among its attributes a location in time, either absolute or in relation to other states or events. Exactly analogous to the problem of picking out discrete individuals from a set of raster data is the problem of identifying discrete events from a sequence of snapshots. The converse problem is that of determining the complete state of the world at different times, given only the information about, say, an initial state and the events that have occurred subsequently.

Once again, the QSS is pre-eminently apt for event-based descriptions, the events being the transitions between different qualitative states. Any QSS-based system can provide a basis for describing possible sequences of events of this kind and perhaps also for using such descriptions for reasoning about events (e.g., for the purposes of prediction or retrodiction). But as in the pure spatial case, there are problems about where the event data are supposed to come from in the first place. Indeed, when we consider that we might want to handle both snapshot-based and event-based temporal data relating to changes in both location-based and object-based spatial data, it is easy to see that a wealth of different possibilities opens up. These do not appear to have been investigated in a systematic and principled way by people working at incorporating time in GIS.

A further possibility might be to develop a unified framework within which the time-dimension is handled exactly on a par with the spatial dimensions, the objects of study being 'object-histories', coherent chunks of space-time corresponding to the life-histories of what on a static view we would normally consider to be complete objects. For a general discussion of some of the issues involved, see Galton (2004). An interesting philosophical perspective on the various ways in which the spatio-temporal aspects of the world can be conceptualised is provided by Zemach (1970); philosophical defences of integrated spatio-temporal theories include Heller (1990) and Sider (2001). In the QSS framework, unified spatio-temporal theories have been proposed by Muller (1998) and Hazarika and Cohn (2001). In a quite different context, something of this sort seems to be envisaged by Massey (1999), who cites the geomorphological modelling work of Raper and Livingstone (1995) as an important attempt to integrate the temporal dimension into a spatial information system in a more thoroughgoing way than is usual. The representations implied by these theories must be considered as very high-level: by its nature, a single object-history encapsulates data that do not co-exist at any one time or in any one place, and thus the passage from raw low-level data to the high-level representation involves a degree of abstraction exceeding anything considered up to now.

10.4 Conclusions

In conclusion, it has to be said that although the QSS approach has attracted a wide following and has produced some interesting and elegant results, its claims to be of practical use to GIS remain to be adequately evaluated. There is a need for research into how QSS-based systems can serve GIS, and at the present time this is more important for GIS than the further elaboration of the QSS paradigm itself. The papers in this book identify a number of crucial concepts for the future development of GIS, and many of these – for example, indeterminacy, multiple representations, and the integration of space and time – are also of specific relevance to the applicability of QSS. There is wide scope here for the fruitful interaction between those working in QSS and researchers in other areas of GIS.

References

Asher, N. and Vieu, L., 1995, Toward a geometry of common sense: A semantics and a complete axiomatisation of mereotopology, In: *Proceedings of the Fourteenth International Joint Conference on Artificial Intelligence (IJCAI'95)*, edited by Mellish, C.S., Morgan Kaufmann, San Mateo, CA, pp. 846–852.

Balbiani, P. and Osmani, A., 2000, A model for reasoning about topological relationships between cyclic intervals, In: *Proceedings of the Seventh International Conference on Principles of Knowledge Representation and Reasoning (KR2000)*, edited by Cohn, A.G., Giunchiglia, F. and Selman, B., Morgan Kaufmann, San Francisco, CA, pp. 378–385.

Clementini, E. and di Felice, P., 1996, An algebraic model for spatial objects with indeterminate boundaries, In: *Geographic Objects with Indeterminate Boundaries*, edited by Burrough, P.A. and Frank, A.U., Taylor and Francis, London, pp. 155–169.

Cohn, A.G. and Gotts, N.M., 1996, The 'egg-yolk' representation of regions with indeterminate boundaries, In: *Geographic Objects with Indeterminate Boundaries*, edited by Burrough, P. and Frank, A.U., Taylor and Francis, London, pp. 171–187.

Cohn, A.G. and Hazarika, S.M., 2001, Qualitative spatial representation and reasoning: an overview, *Fundamenta Informaticae*, **46**, 1–29.

Cohn, A.G. and Varzi, A.C., 1998, Connection relations in mereotopology, In: *Proceedings of the Thirteenth European Conference on Artificial Intelligence (ECAI'98)*, edited by Prade, H., Wiley, Chichester, pp. 150–154.

Cohn, A.G., Bennett, B., Gooday, J. and Gotts, N.M., 1997, Representing and reasoning with qualitative spatial relations about regions, In: *Spatial and Temporal Reasoning*, edited by Stock, O., Kluwer Academic Publishers, Dordrecht, pp. 97–134.

Couclelis, H., 1992, People manipulate objects (but cultivate fields): Beyond the raster-vector debate in GIS, In: *Theories and Methods of Spatio-Temporal Reasoning in Geographic Space*, edited by Frank, A.U., Campari, I. and Formentini, U., Lecture Notes in Computer Science 639, Springer-Verlag, Berlin, pp. 65–77.

Cui, Z., Cohn, A.G. and Randell, D.A., 1992, Qualitative simulation based on a logical formalism of space and time, In: *Proceedings of the Tenth National Conference on Artificial Intelligence (AAAI-92)*, AAAI Press/MIT Press, Menlo Park, CA, pp. 679–684.

Davis, E., 1990, *Representations of Commonsense Knowledge*, Morgan Kaufmann, San Mateo, CA.

Egenhofer, M.J., 1989, A formal definition of binary topological relationships, In: *Third International Conference on Foundations of Data Organisation and Algorithms (FODO)*, edited by Litwin, W. and Schek, H., Lecture Notes in Computer Science 367, Springer-Verlag, New York, NY, pp. 457–472.

Egenhofer, M.J. and Al-Taha, K.K., 1992, Reasoning about gradual changes of topological relationships, In: *Theories and Methods of Spatio-Temporal Reasoning in Geographic Space*, edited by Frank, A.U., Campari, I. and Formentini, U., Lecture Notes in Computer Science 639, Springer-Verlag, Berlin, pp. 196–219.

Egenhofer, M.J. and Franzosa, R.D., 1991, Point-set topological spatial relations, *International Journal of Geographical Information Science*, **5**, 161–174.

Egenhofer, M.J. and Franzosa, R.D., 1995, On the equivalence of topological relations, *International Journal of Geographical Information Science*, **9**, 133–152.

Egenhofer, M.J. and Sharma, J., 1993, Topological relations between regions in R^2 and Z^2, In: *Advances in Spatial Databases – Third International Symposium on Large Spatial Databases SSD'93'*, edited by Abel, D. and Ooi, B.C., Lecture Notes in Computer Science 639, Springer-Verlag, New York, NY, pp. 316–336.

Frank, A.U., 1992, Qualitative spatial reasoning about distances and directions in geographic space, *Journal of Visual Languages and Computing*, **3**, 343–371.

Frank, A.U., 1996, Qualitative spatial reasoning: cardinal directions as an example, *International Journal of Geographical Information Science*, **10**, 269–290.

Freksa, C., 1992, Using orientation information for qualitative spatial reasoning, In: *Theories and Methods of Spatio-Temporal Reasoning in Geographic Space*, edited by Frank, A.U., Campari, I. and Formentini, U., Lecture Notes in Computer Science 639, Springer-Verlag, Berlin, pp. 162–178.

Galton, A.P., 1994, Lines of sight, In: AI and Cognitive Science '94', edited by Keane, M., Cunningham, P., Brady, M. and Byrne, R., Dublin University Press Proceedings of the Seventh Annual Conference, September 8-9, 1994, Trinity College Dublin, pp. 103–113.

Galton, A.P., 1997, Space, time and movement, In: *Spatial and Temporal Reasoning*, edited by Stock, O., Kluwer Academic Publishers, Dordrecht, pp. 321–352.

Galton, A.P., 1998, Modes of overlap, *Journal of Visual Languages and Computing*, **9**, 61–79.

Galton, A.P., 2000a, Continuous motion in discrete space, In: *Proceedings of the Seventh International Conference on Principles of Knowledge Representation and Reasoning (KR2000)*, edited by Cohn, A.G., Giunchiglia, F. and Selman, B., Morgan Kaufmann, San Francisco, CA, pp. 26–37.

Galton, A.P., 2000b, *Qualitative Spatial Change*, Oxford University Press, Oxford.

Galton, A.P., 2001a, Dominance diagrams: A tool for qualitative reasoning about continuous systems, *Fundamenta Informaticae*, **46**, 55–70.

Galton, A.P., 2001b, A formal theory of objects and fields, In: *Spatial Information Theory: Foundations of Geographical Information Science*, edited by Montello, D., Lecture Notes in Computer Science 2205, Springer-Verlag, NY, pp. 458–473.

Galton, A.P., 2004, Fields and objects in space, time, and space-time, *Spatial Cognition and Computation*, **4**, 39–67.

Galton, A.P. and Meathrel, R.C., 1999, Qualitative outline theory. In: *Proceedings of the Sixteenth International Joint Conference on Artificial Intelligence, IJCAI 99*, edited by Thomas Dean, Morgan Kaufmann, San Francisco, CA, pp. 1061–1066

Gottfried, B., 2003, Tripartite line tracks: Qualitative curvature information, In: *Spatial Information Theory: Foundations of Geographic Information Science*, edited by Kuhn, W., Worboys, M. and Timpf, S., Lecture Notes in Computer Science 2825, Springer-Verlag, New York, NY, pp. 101–117.

Hazarika, S.M. and Cohn, A.G., 2001, Qualitative spatio-temporal continuity, In: *Spatial Information Theory: Foundations of Geographical Information Science*, edited by Montello, D., Lecture Notes in Computer Science 2205, Springer-Verlag, NY, pp. 92–107.

Heller, M., 1990, *The Ontology of Physical Objects: Four-dimensional Hunks of Matter*, Cambridge University Press, Cambridge.

Hernández, D., Clementini, E. and Di Felice, P., 1995, Qualitative distances, In: *Spatial Information Theory: A Theoretical Basis for GIS*, edited by A.U. Frank and Kuhn, W., Lecture Notes in Computer Science 988, Springer-Verlag, New York, NY, pp. 45–57.

Isli, A. and Cohn, A.G., 1998, Qualitative spatial reasoning à la Allen: An algebra for cyclic ordering of 2D orientations, In: *Proceedings of the 12th International Workshop on Qualitative Reasoning*, edited by Zhao, F. and Yip, K., AAAI Technical Report WS-98-01, AAAI Press, pp. 65–76.

Kulik, L. and Klippel, A., 1999, Reasoning about cardinal directions using grids as qualitative coordinates, In: *Spatial Information Theory: Cognitive and Computational Foundations of Geographic Science*, edited by Freksa, C. and Mark, D.M., Springer-Verlag, New York, NY, pp. 205–220.

Ligozat, G., 1994, Towards a general characterization of conceptual neighbourhoods in temporal and spatial reasoning, In: Proceedings of the AAAI-94 Workshop on Spatial and Temporal Reasoning, American Association for Artificial Intelligence, pp. 55–59.

Ligozat, G., 1999, Simple models for simple calculi, In: *Spatial Information Theory: Cognitive and Computational Foundations of Geographic Science*, edited by Freksa, C. and Mark, D.M., Springer-Verlag; Berlin, pp. 173–188.

Massey, D., 1999, Space-time, 'science', and the relationship between physical geography and human geography, *Transactions of the Institute of British Geographers*, **NS 24**, 261–276.

Muller, P., 1998, *Eléments d'une théorie du mouvement pour la formalisation du raisonnement spatio-temporel de sens commun*, Ph.D. thesis, Université Paul Sabatier, Toulouse III.

Pawlak, Z., 1982, Rough sets, *International Journal of Computer and Information Sciences*, **11**, pp. 341–356.

Pawlak, Z. 1986. *Rough Sets: Theoretical Aspects of Reasoning about Data*, Kluwer Dordrecht, The Netherlands.

Peuquet, D.J., 1984, A conceptual framework and comparison of spatial data models, *Cartographica*, **21**(4), 66–113.

Pratt, I. and Lemon, O., 1997, Ontologies for plane, polygonal mereotopology, *Notre Dame Journal of Formal Logic*, **38**(2), 225–245.

Randell, D.A. and Cohn, A.G., 1992, Naive topology: modelling the force pump, In: *Advances in Qualitative Physics*, Struss, P. and Faltings, B., MIT Press, Cambridge, MA, pp. 177–192.

Randell, D.A., Cui, Z. and Cohn, A.G., 1992, A spatial logic based on regions and connection, In: *Principles of Knowledge Representation and Reasoning: Proceedings of the Third International Conference (KR'92)*, edited by Nebel, B., Rich, C. and Swartout, W., Morgan Kaufmann, San Francisco, CA, pp. 165–176.

Randell, D., Witkowski, M. and Shanahan, M., 2001, From images to bodies: Modelling and exploiting spatial occlusion and motion parallax, In: *Proceedings of the Seventeenth International Joint Conference on Artificial Intelligence (IJCAI'01)* edited by Nebel, B., Morgan Kaufmann, San Francisco, CA, pp. 57–63.

Raper, J. and Livingstone, D., 1995, Development of a geomorphological spatial model using object-oriented design, *International Journal of Geographical Information Systems*, **9**(4), 359–383.

Renz, J. and Nebel, B., 1999, On the complexity of qualitative spatial reasoning, *Artificial Intelligence*, **108**, 69–123.

Schlieder, C., 1995, Reasoning about ordering, In: *Spatial Information Theory: A Theoretical Basis for GIS*, edited by Frank, A.U. and Kuhn, W., Lecture Notes in Computer Science 988, Springer-Verlag, New York, NY, pp. 341–349.

Schlieder, C., 1996, Qualitative shape representations, In: *Geographic Objects with Indeterminate Boundaries*, edited by Burrough, P.A. and Frank, A.U., Taylor and Francis, London, pp. 123–140.

Sider, T., 2001, *Four-dimensionalism: An Ontology of Persistence and Time*, Clarendon Press, Oxford.

Smith, B., 1993, Ontology and the logistic analysis of reality, In: *Proceedings of the International Workshop on Formal Ontology in Conceptual Analysis and Knowledge Representation*, edited by Guarino, N. and Poli, R., Institute for Systems Theory and Biomedical Engineering of the Italian National Research Council, Padova, Italy, pp. 51–68.

Smith, B., 1994, Topological foundations of cognitive science, In: *Topological Foundations of Cognitive Science*, edited by Eschenbach, C., Habel, C. and Smith, B., Kognitionswissenschaft, Hamburg, pp. 3–22.

Weld, D. S. and de Kleer, J. eds, 1990, *Readings in Qualitative Reasoning about Physical Systems*, Morgan Kaufmann Publishers, Inc., San Mateo, CA.
Wolter, F. and Zakharayaschev, M., 2000, Spatio-temporal representation and reasoning based on RCC-8, In: *Proceedings of the Seventh International Conference on Principles of Knowledge Representation and Reasoning (KR2000)*, edited by Cohn, A.G., Giunchiglia, F. and Selman, B., Morgan Kaufmann, San Francisco, CA, pp. 3–14.
Zemach, E. M., 1970, Four ontologies, *Journal of Philosophy*, **67**, 213–247.

11

Network Geography: Relations, Interactions, Scaling and Spatial Processes in GIS

Michael Batty

11.1 Spatial Relations and GI Science

If we unpack the term 'geographic information science', it is immediately clear that its reference to 'geography' is simplistic. This is the geography of locations not relations. It is the geography of place in an absolute sense, represented by points, lines and polygons which enable attributes to be associated with these geometric objects, attributes that are largely unordered. Moreover, in terms of the word 'system', GIS does not have any of the connotations of general systems theory which is the study of basic elements and their interactions, of their parts and wholes, with the whole being 'more than the sum of the parts' (von Bertalanffy, 1968). Interactions – relations – are also key to systems. The system in GIS does not refer to the geographical system but to the way information is organised and only in this sense can relations be found, based on ways in which raw or processed data are represented in relational fashion: how points, lines, and polygons are ordered geometrically to aid efficient processing and visualisation, rather than implying order to the geography that GIS purports to represent.

These limitations to GIS are perfectly understandable given its history and purpose. But in an era when the 'system' in GIS is being translated into 'science', and in the light of rapid developments in systems theory in many fields which are changing the focus to bottom-up dynamics as the generator of more aggregate spatial order and pattern, the inability of GIS to embrace spatial relations, interactions, and their connectivity is posing

Re-presenting GIS Edited by P. Fisher and D. Unwin

a major barrier to its continued development. The rapidly developing science of networks is predicated not on the representation of static structures *per se* but on processes of change occurring on and within such structures. This problem of an appropriate representation lies at the very heart of GIS. If GIS is to be useful in articulating and operationalising contemporary geographical theory, it must not only incorporate relations but also enable such representation to be focused on processes rather than structures. For the emerging logic of how many kinds of system – from rivers to central places, from natural ecologies to cities as built forms – is one that espouses any form of equilibrium, treating evolution as processes in which relations and interactions are never at rest.

In this chapter, we will outline this network view, drawing examples mainly from cities and their transport but alluding to many other kinds of spatial system. There is currently a sea change occurring in how geographical systems are being articulated. Many systems appear to function from the bottom up in contrast to the top-down ways in which such systems have been traditionally conceived (Holland, 1996). Ideas of how structures emerge in time are central to these theories with structures being conceived as networks of components which exist through their self-organisation rather than being directed by any hidden hand. In the social sciences, cities, societies, economies are all being reworked in these terms. In the physical and biological sciences, ideas about networks and interactions embedded within intrinsically dynamic processes of change are gaining wide currency. These ideas are part of the move to define new theories of complexity but they extend to the treatment of all systems that are far-from-equilibrium which, it turns out, embraces the majority of geographical systems that GIS intends to represent. Therefore, getting such ideas into GIS is an enormous challenge for it is much easier to see GIS as providing some convenient form of visualisation, data storage and manipulation technology rather than a vehicle on which to make contemporary geographical theory applicable and practicable.

We should also note that by identifying networks of interactions and relations as having a central place in contemporary geography, necessarily this involves the temporal dimension. Traditional GIS is largely atemporal, representing locational structures at a single point in time. In so far as time has been involved, such systems simply represent a series of cross-sections in a comparative static manner, with little functionality or science being developed to deal with processes that link these cross-sections through their temporal evolution. The problems posed by these limits on our use of traditional GIS for the study of networks become clear when we begin our discussion of ways in which geographic information scientists have attempted to adapt and extend GIS to embrace interactions. As we shall see, such extensions are invariably *ad hoc*, treating interactions between locations at a snapshot in time, often using GIS simply as a means to visualise resultant relational structures. GIS is not well adapted to treating interactions although its tool box nature does allow various plug-ins to be developed that deal with networks. However, our main purpose here is to illustrate how new views of networks need to be incorporated within GIS and this means new forms of representation in time as well as space.

Thus, we turn to the emerging network paradigm and sketch its key components, illustrating how important the idea of evolution is to our understanding of locations and their relations. There is a strong spatial theme in all this which involves the unifying force of scaling. Many networks scale in time as well as space and this uniformity can be exploited in building a spatial science of networks. Although these ideas can be demonstrated easily in traditional geographic applications, we turn to two popular examples of these developments: i) ideas about the scale of networks in the form of small worlds which have local properties which generate global pattern; ii) notions about how networks mirror ways in which systems, such as the World Wide Web, grow from the bottom up, in cyberspace, which can be mapped onto the Euclidean space (the world that GIS represents). Here we will challenge GI science to deal with network representations that are intrinsically dynamic, scale with both space and time, and map onto different kinds of geometries, other than the Euclidean, suggesting that this is a major challenge in our quest for re-presenting GIS.

11.2 Spatial Interactions at the Cross-Section

All proprietary GIS reach out to represent spatial interactions in some manner although the way such relations are handled is simplistic and *ad hoc*. GIS software, which essentially is concerned with manipulating layers of map objects, is focused almost entirely on representing such map objects layer by layer and any processual simulation that occurs between these objects is largely confined to ways in which layers are related and manipulated. This rather specific view of process tends to discriminate against representations of the spatial world in which objects interact with one another across space (and through time). This is in contrast to how the attributes of any object interact with each other in space which is what map layer and map algebra approaches emphasise.

This problem is seen quite distinctly in the failure of transportation to be represented appropriately in GIS. This area, called GIS-T, has never really taken off. The recent book by Miller and Shaw (2001), which is as good a summary of the state-of-the-art in GIS-T as there is, reveals that GIS, in terms of its software, is largely peripheral to the transportation modelling, forecasting, and planning processes. Although shortest route problems have been configured within some GIS software, these have been mainly used for more partial planning problems such as location-allocation modelling which lies outside the mainstream of transportation planning. GIS has largely been influenced by transportation geographers dealing with problems of facilities location and the software that has been developed is unable to embrace the fully-fledged transportation planning process which still revolves around the four stage modelling process of trip generation, distribution, modal split and assignment. GIS at best is used in a supportive role, for organising data inputs and outputs to and from more mathematically-based software and of course for visualisation. What at first sight appears to be a synthesis of transport and GIS turns out to be a recasting of traditional and new transportation planning technologies under the general banner of GIS in its interpretation as GI science, and by GIS scientists rather than transportation engineers.

The limits to GIS in dealing with transport are seen most widely in problems of visualising interactions. Since GIS software is not well adapted to dealing with anything other than area or point locations, it has proved difficult to develop programs within GIS for displaying flows between locations, for example, which are the stock in trade of transport modelling. Desire line diagrams and assigned flows to networks are rarely a feature of visualisation in GIS. For example, de Jong and van Eck (1996) have developed specific software for enabling such flows to be represented. Their FLOWMAP software plugs into various GIS packages (http://flowmap.geog.uu.nl/); it is focused entirely on a range of problems involving the direction and volume of movement relevant to everything from migration to road traffic, and is designed to exploit standard GIS packages from which the various inputs and outputs used in visualising such flows are taken.

These peripheral attempts to develop GIS in the context of transportation treat such systems as static networks. Other software systems – which deal with networks in cities such as the street oriented analysis based on graph accessibility, a technique that is called space syntax – do not relate to GIS at all. In space syntax, the topology of street segments is used to build up a picture of accessibility at the very local level (Batty *et al.*, 1999). Distance is not a feature of these systems in any explicit fashion for it is topology and porosity of the networks that is the main concern. Space syntax is the only urban-architectural-based graph theoretic system operationalised to the point where software is available. But as land use and related area and point location data are not a central feature of such systems, the visualisation capabilities of GIS are not widely used in applications. There have been attempts to develop such graph theoretic analysis within GIS where indices of street segment access are the focus (Jiang *et al.*, 1999) and where the analysis is linked to viewsheds (Batty and Rana, 2002), but in these examples, GIS is once again peripheral (see http://www.casa.ucl.ac.uk/sanjay/software_isovistanalyst.htm). Such software has been used largely because its plug-ins and extensions enable various kinds of spatial analysis and 3D visualisation to be accomplished with relative ease.

A further example of interaction between networks and GIS is a rather new development, very much in the spirit of other chapters in this book. Okabe (2003) argues that for many spatial problems, the kind of homogeneous spatial landscape where every point and area has the same intrinsic importance is often a gross simplification of the reality under scrutiny. Many features and objects locate with respect to a landscape that is a network. Indeed one of the arguments of space syntax is that streets or lines of sight are more important to understanding how activities locate than areas, plots, or viewsheds. In short, the notion of developing spatial analysis on a network is gaining widespread interest. In such characterisations, what might appear to be a random distribution of activity in Euclidean space is often seen as being highly ordered on a network. What is random on a network is often highly ordered in Euclidean space and there are many other possibilities in between that differentially combine different networks and planar spaces. Okabe's (2003) software – SANET: Spatial Analysis on a NETwork – has been developed for such problems. Again, the software is compatible with GIS for visualisation purposes but stands independent of the representational basis of conventional GIS, which does not have the flexibility to embrace these new kinds of order that rely on the connectivity of point locations, rather than point patterns *per se*.

We should not give the impression that GIS software engineers have been unable to adapt their software to embrace ideas about networks. Much of what exists is dictated by the market place where network applications are considerably less prominent than location-based projects. However, GI scientists have not in general developed their science to embrace the idea of the network. Where this has occurred it has been peripheral to the mainstream and where there is real momentum for network applications as in transport, GIS has not responded. Systems for such planning focus much more on analysis and simulation than on representation. All this, however, is changing. The idea of the network and more particularly the evolution of networks is a science in the making, and this will force us to re-present GIS. Devising a GI science on the basis of a landscape of changing relations is likely to lie at the heart of this field in the coming years.

11.3 The Network Paradigm

In the last quarter century, there has been an enormous shift in the way we explain large scale systems in the physical and social sciences. Classical science proposed a model of the world that was essentially reductionist in form, where it was assumed that systems could be understood from the top down, by gradually disaggregating behaviour to ever finer scales, in the faith that what had already been explained at higher levels was always consistent with new explanations at the lower levels. However, wide experience in many fields suggests that systems do not appear to get any easier to understand through blind application of the reductionist principle and it is now widely regarded that some synthetic, bottom-up characterisation is essential in grasping the way large scale systems function and evolve. In one sense, this break in thinking might be said to be one that shifts the baseline from physics to biology where growth and evolution rather than structure is now the prime concern. But the shift is much wider; as human systems and societies have evolved over the last 50 years, it has become increasingly apparent that societies function from the bottom up and that top-down control can never effectively manage or control the problems that such systems manifest.

At the heart of this shift from a centralised to decentralised view of systems, is the idea that, in Adam Smith's terms, there is no 'hidden hand' guiding evolution. Systems function from the bottom up where local action, often motivated by purely local gain, gives rise to more global order. Indeed complexity theory, which is another dimension of this concern for decentralised explanation, argues that it is this uncoordinated, relatively non-ordered basis for local action that gives rise to the kind of global order that we see around us. Most systems that survive, depend on their functioning at the local level adapting to produce resilient structures which are then reflected in a relative global order. In this sense, the structures that emerge cannot be explained without recourse to the dynamics of local action. Moreover any order that is generated must be maintained and this requires energy. In fact, order is the equilibrium that we see around us although as it requires much energy to sustain it, this order is hardly an equilibrium based on least effort. It is, in fact, an order that is far-from-equilibrium in the traditional terms of classical physics and it is this that makes the focus on processes that reach, maintain and evolve these structures so essential to this new science.

One of the key signatures of complex systems is in the fact that many patterns repeat themselves at different spatial and temporal scales. If we focus on spatial scales for the moment, then simple processes that generate growth at the most local level, when applied uniformly in building up structures from the simplest seed location, generate the same pattern at successively larger scales. Such order across scales, which is marked by spatial self-similarity, is said to be scaling and the structures that result are fractals. One of the best examples is the tree-like structure that shows similarity at all scales from the tiniest branches to the entire network. Dendritic patterns such as rivers and transport systems in cities, which both grow at the local level to 'service' their surrounding space in the most minimalist fashion, provide excellent examples of growth processes that scale, are never in equilibrium in the traditional sense, and require the same continual operation of basic local rules or codes to maintain their structure.

Many models have been developed to simulate such processes, and the current fascination in urban growth modelling with cellular automata mirrors a concern for generating urban patterns from the bottom up. These models are not quite network models in that they are concerned with generating local development in restricted neighbourhoods around already developed sites. But the rules that are used routinely and repetitively tend to give rise to structures which, in their idealised form, are dendritic, and this reflects the transport of energy between developed cells which together define the growing urban structure or city. These patterns fill space at various densities producing spatial order which is fractal, self-similar on all scales or within a restricted range of scales relevant to the phenomena in question. Essentially, these models are based on local diffusion where the diffusion in question is land development configured so that the various developed sites are always connected. Network representations in such models, although implicit, do indicate the way in which formal networks underlie such complexity.

These kinds of process change the representation quite radically once formal relations between their objects are involved. In fact, proper representation of this kind of complexity is through the notion of relations embodied in networks with their interactions formed by processes that operate over such networks. The main focus in network science is no longer on searching for patterns in structures (these have been developed satisfactorily enough in GIS software through well established algorithms that generate shortest routes) but on the way such networks change and evolve. An essential measure of how such systems evolve is through the way various elements are connected. We will sketch two emerging models. The first is a model where the number of objects that form the network remains fixed but the connections between these objects evolve or change. The second model, developed in Section 11.4, is where the number of objects and their connections change through time. The first model does not deal with growth in the size of the system but with the growth or change in the number of links – the connectivity of the system. The interest in this kind of model is the effect of the change in the number of links on how complex – or how connected (how accessible if you like) – the system might become. The second model examines this too but is also concerned with the distribution of objects in terms of the strength or otherwise in their competition for new links, as new objects are introduced into the system. The latter is more general than the former.

We need some formal definitions to make sense of this problem. We first index an object by its node number i. We call each node $n_i = 1$ if it exists and $n_i = 0$ if it does not and thus there are $n = \sum_i n_i$ fixed nodes in the system. An arc between any two nodes is defined as $a_{ij} = 1$ if nodes i and j exist or 0 otherwise with the total number of nodes in the system $a = \sum_{ij} a_{ij}$. From these definitions, the average connectivity of the system is given as a/n which varies from zero, where no nodes are connected, to n where every node is connected to every other. In-degrees (incoming arcs into any node) and out-degrees (outgoing arcs from any node) are defined respectively as $a_i = \sum_j a_{ij}$ and $a_j = \sum_i a_{ij}$. Where an arc is symmetric in that it exists in both directions $a_{ij} = a_{ji}$, then the in-degree of its node is the same as its out-degree. Let us now imagine a situation in which there are a fixed number of n nodes which we can consider as places or locations with a population. Initially, there are no connections between any of these places. This might represent a world in which there is subsistence agriculture where there is no economic cooperation whatsoever. Gradually connections are made between places as farmers come to see that some form of exchange is useful.

Let us then introduce one link at each time period gradually adding links between the farmers until ultimately everyone is connected to everyone else. This is a process that generates significant change that in every sense is surprising. The best way to show this is to take a simple example and in Figure 11.1 we show a process where 12 locations – nodes – become connected, one by one, until all 12 nodes are connected to one another. The way we add links is one at a time and we choose them randomly. In terms of the average connectivity a/n, this increases linearly but if you measure the average travel distance in the network taking each link as a distance of unity, then what happens is that quite suddenly, the average distance drops as nodes become more connected.

Essentially the probability of finding a shortest path between two nodes increases rapidly as more nodes are connected gradually. In short, from a situation where most people have to traverse several nodes to get from i to j, then suddenly it becomes easier to go from i to j by ever shorter routes. In the limit when all nodes are connected to each other, the average distance is unity and everyone can reach everyone else in one step. This might be pictured as the evolution of a transport network where more and more links are built as the society gets richer and people want to travel more. But if those financing the transport network realise that you can achieve the same by simply adding more capacity to the routes, then it is possible that once everyone can reach everyone else either directly or indirectly, no more links would be built and thus a kind of critical level of connectivity might be reached. In this sense, we might picture the evolution of the transport network as being self-organised to this critical level.

In Figure 11.1, we show several stages of network construction for a system composed of $n = 12$ nodes. Figure 11.2 shows a graph of the number of arcs used to connect nodes (out of a total of $a = n(n-1)/2 = 66$ for 12 nodes where we do not count the self connections from i to i) and the average distance travelled. Since up to a given level, some nodes are not reachable from any other, then with a small number of nodes, the average distance in the graph is infinity. To get over this problem, we assume that when we begin, the average distance to go anywhere from anywhere is equal to 12 units. When two nodes are connected, then the arc distance for the nodes in question is equal to one and then the average distance reduces quite rapidly. In fact, what happens is that quite suddenly we reach a threshold where every node is connected to every other directly or indirectly and then the average distance travelled from any node to any other stabilises. This is the critical threshold. After this, adding more links does not change the average distance very much. What happens to our hypothetical 12 node network is shown in Figure 11.2 where we also plot the graph of the average distance. Note how it stabilises when the network reaches 17 links, at which point it becomes strongly connected. In fact the network can have a total of 66 two-way links, the ratio of network density varying in the same way as the number of arcs or connectivity.

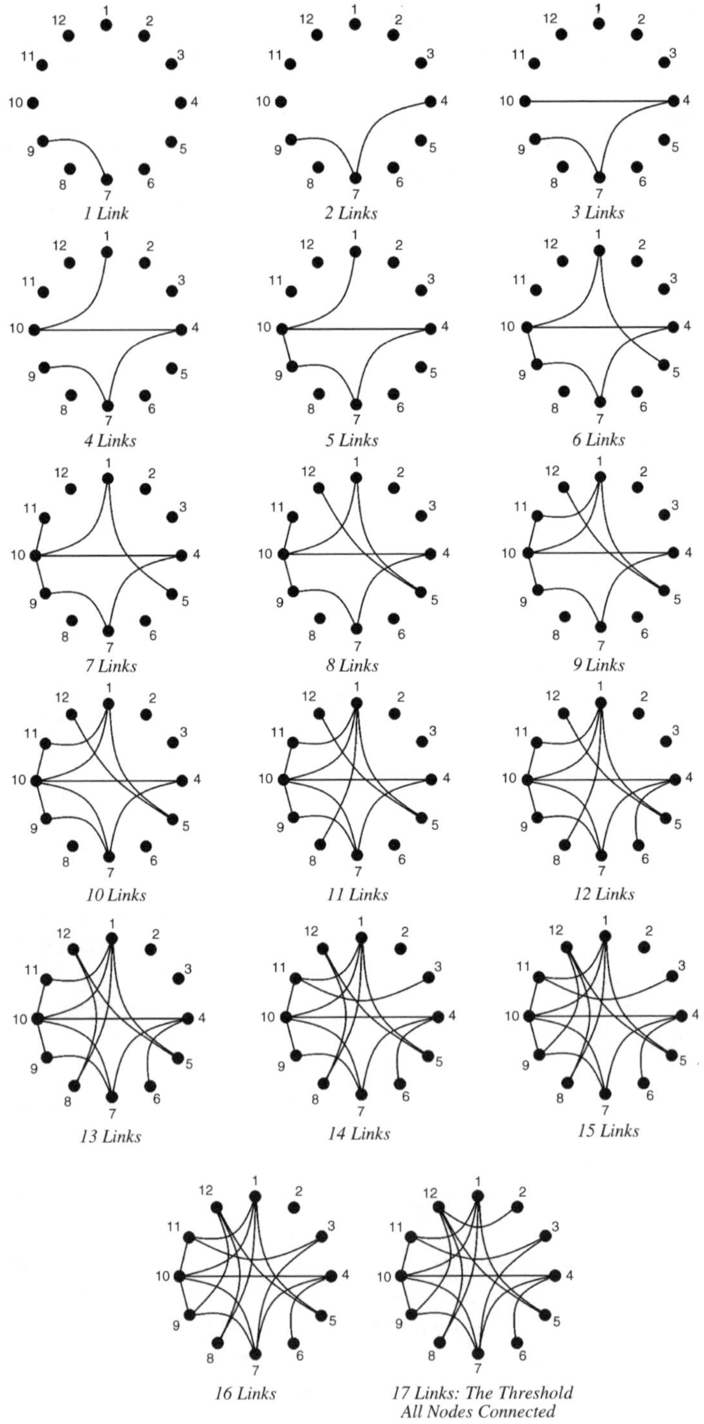

Figure 11.1 *Increasing connectivity in a 12-node graph*

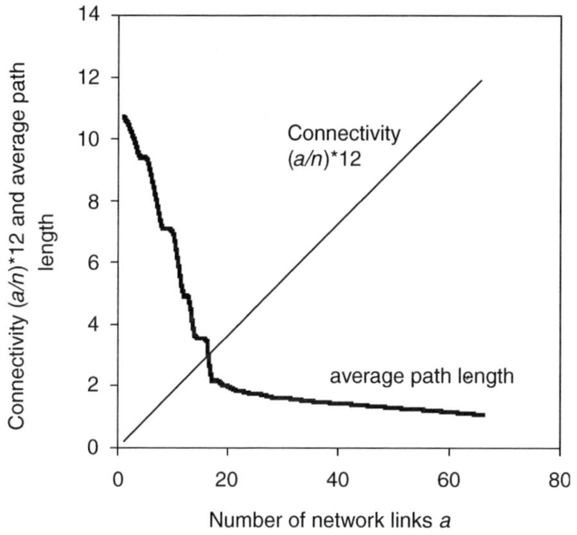

Figure 11.2 *Link density, connectivity and average path length (note the breakpoint threshold at around 17 links)*

What we are seeing here is an increase in complexity where we use not connectivity but average distance as a measure of complexity or order. There comes a critical point like a phase transition where a threshold is crossed and the system becomes connected. This is a qualitative change before which the system is not connected, beyond which it is. Beyond this threshold, average distance does not change very much and there is no further qualitative change. It is the threshold that is of interest and this is often referred to as a state of self-organised criticality. The physical analogy is based on phase transitions, the easiest of which to understand are transitions from ice to water and water to air, from a liquid to a solid to a gas. Freezing or boiling points are thus critical thresholds because once they are reached and passed, the nature of the system changes dramatically. But these states are not self-organised because systems do not remain or maintain themselves on the boundary or at the edge of these transitions. The idea of a self-organised criticality (or self-criticality as Bak (1996), the inventor and populariser of the terms, calls it) is one where the dynamics keeps the system at the critical state in most situations through a combination of local actions which never combine to breach or pass the threshold needed to initiate such a radical change. Such radical changes can in fact take place in self-organised critical systems but then the system would move to another state, another regime, where its fundamental structures and processes are assumed to act rather differently. Technological change sometimes leads to such fundamental transitions and for city transport we might see the changes in connectivity as reflecting such technological shifts. A similar demonstration is given by Kauffman (1995) and summarised by Batten (2000). The problem is also related to percolation in networks where gradual increases in connectivity can suddenly lead to dramatic changes in percolation as the network connects up. Clearly the movement of fluids through porous media is a basic application but so too are diffusion problems such as forest fires.

We need to summarise what all this means. In essence, we are saying that systems become more complex as they accumulate interactions. In a system where interactions take place between objects, as the number of interactions grows as the square of the number of objects, then complexity increases according to this power law. However, critical thinking is all about identifying thresholds and limits to such increasing complexity. Systems do not have undifferentiated complexity as they grow; they add links selectively and they optimise. Another example that illustrates how such growth in complexity self-organises is traffic density based on the number of cars on a highway; as their density increases, after a certain point they are all interacting with each other and queues and jams form whereas when traffic density is light, they barely interact with each other. This shows how accumulated interactions lead to qualitative changes in state; to critical thresholds.

There are some clear examples where this kind of evolution in networks has taken place despite the fact that the model is a somewhat idealised schematic of how real systems develop. It is difficult to think of a world in which there are no connections, for example, although there are situations usually caused by political constraints where well developed networks are separated. A good example is in the division of Berlin between east and west which almost entirely separated the transportation network structure for over 40 years following World War 2. When the Berlin Wall came down, the city was reconnected. Simple accessibility analysis based on the importance of different street segments has been developed by Desyllas (1999) where he shows how the average distance between the two halves of the city changes dramatically in 1990 when people were once again free to move over the whole network. An image of this change based on before and after street accessibility in the centre of Berlin is shown in Figure 11.3 (Colour Plate 1).

A second example is longer term and more qualitative and this relates to social transitions. The word transition has been most widely used in the social sciences to describe the change in the way populations reproduce themselves – the demographic transition – and the way they locate themselves in cities – the urban transition. There is an assumption that such transitions are smooth without any clean breaks or thresholds that disrupt this smoothness. But historically there is a sense in which the breaks have been very sharp – almost bifurcations from earlier established paths, and there is some speculation that the process of urbanisation is composed of a series of abrupt transitions. The movement from nomadic existence to settled agriculture about 10 000 BC is one such transition, the formation of the first cities in approximately 3000 BC is another. The collapse of ancient civilisation and the ensuing dark ages is a reversal while the revival of trade in the early Middle Ages and then the Renaissance yet another. Finally, the Industrial Revolution is the most recent transition with some speculation that the ongoing transition to a Post-Industrial society is another.

In some senses, this kind of punctuated equilibrium is reminiscent of phase transitions. Whether or not the notion that civilisations are held at some sort of threshold – a level of criticality – is speculation but there is some sense in which balance is reached. Quite clearly transitions and criticality are issues that are scale dependent in both space and time, and this is of profound importance in applying these ideas. It is worth quoting from an early paper on these ideas as applied to urban development. Iberall and Soodak (1987),

Berlin 1989 before unification

Berlin 1999 after unification

Figure 11.3 (Plate 1) *Accessibility in Berlin: before and after the wall is demolished. The colours indicate the street segment accessibility illustrating a massive increase in the central area after the wall is demolished. The scale for accessibility varies from high (red) through yellow to and green to low (blue)* referenced in Johnston (2001), describe the process by which Europe underwent a transition

> *... not unlike that between H_2O molecules changing from the fluid state of water to the crystallised state of ice: for centuries the population is liquid and unsettled – and then suddenly a network of towns comes into existence, possessing a stable structure that would persist more or less intact until the next great transformation in the nineteenth century during the rise of the industrial metropolis.*
>
> (Johnston, 2001, pp. 110–111)

11.4 The Unifying Force of Scaling

Our first demonstration of scaling relates to the paths that exist in a graph as its connectivity increases. It is at the critical threshold, when the graph becomes strongly connected, that it first becomes possible to move from any node to any other; paths in the network exist on all scales. Like many mathematical demonstrations, the proof is long winded but a sketch of what happens at this point is easy to make. As the network becomes more and more connected by adding one link at a time, then more and more paths of different lengths exist with shorter paths clearly being more frequent than longer paths. As the threshold is reached, the frequency distribution of these paths implies that paths of different lengths exist on all scales, that is, that the relationship between the number and length of paths decreases according to a power law which is the hallmark of scaling. At this point the system is fractal, but beyond the threshold the frequency distribution becomes degenerate in that when the graph is completely connected, all path lengths are the same – all are of any length and thus the distribution is no longer scaling as an inverse power but is uniform. This kind of characteristic essentially implies that only at the point of criticality – at the threshold – is the system complex. Before that point it has little order but beyond that point it is disordered in an entirely different way with no structure whatsoever. Although we cannot demonstrate this here, different levels of connectivity imply different kinds of order with four classes of order being identified for different levels of connectivity as indicated in Wolfram's (2002) work on cellular automata.

As we have already implied, this model of network evolution is rather artificial. For example, in cities it is most unlikely that a city exists that is anything but ordered. This means that the notion of cities existing that have simplistic order – i.e., cities with connectivity less than that at the threshold, and cities with the kind of uniformity that exists beyond the threshold – must be hypothetical. It might be argued that the world is getting more complex and in some characterisations of urban evolution, it might be useful to establish the current city as the baseline, compare cities in history with this, and thus extract their increasing connectivity and changing order. But what is more likely is that any city always exists at some critical threshold and that this threshold changes qualitatively as technology changes. Mediaeval cities were indeed connected to one another but the connectivity was different in form from the industrial city. The post-industrial city will still have order on all scales but this is likely to have a different form from the industrial city, being based on multiple technologies of communication rather than the few technologies that characterise earlier city forms. Nevertheless, although our first model is limited in direct applicability, it is still a useful metaphor for establishing the idea of evolving networks and it illustrates that connected cities where the connections are just enough to make the city coherent and workable, imply scaling, in contrast to hypothetical cities with too few links or too many which do not display the same levels of complexity.

Our second approach to networks is somewhat different in that our concern is with the overall properties of a growing network rather than the more static view of its connectivity. To explore these properties, we need to focus on the frequency distributions of the size of the nodes that make up these networks, where we will define the size of a node in terms of its in-degrees (or out-degrees). We refer to the frequency of a node in terms of its in-degrees, say, as $N(a)$ where this is the number of nodes – frequency of nodes – with in-degree of size a. If we consider the distribution of these frequencies, in a graph whose arcs have been formed randomly, this distribution is Poisson but in many networks which reflect competition in space and time, there is considerably more order with such distributions following a power law. In short, what we see in the real world are numbers of nodes which vary inversely with the number of links that are associated with them. In other words, we see very many nodes with hardly any links at all but a few, very significant nodes with an extremely large number of links. Thinking now of frequency as probability, then the typical density function for networks in urban space and cyberspace is $N(a) \sim a^{-\lambda}$ where λ is the rate at which the density falls off as size of the node increases. It is often easier to work with cumulative distributions and the one that is favoured here is the complementary or counter cumulative defined as $r = N(A > a) \sim a^{-\lambda+1}$ which is in effect the Pareto distribution. This is the distribution of the number of nodes greater than size a which is the rank (r) of the node(s) in question. The rank-size distribution popularised for city systems and word frequencies by Zipf (1949) is based on a simple manipulation of this as $r \sim a_r^{[1/(1-\lambda)]}$ where a_r is now the in-degree associated with the node ranked r.

This is a classic scaling which has been found in many different kinds of systems where there is differential growth and competition between the system elements. The best known examples of such scaling exist for incomes, city size distributions, and word frequencies, all credited with observations and models first proposed over one hundred or more years ago (by Pareto, Estoup, and Auerbach respectively). In the last decade, there has a been flurry of work associating rank-size scaling with other human systems such as scientific citations, company size, and productivity while quite recently these ideas have been applied directly to networks (Barabasi, 2002). The World Wide Web has been the most significant focus but other kinds of system such as social networks, food webs, cell networks, and the transport of energy, information, and people have all been shown to scale in this way: clear evidence of course for systems that grow from the bottom up.

We will show evidence of such scaling below but our concern is for plausible models of network growth that gives rise to such scaling. Many such models exist, all of them based on simple stochastic processes that generate growth through the addition of new objects that connect to existing objects in some preferential manner. The basic model goes back to Yule in the 1920s but was first properly formalised by Simon (1955). Essentially, when applied to network growth, this model is based on two key assumptions. The process of growth adds one new node for each time period and thus the growth of the system in terms of the total number of nodes n is directly proportional to time t.

When a node is added, there is a probability ρ that it attaches itself to any node in the system and there is complementary probability $1-\rho$ that it attaches itself to nodes in proportion to the number of in-degrees that they already have; that is, this probability is proportional to the product of the number of nodes $N(a)$ and their in-degrees a, $aN(a)$. We can formalise this as follows. An increase in the in-degree of nodes $N(a)$ is defined as $[\rho N(a-1)+(1-\rho)(a-1)N(a-1)/t]$ while there is corresponding decrease in the nodes $[\rho N(a)+(1-\rho)aN(a)/t]$. This occurs for all nodes in the system other than the new node $N(0)$. It generates a differential equation that defines the change in the in-degree of any node as

$$\frac{dN(a)}{dt}=\frac{\rho\,[N(a-1)-N(a)]-(1-\rho)[(a-1)N(a-1)-aN(a)]}{t},\ a\geq 1$$

whereas the appropriate equation for the new node $a \neq 0$ is

$$\frac{dN(0)}{dt}=1-\frac{\rho N(0)}{t}$$

If we now argue that in the steady state, $dN(a)/dt=0$, these two equations can be manipulated accordingly. The recurrence, which is established by the steady state, then implies that the distributions of the in-degrees must satisfy a power law for the steady state to hold. In fact, in terms of the counter cumulative or rank-size, it is easy to show that this is equivalent to $r \sim a^{-[1/(1-\rho)]}$. This treatment follows Mitzenmacher (2003) who provides a clear review of how these 'preferential attachment' models, the term coined by Barabasi (2001), are equivalent to Simon's (1955) model and are strongly related to the other stream of models used to generate the rank-size distribution which are based on multiplicative or proportionate effects. Albert and Barabasi (2002) provide a useful summary of how these models are linked to those we examined in the last section where the number of nodes are fixed and where the focus is on connectivity, and those of the next section where the focus is again on systems where the nodes are fixed but where the interest is on the particular structures that such networks display.

To conclude our discussion, it is worth noting that city size distributions can be generated using at least the metaphor of this network model. It is, however, hard to identify the kinds of connections that would be necessary to show how cities grow in this manner. In terms of physical transport, there are limits on the way cities grow in that transport links are fashioned so that they link many settlements at once. In a sense, although not random, the idea that there are more and more physical links from all cities to those further up the rank hierarchy is simply not feasible in physical terms. To show how cities grow in terms of networks, we probably need to look at information flows which are not physical in the traditional sense. If we were to assemble all the kinds of links that cities make with one another in terms of government, business, social networks and so on, then this is likely to show the requisite scaling. We do not have data here to show this but we can examine some of the scaling of in-degrees and out-degrees between 180 countries with respect to the hits generated through web pages. This is based on measuring the hits made to each of the domain names – country domains where we have excluded US domain names which we cannot unambiguously associate with that country. This, we realise is a crude picture but it does reveal scaling as we illustrate in

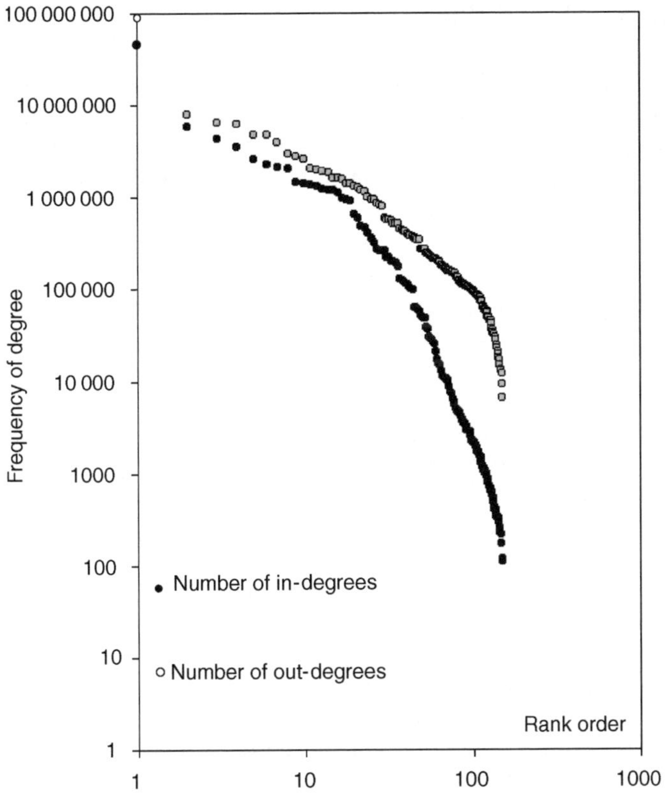

Figure 11.4 *In-degrees and out-degrees associated with country name domains from the Alta Vista search engine, April 1999*

Figure 11.4 for the in-degrees and out-degrees for the countries in our dataset. The data were produced using the AltaVista search engine (Shiode and Batty, 2000).

The relationships in Figure 11.4 deviate from pure scaling suggesting that the underlying probability distribution is more likely to be lognormal than a power function. In fact, this is generally the case for many distributions in that the power function is usually a good approximation for the fat or heavy tail, in this case, the upper left region of the distribution. In fact for distributions where the variance of the frequencies is large, then the power function is a good approximation whereas in this example, where the distributions of in-degrees and out-degrees is clearly still changing rapidly, the variance is not as large as it will ultimately be and the approximation at this stage is less good. This means that we need to be very careful in studying evolving networks which clearly have not reached any steady state. Indeed all scaling models rely on finding a distribution at the steady state and comparing this with an actual distribution which is unlikely to be in the steady state. This is the Achilles heel of network science in that we do not have much idea about how actual networks grow although we are ready to make assumptions about their equilibrium form. Much remains to be done on their dynamics and this is what makes the area so challenging.

11.5 Small Worlds and Wide Webs

So far we have looked at macro properties of evolving networks in terms of connectivity and scaling but we have not looked at structure. Although the rekindling of a scientific interest in networks can be linked to complexity theory and the emergence of decentralised thinking as a major force in science, it is also due to the resurrection of a long standing problem in sociology involving so-called 'small worlds'. The term small world was first used by Milgram (1967) in a popular article in *Psychology Today* in which he reported the results of an experiment of sending letters to unknown targets in very different geographical locations in terms of the number of intermediaries needed to pass from the points where the letters originated to the targets. He found that the average number of individuals through which such letters would need to pass before they reached their target was around six. From this, he speculated that the average number of links in the social network to reach anyone from anywhere was of this order. To enable this, he directed the sender to a particular target they did not know by asking them to send the letter to someone they thought was closer to the target than they were, with the instruction that that person was to send it on to an even closer target, and so on until the letter arrived. The fact that it only took six steps was evidence of the fact that the world was much 'smaller' than had been imagined, although it has subsequently been noted that a person six steps removed from one, still might be a lifetime and a continent away (Watts, 2003).

What is so surprising about these kinds of network is that at the most local level, we know they must be based on rather dense clusters of friendship ties but at the aggregate level it is still possible to reach anyone from anywhere in an average of six steps. Thus there is a high level of local clustering, meaning short local distances but also short overall network distances, the best of both worlds. Watts and Strogatz (1998) were the first to formalise this problem in a way that articulated this local-global nexus. They showed that graphs with low average distances, that is, shorter paths, could be formed by randomly selected arcs between a set of nodes, as we did in our first example (p. 157) where we illustrated how distances reduced massively at the critical threshold where the graph became strongly connected. This is one extreme form where there is little local clustering but low average path lengths. At the other extreme, there are what Watts and Strogatz (1998) call 'cave-man worlds' where there are dense clusters strung together but these have high average path distances because one has to pass through each cluster to visit any other. Small worlds lie somewhere in between. Watts and Strogatz (1998) show that by starting from a cave-man world of dense clusters and requiring a small fraction of links to bypass the local clusters leads very quickly to a small world with a much lower average distance between the nodes.

We illustrate a picture of these kinds of network in Figure 11.5 where we show a clustered graph (cave-man world) and a small world which results by rewiring the clustered graph through changing only three links. We have not computed the path lengths or the cluster values but the example is obvious enough in making the point that real networks often show a small world quality simply to enable efficient movement. The resurrection of this problem has led to a flurry of work. All kind of networks appear to have a small world quality, the World Wide Web being the most obvious one

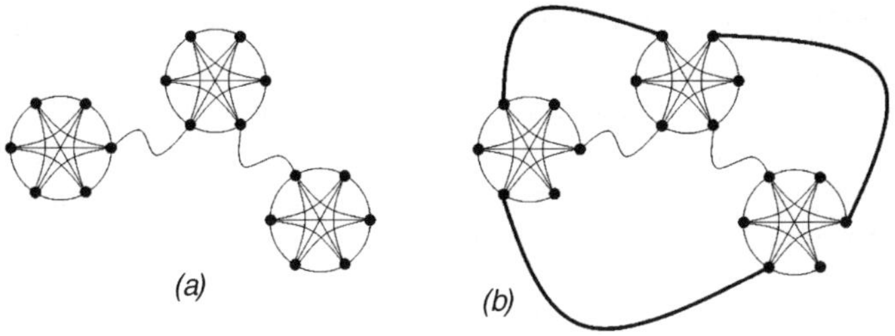

Figure 11.5 *Creating a small world: (a) the original 'Cave Man World' based on weakly connected clusters; and (b) rewiring to add strategic links retaining the high degree of clustering but enabling a massive reduction in the average path length between all nodes*

(Barabasi, 2002) but so do energy networks, nervous systems, chemical bonds, and social networks which spread everything from friendship to disease (Watts, 2003). Moreover small worlds can also be consistent with the kind of scaling that we illustrated in Section 11.4 although many networks that scale are not small worlds. We do not have the space to describe the ways in which connectivity, scaling and structure in the various types of networks we have used here can all be integrated through the new science of networks, but this is occurring rapidly at present and readers are referred to the work of Albert and Barabasi (2002) for a comprehensive survey.

We will conclude our speculations with an example that is suggestive rather than definitive but does pose the key challenge for network geography and for the way this might be incorporated into GI science and GIS. The network of streets that is the base on which cities grow and change is clearly not a small world. Street networks are planar graphs composed of junctions and street segments with junctions usually having a small number of in-degrees and out-degrees, usually no more than six with a mean around four. In fact, the distribution of in-degrees and out-degrees in street systems is likely to be Poisson. Only in countries such as Britain, where the roundabout is used widely for street intersections, are there many examples where more than four ways intersect and even there, five or six way intersections are very much a minority. However, physical networks to move people in cities are small worlds if all the various networks corresponding to different modes of travel are considered. Let us consider the growth of the modern city in these terms as a metaphor for how small worlds emerge. If we go back to the medieval city then the street system was simplistic, although there were arrangements to control some streets for certain restricted purposes. In the industrial era all this changed, for as cities began to grow new forms of transportation technology were needed to enable people to move greater distances at faster speeds. For example, freight was moved using canal systems while the emergence of the railways in the early nineteenth century in western Europe enabled people to move into peripheral locations which eventually were called suburbs. The street car or tram system also developed a little later and in these cases a new transport system with much restricted nodal access to the street system was layered on top of the existing roadways. In short, people would use the train with its limited stops to access the local clusters. In London, for example, which was expanding to embrace local villages in its hinterland, this was a way of connecting previously remotely linked clusters.

In the twentieth century, all this was added to by the development of the automobile and new high speed road systems. In the 1920s, the ring road and the bypass were examples of such roads with relatively restricted access compared with the basic street system. But it was the development in the 1950s and 1960s of freeways with highly restricted access that really produced new layers of segregated movement, akin to the railways, and that reinforced the small world quality of urban networks. As large cities have become global, these networks have extended to the airlines, and more recently to cyberspace, to information flows that really do produce dramatic changes in the way we can communicate over very large distances. In one sense, one can see this process as adding new layers of transportation, which involves a limited amount of rewiring of the old, but more new wiring for new forms of movement technology with higher speeds and the need for restricted access to enable such networks to function as intended.

To enable cities to grow and continue to function in moving ever more people about between their parts, we need new networks based on new technologies, and this implies a process of co-evolution between population growth and changes in technology. In a sense, growth would not have been possible without the existence of these new technologies, not the other way around, so population growth and technological innovation have gone hand-in-hand, best seen in the way cities have made innovations to their transport systems. We are not able to illustrate this with specific examples of where the small world idea has been applied, for as yet there are none, but we can show how new networks have been formed and provide some snapshots of how cities have evolved in terms of linking new transport nets to the existing structure. A particularly good example of the way two different networks – the street and the subway system – have co-evolved can be illustrated with respect to London, although over the last 200 years, the suburban railway system and more recently the freeway system have also reinforced the 'small worldliness' of the city. Nevertheless, the street and subway system is a good example of how it is necessary to move people around more efficiently as the city has grown, in that the subways interface with the street system at very specific points. Indeed many people who work and live in London often find it difficulty to provide an integrated picture of how these systems interact in that the subway system implies a topology quite different from the street system, which is even encoded in the maps that people use.

In Figure 11.6 (Colour Plate 2), we show Sam Rich's map of the topology of the subway system as mapped onto the actual street pattern for the central area. His website http://www.fourthway.co.uk/ illustrates how the real subway map can be mapped back onto Beck's original topology which forms the basis of the current non-Euclidean map used by most travellers to navigate their way around the subway. Although we are not able to say more about this process of moving from the topological to the Euclidean and back, which is another aspect of the way we might represent networks (Cox, 2002), such transformations make it clear how small worlds can never faithfully represent movement in terms of real maps. From the real map, one can see directly how one can travel quickly between local neighbourhoods using this mapping. If one adds the suburban and mainline rail network to this, then we begin to get a more comprehensive picture of how useful the

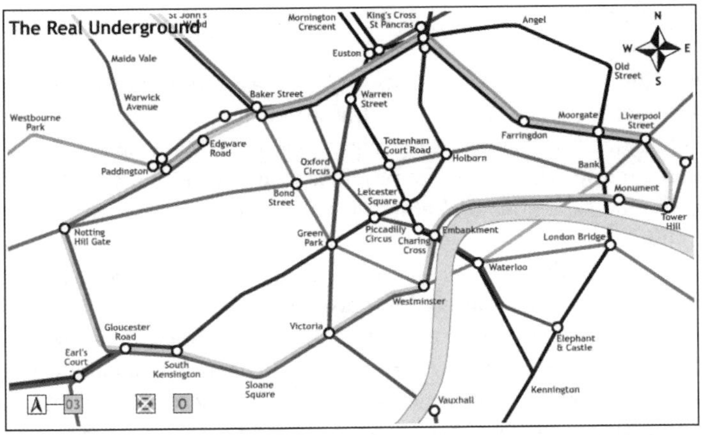

(a)

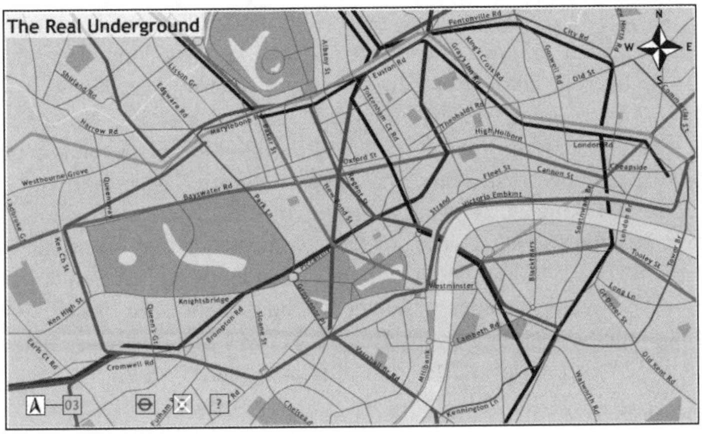

(b)

Figure 11.6 (Plate 2) *The way the London subway map connects to the real street pattern. (a) Morphing the topological subway map to the real geography. (b) Connecting the subway topology to the Euclidean pattern. A dynamic illustration of this morphing is at http:// www.fourthway.co.uk/ (by permission of Sam Rich and Transport for London)*

small world idea can be. Moreover, we can begin to think about measuring its connectivity and clustering, the distribution of its path lengths, and then consider how these can be used to compute its efficiency. In fact, it could be argued that in London, the current difficulties of travelling in the city are due to lack of investment in keeping the small world quality of the network intact, given the underinvestment in freeway and the subway systems during the last 50 years. To this picture, we must also add cyberspace and all the other networks that exist in transporting people, information, and energy, but this will remain another challenge until we perfect our measurement systems for identifying such comparatively invisible interactions.

11.6 Next Steps

Given the nature of this argument, readers might have expected some stronger guidelines as to how contemporary GIS might be transformed to take account of the dynamics of network representation that we have emphasised here. We would counter that it is first necessary to explore network geography much further before we are in any position to develop new representations and new software which must necessarily have routinised and generalised applications in mind. Our concern is more with GI science, which has taken the place of spatial analysis specifically and quantitative geography more generally, and it is in this domain that we need to grasp the dynamics of networks. Okabe's (2003) proposal is a start, notwithstanding his focus on point patterns on a network rather than relationships and interactions *per se*. But we need new research on how network systems evolve and in this regard, developments in the physical world (for example, for river and related hydrological systems), in the economic world based on trade flows and exchange, and in the social world where there is a renaissance in ideas about evolving social networks, all point the way.

GIS is concerned very largely with representation but here we have argued that this is not enough. We need to move beyond representation to dynamics and change, and this means that we need to involve process. In a sense, these ideas await further unification but all the seeds are there already in spatial analysis, which is founded on stochastic processes in which time is clearly implicit. It is this kind of development that must go hand-in-hand with new ideas about the representation of networks within GIS. In this way, processes working on networks that interact with ways in which such networks evolve must provide the new focus. In this, we consider that GI science must become more substantively based, that methodologies embracing processes that have substantive content must be developed. In this, the network idea is intrinsic to spatial systems where exchange and interaction are key constituents of the way in which spatial structures can be explained, represented, modelled and ultimately understood.

Acknowledgements

The auther wishes to thank the ESRC Nexsus Project (L326-25-3048) for support. Jake Desyllas provided the images of Berlin from his PhD thesis. Sam Rich gave permission for the use of two screen shots from his London Underground web pages at http://www.fourthway.co.uk/.

References

Albert, R. and Barabasi, A.L., 2002, Statistical mechanics of complex networks, *Reviews of Modern Physics*, **74**, 47–109.

Bak, P., 1996, *How Nature Works: The Science of Self-Organized Criticality*, Springer-Verlag, New York.

Barabasi, A.L., 2001, The physics of the Web, *Physics World*, **14**, 33–42.

Barabasi, A.L., 2002, *Linked: The New Science of Networks*, Perseus Publishing, New York.

Batten, D.F., 2000, Complex landscapes of spatial interaction, in Reggiani, A. (editor), *Spatial Economic Science*, Springer-Verlag, Berlin and New York, 51–74.

Batty, M., Dodge, M., Jiang, B. and Smith, A., 1999, Geographical information systems and urban design, In: *Geographical Information and Planning*, edited by Stillwell, J., Geertman, S. and Openshaw, S., Springer-Verlag, Heidelberg, Germany, 43–65.

Batty, M. and Rana, S., 2002, Reformulating space syntax: The automatic definition and generation of axial lines and axial maps, *Working Paper 58*, Centre for Advanced Spatial Analysis, University College London, at http://www.casa.ucl.ac.uk/working_papers/paper58.pdf.

Bertalanffy, L. von, 1968, *General System Theory: Foundations, Development, Applications*, George Braziller, New York.

Cox, L., 2002, Investigating the cartography of the London Underground map, MSc Thesis, Department of Geomatic Engineering, University College London, Gower Street, London.

de Jong, T. and van Eck, J.R., 1996, Location profile based measures as an improvement on accessibility modelling in GIS, *Computers, Environment and Urban Systems*, **20**, 181–190, see http://flowmap.geog.uu.nl/.

Desyllas, J., 1999, The relationship between urban street configuration and office rent patterns in berlin, *PhD Thesis*, The Bartlett, University College London, Gower Street, London.

Holland, J.H., 1996, *Hidden Order: How Adaptation Builds Complexity*, Perseus Books, Cambridge, MA.

Iberall, A. and Soodak, H., 1987, A physics for complex systems, in Yates, F.E., (editor), *Self-Organizing Systems: The Emergence of Order*, Plenum Press, New York, 499–520.

Jiang, B., Claramunt, C. and Batty, M., 1999, Geometric accessibility and geographic information: extending desktop GIS to space syntax, *Computers, Environments and Urban Systems*, **23**, 127–146.

Johnston, S., 2001, *Emergence: The Connected Lives of Ants, Brains, Cities, and Software*, Scribner, New York.

Kauffman, S., 1995, *At Home in the Universe*, Oxford University Press, New York.

Milgram, S., 1967, The small world problem, *Psychology Today*, **1**, 61–67.

Miller, H.J. and Shaw, S-L., 2001, *Geographic Information Systems for Transportation: Principles and Applications*, Oxford University Press, New York.

Mitzenmacher, M., 2003, A brief history of generative models for power law and lognormal distributions, *Internet Mathematics*, **1**, 226–251.

Okabe, A., 2003, A new approach to spatial analysis: spatial analysis on a network with SANET, *CUPUM '03: Proceedings of the 8th International Conference on Computers in Urban Planning and Urban Management*, Center for Northeast Asian Studies, Tohoku University Sendai, Japan, 9–20, see http://okabe.t.u-tokyo.ac.jp/okabelab/atsu/sanet/sanet-index.html.

Shiode, N. and Batty, M., 2000, Power law distributions in real and virtual worlds, Working Paper 19, Centre for Advanced Spatial Analysis, University College London, at http://www.casa. ucl.ac.uk/powerlaw.pdf.

Simon, H.A., 1955, On a class of skew distribution functions, *Biometrika*, **42**, 425–440.

Watts, D.J., 2003, *Six Degrees: The Science of a Connected Age*, Norton, W.W. and Company, New York.

Watts, D.J. and Strogatz, S.H., 1998, Collective dynamics of 'small world' networks, *Nature*, **393**, 440–443.

Wolfram, S., 2002, *A New Kind of Science*, Wolfram Media Incorporated, Urbana, IL.

Zipf, G.K., 1949, *Human Behavior and the Principle of Least Effort*, Addison-Wesley, Cambridge, MA.

12

The Nature of Everyday Experience: Examples from the Study of Visual Space

Marcos Llobera

Learning is experience. Everything else is just information.

Albert Einstein

12.1 Introduction

The discussion in this essay concerns a huge topic: how do people perceive and experience the world *within the context of everyday life*? Indeed, this is a topic that spreads beyond any single discipline or area of expertise and cannot be fully covered in this paper. But more specifically, it is concerned with the place that Geographic Information Science (GISc) may have in dealing with such a topic. Can GISc bring us any closer to how people experience the world? As I hope to demonstrate, I believe the answer is that it can, or at least that exploring such a possibility is well worth a try.

To address this question I will briefly review, the main precepts that underlined several important GISc initiatives, e.g. NCGIA research initiative #21, *Varenius Project* (Mark and Freundschuh, 1995, Mark and Egenhofer, 1995, Mark and Frank, 1996, Mark *et al.*, 1997, Mark *et al.*, 1999) concerned with spatial cognition, which will be called GISc-cognitive henceforth. These initiatives have claimed the possibility of linking spatial tools with human spatial cognition in order to promote the design of better tools and, conversely, to better understand everyday human interaction with the world.

Re-presenting GIS Edited by P. Fisher and D. Unwin

In recent years, discussions on the experiential nature of the world have been rather restricted in GISc especially when compared to the wealth and variety offered by 'post-modern' works in human geography and other areas such as anthropology and archaeology. One could claim that this is due to the rather elevated, highly formal, output the former seeks to attain, in the way of axiomatic principles and rules, as opposed to the less strict and more open ended narratives produced by the latter. As I will try to argue in the following sections, many of the restrictions and limitations found in some of the GISc-cognitive initiatives are self-imposed, and can be ultimately traced back to the *classic* cognitive science view (Clark, 1997) to which these subscribe.

To expose these limitations, I will rely quite heavily on current anthropological work on perception, more specifically, on Ingold's critical examination of the evolution of theories of perception in anthropology as found in his most recent work, *The Perception of the Environment* (Ingold, 2000). In spite of its anthropological overtones Ingold's exposition resonates with similar debates found in many other fields (e.g. Thrift, 1996, 2000), and provides a clear and overall synthesis of many of the main ideas surrounding our understanding of human perception. Given his background, as a social anthropologist, Ingold has striven to understand how people perceive the world outside the context of western culture. This process has served to reveal many of the western biases that are built into our understanding and descriptions of how people perceive the world. Many of these biases can be traced back to the influences that predominant scientific currents exert on other fields of enquiry.

12.2 Dis-embodied Mind

In his chapter on theories of perception (Ingold, 2000, Chapter 9), Ingold uses the following question as his starting point: 'why should people from different cultural backgrounds perceive the world in different ways?', a question that within Anthropology could not be easily divorced, at least until recently, from a more general discussion on the nature of culture and the interrelationship between the individual and society.

Anthropologists in the 19th century, such as Emile Durkheim, did not consider an individual's psychology to be a matter of anthropological enquiry. Anthropology was to be concerned with the mind of the collective. To describe the relationship between the consciousness of the individual and that of the collective, he made reference to another opposition, the one that according to him existed between sensation and representation. Durkheim explained this distinction by making reference to some of the properties of sensations. First, sensations are not durable; they change from one moment to the next. Hence the only way we can retain the stream of sensations in our consciousness is by capturing them using a more stable and durable system of concepts. Second, sensations can only be experienced individually, therefore in order to be able to share them among others they need to be captured by symbols, for example, words.

Fifty years later (early 1960s), British anthropologists, such as Leach and Douglas, were still using the same recipe to describe the process of perception. Life is partitioned and matched against culturally transmittable concepts that order experience. Many of these concepts are embedded in language categories that members of a same society use to communicate with each other, hence they only make sense in relation to a collective. American anthropologists, like Boas and later Goodenough, dissented from this view and maintained that culture, defined as a *system of habits, beliefs and dispositions*, could only be understood in relation to the individual and that it was, therefore, psychological in nature. The stress was on culture as a set of transmissible and internalized rules (i.e. within the individual) rather than on any other sort of observable patterns whether material or behavioral. This separation between internal and external aspects of culture is an important issue that later developments in many branches of social theory would aim to re-evaluate. Not all American anthropologists (e.g. Geertz) shared Boaz and Goodenough's ideas on the individual aspects of culture; for them, as with the British school, culture was to be studied as a set of public and social symbols that imposed meaning upon experience.

By the late 1960s, based on Goodenough's view of culture as a set of shared conceptual *schemata* that reside in the mind of the individual, several anthropologists established what was to be known as Cognitive Anthropology. Their aim was to discover how people in other cultures structured their experience of the world by using, what they believed, was a set of finite, hierarchical ordered classes. Their early attempts were characterized by the use of various sorts of formal semantic analyses, aimed to extract and identify the nature and structure of taxonomies for different knowledge domains (e.g. kinship terms). However, these classification *schemata* seemed to be more artifacts of the investigation, resulting from unnatural and controlled contexts of elicitation, than part of the cognitive organization people employed in their everyday life. They did not bring us any closer to understanding how people negotiate their relationships with others and/or their environment. This could only be achieved, according to Ingold, by allowing a certain flexibility in the use of the concept or category elicited and a sensitivity towards the context in which it is being used, precisely what formal methodologies are meant to disregard. It also presupposes the primacy of propositional information, i.e. information that can be represented in some way, say by words or any other sort of representation, over non-representational information.

Later developments in cognitive anthropology sought to substitute early (simpler) taxonomies with more complex structures of representation, or *cultural models*. These were superior as they incorporated several features: a description of the world as a set of interconnected propositions where objects, events and situations constitute prototypes; the acknowledgement that their relation with cultural knowledge is indirect and can only be retrieved through 'the analysis of richly textured material of ordinary discourse' Merleau-Ponty (1962); and finally, their role, once internalized, in directing future behavior.

Ingold maintains that ultimately, cognitive and early anthropological views were/are flawed as they are deeply rooted in Cartesian ontology whereby the mind is kept separated from the body (and the world) and given precedence over the latter. According to this view, the mind is solely responsible for any cognitive (intelligent) process while the body behaves as a passive transducer of outside stimuli. Inevitably, perception can only be understood as a two-stage process that involves first, receiving stimuli from the outside world through the senses in the body and second, 'matching' these against some kind of stable, culturally transmitted, information pattern or *schemata*, so as to generate an ordered experience. Perception is reduced to a faculty strictly of the mind. People of a similar cultural background share the same perceptions, insofar as they share the same *schemata* established through some sort of consensus. This separation is also responsible for generating an unbalanced, one-way, relationship between the individual and the world, in which the world is seen as a domain full of problems and the background where the latter are worked out, rather than as a resource for problem solving.

12.2.1 GISc – Cognition Agenda

At this point, it is useful to recall the main objective behind most GISc-cognitive initiatives: to provide the theoretical underpinnings (as derived from human cognition) that will inform the design of new, more intuitive, interfaces allowing, ultimately, a wider range of people (non-specialists) to use and manipulate spatial information. Given the predominant role of spatial information systems as systems of representation, this objective sets out to uncover/generate formal categories and principles of human spatial cognition that are generic and useful at the same time. It is no surprise then that these initiatives would favor areas in cognitive science that would have similar objectives as mentioned above.

Overall, these studies subscribe to an early and quite restricted view in cognitive science, known as the *classic*, *symbolic*, '*rule-and-symbol*' style (Thagard, 1996; Clark, 1997). It is a style in which intelligence is equated directly with the manipulation of mental symbols (explicit representations of some sort) that follow specific rules. It stands opposed to a later version known as *connectionism* that emerged in the 1980s. This style promotes the idea of a distributed representation: one in which meaning is not captured by a single symbolic unit but rather arises from the interaction of a set of units, that interact within some sort of network (Churchland and Sejnowski, 1992). In their original forms, both movements are similar in that they prioritize the working and properties of the mind over those of the body and the world. They differ in the way they explain how the mind operates internally, but the source of intelligence is still restricted to the functioning of the brain alone. In recent years, the idea of intelligence as being restricted exclusively to the workings of the mind has been challenged by several works in artificial intelligence (AI) (e.g. Brooks, 1991). These works promote the understanding of intelligence as the product of the interaction between the mind, the body and the world. In doing so, they come close to many of the ideas that were earlier put forward by philosophers (e.g. Merleau-Ponty) and social theorists (e.g. Bourdieu, see next section).

An example of the classic approach on which GISc-cognitive initiatives are based is the *CYC project*. This project started in 1984 with the aim of instilling commonsense understanding into a computer, i.e. to deliver a sort of intelligent machine/mechanism capable of reading and directly assimilating written texts, and that could, ultimately, derive the rest of its knowledge-base internally. The entire project is totally reliant on *explicit* symbolic representations. However, as Clark has rightly pointed out:

> *[By now] the commonsense database it now encodes will doubtless be of great practical use as a resource for the development of better expert systems. But we should distinguish two possible goals for CYC. One would be to provide the best simulacrum of commonsense understanding possible within a fundamentally unthinking computer system. The other would be to create, courtesy of the CYC knowledge base, the first genuine artificial mind.*
>
> *Nothing in the performance of CYC to date suggests that the latter is in the cards.*
>
> (Clark, 1997: 3)

One major argument against the idea that we operate solely using internal representations comes from a purely pragmatic point of view. If our performance in the world is based on matching 'noisy' outside information with some sort of internal 'generic' model, we could not operate in real-time (Clarke, 1997: 21). For instance, we would not be able to manage in time the proper adjustments needed to catch a moving train by stretching out our arm and grasping its side-rails with our hand. It is not surprising to find among GISc-Cognitive initiatives examples that illustrate several of the limitations Ingold has associated with cognitive anthropology. These can be found in the works of Mark and Frank, (1996) and Mark *et al.,* (1997). In the first of these studies, the authors make reference to two definitions of mental *schemata*, neither of which are particularly informative or helpful:

> *A schema is that portion of the entire perceptual cycle which is internal to the perceiver, modifiable by experience, and somehow specific to what is being perceived. The schema accepts information as it becomes available at sensory surfaces and is changed by that information; it directs movements and exploratory activities that make more information available, by which it is further modified.*
>
> (Mark and Frank, 1996; citing Neisser, 1976: 54)

> *A schema consists of a small number of parts and relations, by virtue of which it can structure indefinitely many perceptions, images and events.*
>
> (Mark and Frank, 1996; citing Johnson, 1987: 29)

On the one hand, these definitions make reference to complex data structures that try to avoid the rigidity of earlier taxonomies which have no way of incorporating contextual information needed in order to apply them successfully (how do we assign membership to a certain schemata?). But they do so at the cost of adding so much vagueness that renders the definitions useless (what are they really?). On the other hand, and rather ironically, the end result does not seem to be any different from a set of well-defined categories (e.g. container, path, Mark and Frank, 1996). Another important assumption within the classic or symbolic view is the idea that internal representations (whatever they might be) or schemata are readily accessible and/or intelligible. This is still open to debate but it appears increasingly clear that whatever it is that we might use internally to operate our mind, call it some sort of 'representations', it is closer to a kind of pattern of activation as described by connectionists, e.g. weights in a *neural-network* (Clark, 1997: 233, note 12), than it is to some sort of symbol that might be described though verbal description.

GISc-cognitive initiatives claim the possibility of retrieving a corpus of rules and principles that govern our everyday experience of the world, that they refer to as *naïve* or *commonsense geography.* Putting aside the validity of their framework, which as has been described above can be questioned, the possibility of being able to retrieve a 'commonsense' geography following some top to bottom approach begs some serious questions (e.g. whose commonsense?). Furthermore, the likelihood of being able to formalize any sort of commonsense into a set of identifiable, useful categories and rules appears, on simple reflection, contradictory as 25 years of discussions in philosophy, sociology and anthropology and more recent debates in psychology and AI have shown! For instance, in Mark *et al*., (1997) the authors define naïve geography as being 'based on high level *expert* understanding on how the world works' (my emphasis). If the term 'expert' is to be understood as a synonym to skilful, then it is contrary (see discussion below) to our current understanding on how a skilful operator engages in his/her activity. That is, an essential feature of any skilful action is the fact that it unfolds without the guidance of any kind of conscious rules, but instead through some sense of what 'feels right'. If on the other hand, 'expert' refers to a professional in some field (e.g. cartographer) then it cannot apply to the general public.

There are several problems surrounding these studies, the most important among them being the fact that the aim set out at the beginning, i.e. generating new interfaces to handle spatial information, has taken precedence over the study of how people perceive and operate in the world. As a consequence, what are presented as general propositions about the nature of human spatial cognition and perception are, in fact, highly contested and biased in nature. This does not mean that the results obtained through these exercises are not useful, as for instance any insights obtained from studying the role of natural language and other forms of spatial representations in western (mostly Anglo-American) societies. It simply means that rather than having general validity they are restricted to western societies. Moreover, they do not explain how people interact or perceive their environment but mostly how they employ and use certain forms of spatial representations. Any observations and principles obtained through these studies cannot form the basis for generalized rules on human spatial cognition and perception, either for non-western or western cultures.

12.3 Dwelling Perspective

Within an anthropological context, the long-standing cognitive view on perception was finally challenged. Among its main critics, the work of French anthropologist Pierre Bourdieu (1977) stands out. He disputed the idea of reducing culture (knowledge in general) to a mental construct that exists *a priori* and that is later applied to the world. Instead he proposed that culture is generated in the context (i.e. coupling mind-body-world) of activities carried out everyday repeatedly. It is through these daily routines, which are socially structured, that people 'acquire the specific dispositions and sensibilities that lead them to orient themselves in relation to their environment and to attend to its features in the particular ways they do' (Ingold, 2000: 162). These dispositions and sensibilities constitute what Bourdieu calls *habitus*. *Habitus* could, to a limited extent, be equated to the earlier cultural (mental) models put forward by cognitive anthropologists (they are the result of, and direct, social activity), except that it does not exist prior to the engagement with the world in the way cultural models do. The idea of *habitus* is analogous to that of a skill, a type of knowledge that escapes any form of description or representation, obtained through the daily repetition of activities and not through formal instruction. It is in the context of these acquired abilities that perception and spatial mastery must be understood. People perceive and share similar knowledge of the world because they share the context in which this knowledge is generated but not necessarily the content itself.

Bourdieu's work intertwines with that of other thinkers to create what Ingold calls the *dwelling perspective*: '. . . a perspective that treats the immersion of the organism-person in an environment or lifeworld as an inescapable condition of existence' (Bourdieu, 1977: 153). This perspective shares many of the features that Thrift (1996) identifies as part of his *non-representational theory* and which are spelled out and commented on in the following paragraphs:

> *First [...] non-representational thinking throws a critical light on theories that claim to re-present some naturally present reality [...] Instead it argues that practices constitute our sense of the real.*
>
> (Thrift, 1996: 7)

> *Second [...] it valorises practical expertise. That is, it is concerned with thought-in-action, with presentation rather than representation.*
>
> (Thrift, 1996: 7)

This second point brings to our attention the importance of understanding the stream of everyday life earlier discussed by Bourdieu. For the most part, we operate in the world using a kind of knowledge (*practical knowledge* or *habitus*) which is best understood as the type we would associate with a skill. This is knowledge that has been forged through the repeated performance of routine daily activities and learnt through trial-and-error. It is the type of knowledge that we employ in our daily tasks, which by and large is not consciously represented in our heads (resisting any sort of propositional form), or communicated through any sort of formal instruction.

> *Third, this valorisation of thought-in-action emphasises the particular moment, in that it suggests that representation is always part of presentation, laid out in a specific context which invites only particular kinds of presencing practices.*
>
> (Thrift, 1996: 7)

> *Fourth, it is concerned with thinking with the entire body.*
>
> (Thrift, 1996: 7)

The fourth point refers to the necessity of re-instating the entire body as an active and essential element in perception. A concern that is best understood when we consider the importance of its role during the process of skill acquisition. This process is superbly described by Dreyfus in his article 'The *Current relevance of Merleau-Ponty's Phenomenology to Embodiment*' (1995). In this article, Dreyfus makes reference to two concepts introduced by Merleau-Ponty in his *Phenomenology of Perception* (1962) – the *intentional arc* and the *maximal grip* – to explain the process by which how people become skilful. According to his interpretation, Merleau-Ponty's *intentional arc* refers to the tight connection between the agent and the world, and the fact that as the agent acquires more skills these are retained, 'not as representations in the mind, but as dispositions to respond to the solicitations of the world.' *Maximal grip* on the other hand, '... names the body's tendency to respond to these solicitations in such a way as to bring the current situation closer to the agent's sense of optimal gestalt'. As we become more skilful in our everyday chores, the world reveals itself each time with less ambiguity and in more detail. In a very simplified form this is analogous, according to Dreyfus, to the initial set of values (i.e. weights) that is present in a *neural network* at the time it is presented with a new input. It is precisely this initial pattern of activation that, according to Dreyfus, provides the neural basis for Merleau-Ponty's *intentional arc*. Explaining the process of learning, on the other hand, is a bit more complicated. It requires a way of explaining how is it that the body recognizes a certain input directly as a deviation from a prototypical one (as opposed to matching the same input with some internal representation in our mind). In order to do so, it is necessary to understand our capacity for generalizing, and why the number of generalizations we produce is restricted. Dreyfus, using Merleau-Ponty as his source of inspiration, argues that our body is responsible for constraining the space of possible generalizations in three ways:

- because of our brain architecture;
- by dictating the order and frequency of the inputs (how, what and when we encounter the world) given the structure of the body and that of the world;
- by determining what counts as success.

It is precisely in relation to this last point that the concept of *maximal grip* becomes helpful. As it stands, it would seem that in order to learn from one's successes and failures, one needs a representation of one's goal beforehand. Looked at in a different way, this question can be translated into another one; whether the intentional content (i.e. conditions of satisfaction, in this case, what is the goal?) that governs an action must be represented in the mind. According to Merleau-Ponty, this is not necessary as an action can conform to such conditions without the agent having these conditions in the mind as a goal. Acting is experienced as a steady flow of skilful activity in response to one's sense of the situation. Part of this experience is a sense that when one's situation deviates from some optimal body-environment relationship, the agent tends to alleviate this tension by means of his/her activity, without knowing or being able to put into words what that optimum is. One way of thinking of this mechanism is by using the concept of attractors found in non-linear dynamic systems and Walter Freeman's *attractor theory* in neuroscience.

> *Past experience has set up the neuron connections so that the current perceptual input, which is similar to some past input but never exactly like it, puts the brain area that controls the movement into a specific energy landscape. Once that brain area is in that landscape, movements are caused that tend to move the brain state closer to the bottom of the nearest basin of attraction.*
>
> (Dreyfus, 1995)

Returning to Thrift's main points of his non-representational theory:

> *Fifth, and relatedly, it invites a degree of scepticism about the 'linguistic turn' in the social sciences and humanities, suggesting that this turn has too often cut us off from much that is most interesting about human practices, most especially their embodied and situated nature, by stressing certain aspects of the verbal cum-visual as 'the only home of social knowledge' (Curt 1994, p. 139) at the expense of the haptic, acoustic, the kinesthetic and the iconic. (Claasen 1993; Serres 1986)*
>
> (Thrift, 1996: 7)

> *Sixth, it is concerned with a rather different notion of 'explanation' which is probably best likened to understanding a person [. . .].*
>
> (Thrift, 1996: 7)

This last point briefly makes reference to the nature of what constitutes an explanation, and the possibility that such might be better understood as being less of an ordered description and more of an empathic understanding: an understanding obtained by re-enacting or sharing the context in which knowledge, or perception, is formulated.

This is the type of understanding that Ingold (2000: 167) suggests can be obtained through ethnographic work, as summarized in the following paragraph:

> *When I was a child my father, who is a botanist, used to take me for walks, in the countryside, pointing out on the way all the plants and the fungi – especially the fungi – that grew here and there. Sometimes he would get me to smell them. Or to try out their distinctive tastes. His manner of teaching was to show me things, literally to point them out. If I would but notice the things to which he directed my attention, and recognise, the sights the smells and tastes that he wanted me to experience because they were so dear to him, then I would discover for myself much of what he already knew.*
>
> (Ingold, 2000: 20)

Most authors agree that concern with understanding the world, by attending to everyday practices rather than to thought alone (on which *non-representational* theories and approaches such as Ingold's are based), originated with Heidegger's phenomenology and was carried through by following thinkers. Among these, for instance, was Merleau-Ponty who discussed in depth the nature of perception and the primordial role that the human body plays within it. To review the philosophical background is beyond the scope of this chapter (Langer, 1989, Dreyfus, 1992). However, a few lines on how Heidegger's phenomenology breaks away from the Cartesian dualism inherent in the approaches discussed in the previous section might be useful.

In Martin Heidegger's work, knowledge shifts from being an epistemological issue to an ontological one. In order to understand the world, we need to take as our point of departure the fact that we are part of it as a being-in-the world. No matter what, we cannot escape from this contention. This is well captured in Heidegger's discussion on the different modalities on how the world reveals itself to us (Dreyfus, 1992). Basically, the world has two ways in which it can 'show up', present itself, to a being that is active within it: *availableness* and *occurentness*. The former applies to the way in which it is taken for granted, as we go about doing our everyday tasks without really paying attention to how we proceed. For instance, when writing notes we do not pay attention to the tilt or the pressure applied on the pen, or the angle our arm makes with the paper block. The latter refers to the way in which the world appears when we self-consciously detach ourselves from the stream of everyday action and adopt a contemplative and reflective stance. Cartesian ontology assumes the latter view as its point of departure, so that objects appear as occurent entities ready to be categorized and to which meaning or function must be assigned before they are made available to use. Heidegger reverses this order of priority arguing that given our inescapable condition as beings-in-the-world, objects reveal themselves first as available (in their *availableness*). They derive their meaning from their uses within the context of daily routines. To arrive at their *occurentness* is to strip them away of their true meaning, as objects-in-the-world (Dreyfus, 1992: 169). The load of daily life relies heavily on *practical knowledge*; we only bring conscious thought into play when something unexpected arises.

To conclude this section, I want to briefly describe the work of another 'non-representational' author, psychologist J.J. Gibson (1950, 1966, 1986). Among the contentions put forward by Gibson was that perception cannot be studied in isolation. It makes no sense to study how an organism perceives without making reference to its environment. For Gibson, the act of perceiving was not restricted to the mind. Rather, it is the product of the intentional movement of the being (body and mind) in the environment.

Movement here does not refer to displacement only, though this is the one usually emphasized, but extends to other motor skills (e.g. the rotation of the eyes). Gibson believed that sensations are not the right unit to understand perception. Instead, what the organism perceives are the *invariants* (the 'constancies') 'underlying the continuous modulations of the sensory array' (Gibson, 1986: 166) as it moves from one place to another. This makes more sense when discussing visual perception (mostly, the area on which he focussed). Gibson defined the *ambient optic array* as those patterns of light reflected from the surfaces that make up an environment and that reach our eyes at each position of our body. As we move around, these patterns transform, but certain *invariants* underlie these transformations. These invariants provide *direct* information to the perceiver about the structure of the environment (this is known as *direct perception*).

The implications of Gibson's work are many, and increasingly accepted. Perhaps his single most important contribution is the fact that he acknowledged the necessity of studying perception as the purposeful interaction between the individual and the environment. Given that perception requires movement, and movement is an action, perception cannot be a prerequisite for action. Perception is an exploratory activity by nature. We perceive different 'environmental' information as we engage in different types of actions, or activities. An illustration of this last point may be found in Ramussen's observations about the *piazza* in front of *S. Maria Maggiore* in Rome:

> *The many tourists that are brought to the church on sight-seeing tours hardly notice the unique character of the surroundings. They simply check off one of the starred numbers in their guide-books and hasten on to the next one. But they do not experience the place in the way some boys I saw there a few years ago did. I imagine they were pupils from a nearby monastery school. They had a recess at eleven o'clock and employed the time playing a very special kind of ball game on the broad terrace at the top of the stairs. It was apparently a kind of football but they also utilized the wall in the game, as in squash – a curved wall. Which they played against with great virtuosity. When the ball was out, it was most decidedly out, bouncing down all the steps and rolling several hundred feet further on with an eager boy rushing after it, in and out among motor cars and Vespas down near the great obelisk.*
>
> *I do not claim that these Italian youngsters learned more about architecture than the tourist did. But quite unconsciously they experienced certain basic elements of architecture: the horizontal planes and the vertical walls above the slopes. And they learned to play on these elements. As I sat in the shade watching them, I sensed the whole three-dimensional composition as never before. At a quarter past eleven the boys dashed off, shouting and laughing. The great basilica stood once more in silent grandeur. In similar fashion the child familiarizes himself with all sorts of playthings which increase his opportunities to experience his surroundings.*
>
> (Ramussen, 1997 [1957]: 17–18)

With every purposeful activity we become sensitized, or attuned, to certain types of information, e.g. the geometry of the piazza by bouncing the ball against its walls (see also the discussion on *visualscapes* below). This information, practical by nature, is what Gibson calls *affordances*. Affordances are properties of an environment that provide behavioral opportunities to an agent in the course of his/her direct perception. The possibility for new affordances is inexhaustible as we are constantly 'tuning our body' to new kinds of information, however, as we 'inherit' certain practices, certain ways of engaging with the world, similar affordances tend to be reproduced. New information comes in the way of purposeful acts of revelation rather than imaging. As a corollary, people perceive in a manner appropriate to a culture, not by reference to complex schemata but by 'training in everyday tasks whose successful fulfilment requires a practised ability to notice and to respond fluently to salient aspect of the environment' (Ingold, 2000: 167). Learning is not about the transmission of knowledge but about the 'education of attention'.

12.4 Comments and Implications

The intention of the previous section was to put forward, in broad terms, an alternative view to the one subscribed to by current GISc-cognitive studies, and used to explain how people perceive and operate in the world, particularly in the context of everyday life. It is a view that seeks inspiration in works on critical theory found in human geography, philosophy, and anthropology where many of these works (mainly in the social sciences and humanities) do offer ample insight into how people engage with the world, in the way of large narratives, but at the expense of methodological concerns. On the other hand critical re-evaluation in other fields like computer science (e.g. AI), mathematics (e.g. geometry), theoretical physics (e.g. non-linear dynamic systems) are developing new theoretical and methodological advances. The purpose is not to deny the usefulness of studying the possible best way to represent space to facilitate the handling and conducting of spatial operations (on these representations, that is), but to negate the fact that this might bring us closer to finding out how people become spatially competent or experience space. This also rejects the possibility of being able to generate a unified single framework that will capture the immense complexities and nuances of commonsense knowledge, given its non-discursive nature and its space-history-culture contingency. Ironically the possibility of frameworks, like those described under banners such as *commonsense geography*, seemed to be much more appropriate for non-commonsense scenarios (non-dwelling spaces) than for those in which we operate on a daily basis. The use of representations in model building, and their existence, is not disputed here but rather their correspondence to genuine cognitive constructs.

In the following section, my aim is to provide an example, admittedly a very simple one, of how GISc methods, particularly GIS, may be employed to help us understand how perceptions are constructed out of the interaction between the body and the environment. That is, to explore how, within each particular context (i.e. body-action-environment coupling), this process of interaction results in the emergence of salient features and/or properties of the environment. The stress is on employing GIS as a heuristic, an exploratory tool, rather than as a prescriptive one. In doing so it is important to tread with care, and to be aware of the limitations of using what is essentially a spatial tool to understand a spatio-temporal process.

To this day, most spatial analytical techniques depend on a single fixed frame of reference and two-dimensional spatial representations. But no one will deny that our experience in the world cannot be captured under such limited restrictions (Llobera, 1996). Our perception is constantly shifting as we move about. We do not engage with our surroundings by starting at them from above or, for that matter, by gliding about in a fashionable *fly-thru*. Instead, the world reveals itself through a sequence of partial encounters. This is true even for those places where we normally feel we can 'grasp the world all at once', e.g. places usually associated with panoramic views. The impact of panoramic views owes as much to the visual quality experienced at that location as it does to the fact that in arriving at these locations we traverse other locations that do not possess such visual quality.

But the world we meet during our daily chores is not random. It has a certain structure that emerges as we move about and re-experience every time we re-visit a place. This structure is an active element in forming our experiences in the world. Human agents are constantly transforming such structure and in doing so they are transforming the nature of the tasks they encounter. As AI has demonstrated (Clark, 1997: 66), this makes sense from a practical point of view: individuals actively structure their environment in order to help them cope with daily problem solving. This structure is socially charged:

> *Every social order systematically takes advantage of the disposition of the body and language to function as depositories of deferred thoughts that can be triggered off at a distance in space and time by the simple effect of re-placing the body in an overall posture which recalls associated thoughts and feelings, in one of the inductive states of the body which as actors give rise to states of mind.*
>
> (Bourdieu, 1990: 90)

We have already seen that such a way of thinking goes against the *classical* cognitive agenda that denies the pervasive and active tendency humans have to structure their environments (Clark, 1997: 150). Moreover, this structure can only be understood in relation to the physical properties of the body. This means that the incorporation of the body, in any possible measure, cannot be seen as being peripheral, as claimed by Mark *et al.* (1999: 757), but as central to our exploration of space. This structure, whilst widely experienced by all of us, is mostly unknown and could be brought to light through the development of new methodology within GISc. Development towards a *body-centered* 'geography' would undoubtedly help reduce the existing divide between GISc and contemporary spatial theory and precipitate many new insights and challenges. But this requires a shift in emphasis, from being concerned with spatial representations and categories to being interested in spatial processes. It is worth mentioning that it is an endeavor that must be pursued in association with other subject areas particularly those such as ecological psychology and ecological physics.

In the following section, I illustrate some of the possibilities by providing examples that refer to the study of visual space.

12.5 Encountering Order: Two Case-studies from Visual Space

Without prior knowledge, it is difficult to know what the designers of the first GIS had in mind when they generated their initial *viewshed*. We do not know whether a *viewshed* was meant to describe what an individual could see or where he/she could be seen from, and whether it was considered to be a good enough expression. Whatever the reasons, the truth of the matter is that since its inception people have been using it, with more or less awareness of its limitations, to model human vision in landscapes.

No review is provided here on the limitations surrounding current viewshed routines as these have already been discussed elsewhere (Fisher, 1991, 1992, 1993, 1996; Gillings and Wheatley, 2000). Many of these limitations are not only computational but conceptual as well, e.g. how useful are lines-of-sight (LoS) to explore visibility? People's vision is not restricted to single points but to entire visual fields. Still, a lot of the information needed to overcome many of these deficiencies may already be available in other subject fields. Among these it is worth mentioning: computer graphics (especially the area of computer vision) and game programming, which offer important insights regarding the computational aspects of visibility. Unfortunately, these areas are not interested in exploring the spatial dimension of visual space.

12.5.1 The Structure of Visual Space: Visualscapes

Among many other concepts, J.J. Gibson's work on ecological psychology introduced the notion of *ambient optic array* as a way of describing how the environment structured human vision, i.e. how the surrounding patterns of light arrive at the eye from the environment and the potential information that these may carry (Gordon, 1989: 155).

Elsewhere (Llobera, 2003), I introduced the concept of a *visualscape*, as an operational concept within GISc that could be employed to explore such structure.

> *A visualscape is defined . . . as the spatial representation of any visual property generated by, or associated with, a spatial configuration.*
>
> (Llobera, 2003: 30)

We can think of visualscapes as different ways of breaking down and representing the spatial aspect of the *ambient optic array.* Most GIS users are familiar with simple visualscapes such as viewsheds, cumulative viewsheds (Wheatley, 1995) and *total viewsheds* (Llobera, 2003). But more informative versions can be generated, and their properties further explored, by mapping in space other visual parameters (see below) and/or incorporating other forms of spatial representation to store visual information in space.

For instance, let us consider the visual structure associated with an entire landscape (in this case, with its physical topography as described by a DEM (Digital Elevation Model) or DTM (Digital Terrain Model; Figure 12.1). One way of obtaining a coarse description of such structure is by calculating the total viewshed (*dominance viewgrid* in Lee and Stucky, 1998). Total viewsheds require lots of processing time but are currently becoming much more accessible, especially when standard GIS capabilities (e.g. the standard

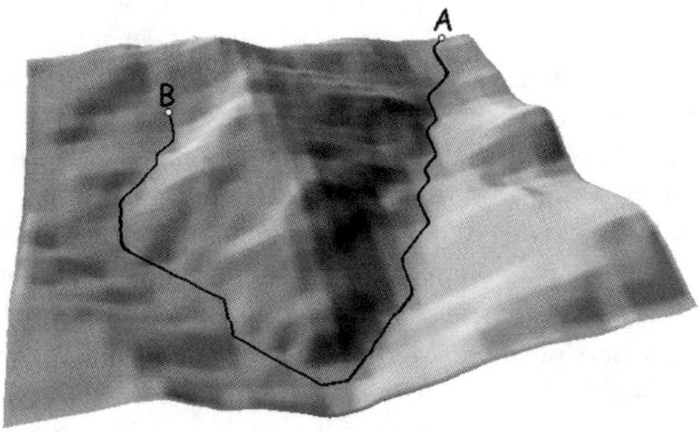

Figure 12.1 *DEM in 3D with path*

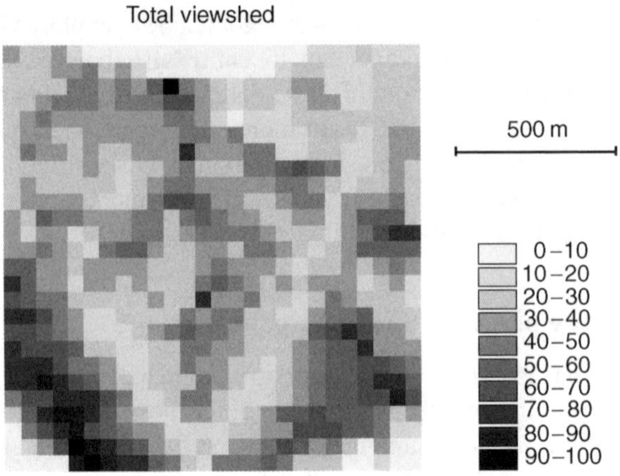

Figure 12.2 *Total viewshed for sample* DEM. *Lighter means a higher value (in this case, more visible locations)*

viewshed routine) are used in combination with grid-computing (Figure 12.2). Approximate descriptions of such a structure (sufficient perhaps to compare different terrains) may be computed in a fraction of the time that is needed to compute a total viewshed by using a surface network to represent the terrain (Morley and Rana, 2002).

Total viewsheds provide useful information at a glance, e.g. locations with corridor views or visually enclosed can be easily detected. They can be further explored using standard image processing techniques among others to derive additional information, e.g. *visual prominence* (Llobera, 2003), and are subject to the same limitations as traditional viewsheds (Gillings and Wheatley, 2000). However, total viewsheds are still relatively crude descriptions of visibility in a landscape as they only describe how many locations can be seen from every location in the landscape. If we consider each grid cell not as a single point but as an extension of land, a plane, (with a maximum size equal to the resolution of the grid cell), viewsheds and by extension total viewsheds cannot distinguish how much of the plane can be seen. We can improve this situation by computing another visualscape: the *(total) visual exposure*.

Visual exposure describes, by means of a *vector field*, how much (in this case of an entire terrain) can be seen at each location. In order to calculate the visual exposure, the same procedure is repeated for every viewpoint–target pair possible. It is possible to vary the sample interval (in this case the same is used for viewpoints and target points) in order to obtain results with different resolutions. The visual exposure for any target location can be computed by first calculating an orthonormal vector (i.e. perpendicular vector with a magnitude of one) to the surface at the target location, and a (normalized) LoS vector with its origin on the target location, and sense and direction pointing towards the current viewpoint. The visual exposure for a viewpoint–target pairing is the vector obtained by projecting the surface orthonormal vector onto the LoS vector, normalized by the distance between the viewpoint and the target point (the farther away, the less you see). By adding the magnitude of each of the vectors obtained by keeping a target location fixed and repeating the same operation

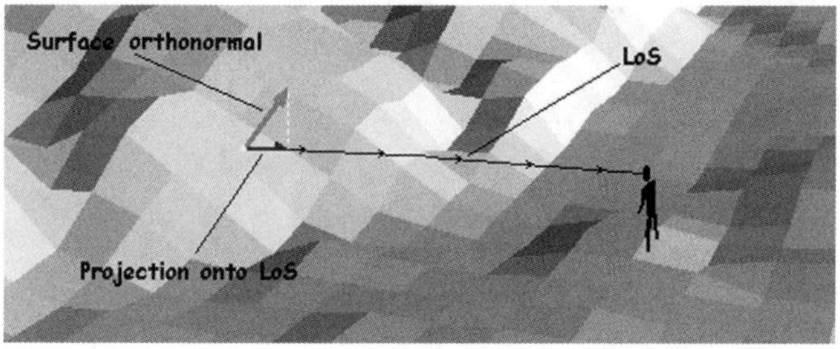

Figure 12.3 (Plate 3) *The vector nature of visual exposure*

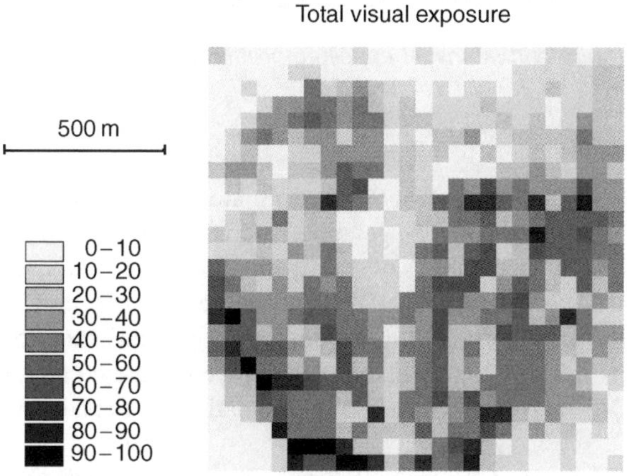

Figure 12.4 *The magnitude of visual exposure for* DEM. *Lighter mean a higher value*

for every viewpoint, we get a final vector that describes the (total) visual exposure at a target point (Figure 12.3, Colour Plate 3). Figure 12.4 shows the result of mapping this value.

Visual exposure is an improvement over the total viewshed (the latter could be generated as a by-product resulting from calculating the former). It may be further processed in order to derive new information, e.g. the result of dividing the total visual exposure by the total viewshed (in this case the *dominance viewgrid* in Lee and Stucky, 1998) allows us to distinguish between locations that are very visible because they can be seen from many locations (in spite of the fact that at each location we may see little) as opposed to being seen much more from fewer locations. As it stands, Figure 12.4 only describes part of the information contained in the visual exposure. Taken as such, this information (as with a total viewshed, Figure 12.2) cannot be used to describe accurately the nature of the visual changes (of the entire landscape) that a viewer experiences as he/she moves in a certain sense and direction (for instance, given by the path in Figure 12.1). This is because neither of these representations includes directional information. Figure 12.5(a) shows a profile obtained by extracting the values of the visual exposure along the path in Figure 12.1. To consider this profile as a faithful description of changes

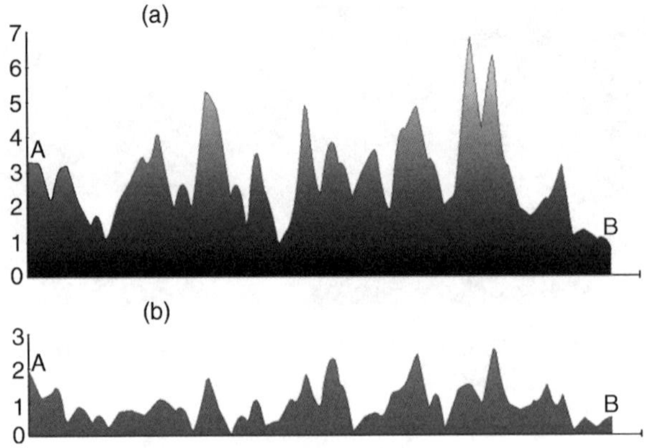

Figure 12.5 (a) Unidirectional visual exposure profile along path and (b) directional visual exposure profile along path

in landscape visibility occurring when travelling along the path from A to B (Lee and Stucky, 1998; Fisher, 1995) would be genuinely wrong, given that the sense and direction of movement (i.e. the orientation of the body following the path) have not been considered. This can be easily fixed. We can sample the path into a set of points and use each point as the origin of a vector that describes the sense and direction that the path follows in 3D. This information can then be employed to create a more accurate profile (Figure 12.5(b)); one that incorporates body orientation as dictated by the orientation of the path.

Still, we must be aware of the many limitations surrounding visual exposure. Four of these are now discussed. First, visual exposure is not immune to many of the criticisms that apply to viewsheds (e.g. no atmospheric attenuation has been included). Second, it is calculated using a very crude model of the individual (with essentially two parameters: *person's height* and *body orientation*) that does not change while moving, i.e. the body is always upright. Third, the effect of light has not been considered (essential in Gibson's discussions), although this could be easily catered for. Fourth, a constant speed of travel is implied. This limitation has to do more with using traditional spatial representations, such as a raster (which assumes a fixed sampling interval), to display the information, rather than to any limitations in computing such information, i.e. sampling could be adjusted to speed of travel. Provided that we have information that describes sense and direction of movement (given by a path or some other spatial description, e.g. cost surface), we can apply standard calculus to map out rates of visual change and the nature of such change (Llobera, 2003). Having accepted the above limitations as part of our model, can we actually use this information as a way of describing the visual experience of someone in space?

This is somewhat debatable. Although we can start thinking of mapping out where visual changes occur and their nature, this does not mean that we necessarily perceive these changes instantaneously. Let me explain with a simple example; Figure 12.6 represents a corridor seen from above. Let's imagine that we are able to derive a measure of *visual enclosure* at each location (Llobera, 1999). Location x would have a single

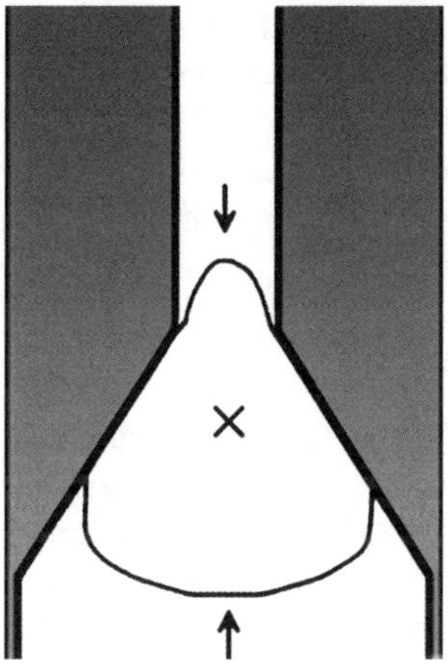

Figure 12.6 The feeling of being visually enclosed is path dependent

value but the feeling of enclosure at that point would be very different, depending on the direction from which we were coming. Thus, to get to the actual experience at a certain location/moment in time, requires that we study how these changes accumulate through time in a meaningful way. Returning to our profile in Figure 12.5(b), this means that if we wanted to fully characterize any point along this profile, we need to consider the values of all those locations that precede it (within a certain time lag?). This has further implications; given that the visual properties at any location depend on the path we use to arrive there, to characterize each point in space we would need, in theory, to consider every possible path that leads to that location (Figure 12.7). In practice, this may

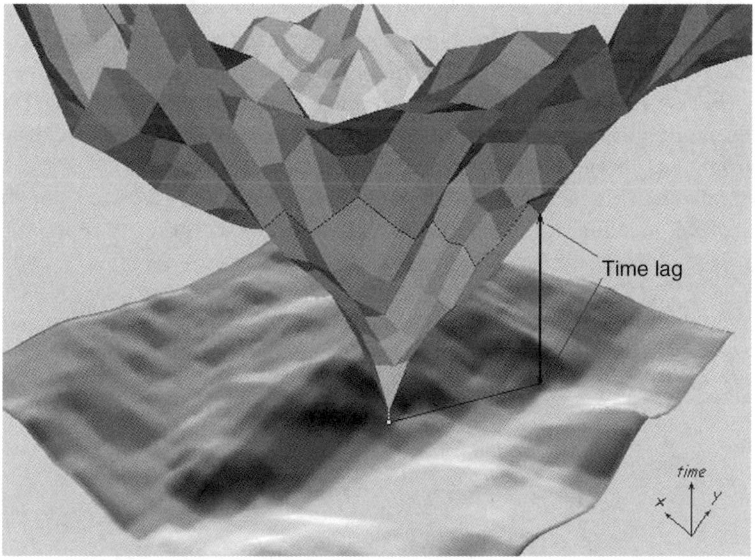

Figure 12.7 The value of visual enclosure along all paths leading to a single location

not be necessary for it is not necessarily true that the probability of moving in all directions is the same. Hence, modeling patterns of movements becomes an essential topic (Helbing and Molnár, 1995, Llobera, 2000). Several interesting questions arise from these observations: e.g. how do perceptions emerge? If in order to understand a property at a particular location, we need to consider the value of that property at previous locations, how far back in time-space do we need to go?

12.5.2 Education of Attention

I will conclude this chapter with another simple example concerning visualscapes. In the previous section I showed how different visualscapes can be used to explore the visual structure of space. In doing so, the spatial configuration used was the entire environment (in our case the entire DEM or DTM). However, I have already mentioned that this does not necessarily need to be the case; we can calculate a visualscape for a relevant subset of the environment. By allowing this possibility, we can incorporate another of Gibson's important contributions: perception as the education of attention (Gibson 1986: 149–150).

For instance, as a landscape evolves through time, so does the relevance of features in it (whether natural or built). During some periods of time, certain features become more salient than others. They might become prominent simply by the mere fact that they were built on a previously barren location. Identifying which features are mostly responsible for anchoring human activity in space during a certain period, is an aspect that landscape historians and archaeologists are trained to recognize and on which they base their interpretations (Barrett *et al.*, 1991; Barrett, 1994). Examples can be found for each period of time, from mortuary sites during prehistoric times (e.g. Neolithic) to abbeys and monasteries during medieval ones. Moreover, by virtue of their location in space and their architecture, we can say that these features act on (structure) the visual structure already extant in a landscape. I believe that with this simple idea in mind we can study how the visual structure of a landscape evolves through time, by examining the *visualscapes* generated by relevant salient features during a certain period of time.

A simple example will suffice to illustrate this point. Figure 12.8 shows the previous landscape (Figure 12.1) with three additional linear features, which could represent some sort of telecommunication antennae (this example could be extended to consider other built environments through 3D solid models).

The next figure (Figure 12.9) provides a 3D view of the magnitude of the visual exposure at each location in the landscape. The value at each grid cell represents, in this

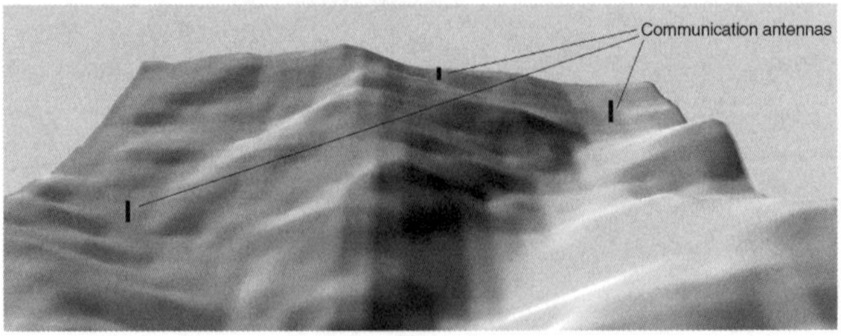

Figure 12.8 *Landscape with antenna*

Figure 12.9 Visual exposure of antenna

case, the magnitude of the sum of three vectors (one vector per antenna). The vector has its origin at each viewpoint, a direction given by the LoS uniting the viewpoint with the base of the visible portion of the antenna, and a magnitude equivalent to the angle describing the visible portion of the antenna.

It is easy to imagine how we could repeat the same calculation for other features, and then extend our study to find out, for instance, how much the visibility of the antennae changes along the path in Figure 12.1 and at which locations the impact of the antennae is greatest. These types of question are the ones on which human perception and experiences of space are built on; the type of question that methods in GISc can help us address and fully explore.

12.6 Final Remarks

My intention with this chapter is to show how methods developed within GISc may be used to throw some light on the way people perceive and experience space. It criticizes the underlying assumptions on which current GISc-cognitive approaches are based, as being rooted in old-fashioned and rather limited views on cognition. It presents an alternative way of understanding these processes based on anthropology, ecological psychology, artificial intelligence and phenomenology.

In a sense, this chapter represents a shift from viewing perception as a simple mechanism where external stimuli is matched to internal (highly complex) representations to viewing perception as a complex process in which people become 'tuned' to certain salient features in their environment (Ingold, 2000). In turn, this represents another shift, from being concerned with categories and representations (what? e.g. ontologies) to being concerned with processes (how?), and from imposing a certain *a priori* set of categories to building understanding in a bottom-up fashion.

As shown at the end of the previous section, understanding and labelling how a single property is experienced in a certain location/moment in time is far more complex than initially considered. A possible starting point is to assume the existence of certain basic properties of space as experienced from the perspective of someone *in* space (surrounded by it). Does it make sense to describe properties when a static, fixed frame of reference (as the body) is considered? Do these properties make sense under translation? Are there properties that can *only* be understood through translation? As the work of Koenderink *et al.* (2002) has pointed out, there is much to be discovered about the structure of perceptual space. Other possibilities (much more challenging) are to simulate how spatial properties arise through interaction with the environment (through a learning process) and how these become biased by constraints on such interaction. What is undeniable is the fact that the human body (with all its characteristics) is central to all of these questions.

Understanding how perceptions and experiences are generated can only be productive if it is attempted in an interdisciplinary fashion, hence, any developments towards this aim within GISc will need to be constantly informed and contrasted with studies in other fields. Finally, I am very hopeful about the possibilities that these explorations will precipitate new challenges and insights, in the way of technical and theoretical developments. But I am also aware that in spite of all of our efforts the richness of human experience will never be reduced to a set of models and simulations (not that we would want this), in the same way it resists itself from being universally described by any system of categories.

Acknowledgments

I would like to thank Dave Unwin for his kind invitation to participate in the meeting for which this work was generated. I would also like to acknowledge the other participants, particularly Nigel Thrift for his encouraging words and Peter Bibby, whose generous enthusiasm and faith convinced me that perhaps I am not doing that badly after all! I will also like to thank Helen Walkington for useful comments and corrections.

References

Barrett, J.C., 1994, *Fragments from Antiquity – An Archaeology of Social Life in Britain*, 2900–1200 BC, Oxford: Blackwell Publishers.

Barrett, J.C., Bradley, R. and Green, M., 1991, *Landscape, Monuments and Society: The Prehistory of Cranborne Chase*, Cambridge: Cambridge University Press.

Brooks, R., 1991, Intelligence without representation, *Artificial Intelligence*, **47**, pp. 139–159.

Bourdieu, P., 1977, *Outline of a Theory of Practice*, Cambridge: Cambridge University Press.

Bourdieu, P., 1990, *The Logic of Practice*, Cambridge: Polity Press.

Churchland, P.S. and Sejnowski, T., 1992, *The Computational Brain*, Cambridge, MA: MIT Press.

Clark, A., 1997, *Being There: Putting Brain Body, and World Together Again*, Cambridge, MA: MIT Press.

Classen, C., 1993, *Worlds of Sense*, London: Routledge.

Curt, B., 1994, *Textuality and Tectonics. Troubling Social and Psychological Science*, Milton Keynes: Open University Press.

Dreyfus, H.L., 1992, *Being-in-the-World: A Commentary on Heidegger's Being and Time, Division I*, Cambridge, MA: MIT Press.

Dreyfus, H.L., 1995, The Current Relevance of Merleau-Ponty's Phenomenology of Embodiment, *Filozofska Istrazivanja*, **15**(3), pp. 385–399.

Fisher, P., 1991, First experiments in viewshed uncertainty: the accuracy of the viewshed area, *Photogrammetric Engineering and Remote Sensing*, **57**(10), pp. 1321–1327.

Fisher, P., 1992, First experiments in viewshed uncertainty: simulating fuzzy viewsheds, *Photogrammetric Engineering and Remote Sensing*, **58**(3), pp. 345–327.

Fisher, P., 1993, Algorithm and implementation uncertainty in viewshed analysis, *International Journal of Geographical Information Systems*, **7**(4), pp. 331–347.
Fisher, P., 1995, An exploration of probable viewsheds in landscape planning, *Environment and Planning B: Planning and Design*, **22**(4), pp. 527–546..
Fisher, P., 1996, Extending the applicability of viewsheds in landscape planning, *Photogrammetric Engineering and Remote Sensing*, **62**(11), pp. 1297–1302.
Gibson, J.J., 1950, *The Perception of the Visual World*, Boston: Houghton Mifflin.
Gibson, J.J., 1996, *The Senses Considered as Perceptual Systems*, Boston: Houghton Mifflin.
Gibson, J.J., 1986, *The Ecological Approach to Visual Perception*, New Jersey: Lawrence Erlbaum Associates Inc.
Gillings, M. and Wheatley, D., 2000, Vision, perception and GIS: developing enriched approaches to the study of archaeological visibility. In: *Beyond the Map*, edited by Lock, G., Amsterdam: IOS Press, pp. 1–28.
Gordon, I.E., 1989, *Theories of Visual Perception*. Chichester: John Wiley and Sons, Ltd.
Helbing, D. and Molnár, P., 1995, Social force for pedestrian dynamics. *Physical Review E*, **51**(5), pp. 4282–4286.
Ingold, T., 2000, *The Perception of the Environment. Essays in Livehood, Dwelling and Skill*, London: Routledge.
Koenderink, J.J., van Doorn, A.J., Kappers, A.M.L. and Todd, J.T., 2002, Pappus in Optical Space. *Perceptions and Psychophysics*, **3**(64), pp. 380–391.
Langer, M.K., 1989, *Merleau-Ponty's Phenomenology of Perception*, A Guide and Commentary, London: MacMillan.
Lee, J. and Stucky, D., 1998, On applying viewshed analysis for determining least-cost-paths on digital elevation models, *International Journal of Geographic Information Systems*, **12**, 891–905.
Llobera, M., 1996, Exploring the topography of mind: GIS, social space and archaeology, *Antiquity*, **70**, pp. 612–622.
Llobera, M., 1999, *Landscapes of Experiences in Stone: Notes on a Humanistic Use of a GIS to Study Ancient Landscapes*, Unpublished DPhil Thesis, Oxford: Oxford University.
Llobera, M., 2000, Understanding movement: a pilot model towards the sociology of movement, In: *Beyond the Map*, edited by Lock, G., Amsterdam: IOS Press, pp. 65–85.
Llobera, M., 2003, Extending GIS-based visual analysis: the concept of visualscapes, *International Journal of Geographic Information Science*, **17**(1), pp. 25–48.
Mark, D.M. and Egenhofer, M.J., 1995, Naïve geography, In: *Spatial Information Theory: A Theoretical Basis for GIS*, edited by Frank. F.U. and Kuhn, W., Lecture Notes in Computer Science No 988. Berlin: Springer-Verlag, pp. 1–15.
Mark, D.M., Egenhofer, M.J. and Hornsby, K., 1997, Formal Models of Commonsense Geographic Worlds. Report on the Specialist Meeting of Research Initiative 21. NCGIA Technical Report 97–2.
Mark, D.M. and Frank, F.U., 1996, Experiential and formal models of geographic space. *Environment and Planing B*, **23**, pp. 3–24.
Mark, D.M., Freska, C., Hirtle, S.C., Lloyd, R. and Tversky, B., 1999, Cognitive models of geographical space. *International Journal of Information Science*, **13**(8), pp. 747–774.
Mark, D.M. and Freundschuh, S.M., 1995, Spatial concepts and cognitive models for geographic information use. In: *Cognitive Aspects of Human-Computer Interaction for Geographic Information Systems*, edited by Nyerges, T.L., Mark D.M., Laurini, R. and Engenhofer, M., Dordrecht: Kluwer Academic Publishers, pp. 21–28.
Merleau-Ponty, M., 1962, *The Phenomenology of Perception*, English translation, London: Routledge.
Morley, J. and Rana, S., 2002, Optimising visibility analyses using topographic features on the terrain. CASA *Working Paper Series* 44, London: University College.
Ramussen, S.E., 1997, *Experiencing Architecture*, Reprinting of 1959 edition, Cambridge, MA: MIT Press.
Sarres, M., 1986, *Les Ging Sens* Paris: Seuil.
Thagard, P., 1996, *Mind: Introduction to Cognitive Science*, Cambridge, MA: MIT Press.
Thrift, N., 1996, *Spatial Formations*, London: Sage.

Thrift, N., 2000, Afterwords. *Environment and Planning D*, 18, pp. 213–255.

Wheatley, D., 1995, Cumulative viewshed analysis: a GIS-based method for investigating intervisibility, and its archaeological application. In: *Archaeology and Geographic Information Systems: A European Perspective*, edited by Lock, G. and Stančič, Z., London: Taylor and Francis, pp. 171–186.

Part III

Time As Well

13

Time As Well: An Introduction

Jonathan Raper, Harvey J. Miller, Subhrajit Guhathakurta, Robert Muetzelfeldt and Tao Cheng

Time is seen by many as the 'other' in representational terms. As a concern it lies beyond objects and space, but for representational completeness it is essential that time should be explored *as well*.

Spatio-temporal representation is highly challenging but has huge potential. Representing the spatio-temporal requires theoretical, methodological and computing advances to realise the insights that can come from a better understanding of time integrated with space. The inclusion of time in our representations also poses psychological difficulties as we can only 'see' the spatio-temporal when motion or rapid change occurs. Gradual change and evolution of large scale phenomena such as sand dunes, vegetation cover or crowds are hard to imagine and to understand. Yet these phenomena and processes of the spatio-temporal are often important and constitutive, demanding further attention.

Space and time create and constrain the fabric of human interaction (Raper, 2000). Virtually all human activities require time and space. Individuals differ markedly with respect to location and timing of the key activities at home and work that anchor their daily existence. Individuals also vary with respect to their discretionary time and transport/telecommunication resources available to overcome space and time in order to access resources and participate in activities in particular locations for limited durations. Societies have created vast and complex cities, transportation and telecommunications systems to reduce the time required to move people, material and information among individuals and entities. In turn, these systems shape human activities by altering space and time. Insights and perhaps even solutions to vexing problems that face modern societies can be gained by considering space and time as an integrated framework for human behaviour and organisation.

Re-presenting GIS Edited by P. Fisher and D. Unwin

Many geographic information system software and tools linked to these software artificially separate space and time in the modelling and analysis of spatio-temporal phenomena such as cities, transportation and telecommunication systems. Static, 'place-based' GIS tools do not exploit the increasing power of computational platforms and information technologies for capturing, analysing and communicating geographic information. Consequently, urban and transportation GIS tools are increasingly irrelevant, perhaps even harmful, in a shrinking but shrivelling and fragmenting world where individuals vary substantially in their ability to overcome space and time and where activities are increasingly disconnected from place due to information technologies.

There has been a range of different strategies to represent the spatio-temporal, including:

- transformations from 4D to 2D, 2D plus time, or 1D plus time;
- addition of dynamic behaviour to spatial representations;
- animation of change;
- formalisation of qualitative spatio-temporal change.

But more work is needed on:

- emerging phenomena from attributes;
- reasoning about change;
- simulating change;
- telling us about change.

Part III of this book contains five papers that address these challenges. In Chapter 14, Jonathan Raper explores the ontological basis of the spatio-temporal in an attempt to rethink the foundations of geo-phenomena representation. He explores concepts of spatio-temporal identity as a starting point for the recognition of geo-phenomena, suggesting that more explicit procedures and richer methodologies are needed for the identification of difference. By tracing the influence of space-time models on entification, the interaction of causality and contingency on the spatio-temporal trajectories of geo-phenomena and the role of cognitive aspects of perception, he argues the case for the construction of identity on explicit grounds. The paper goes on to explore the way that studies of identity link to metadata and space-time data structures.

In Chapter 15, Tao Cheng addresses some key issues of spatio-temporal reasoning. This chapter will first discuss the perceptions of time that could be used to represent the spatio-temporal information. It is considered as one of the bases to building the spatio-temporal structures of the reasoning. Then two spatio-temporal structures of reasoning are presented as examples. One uses the linear time structure to analyse the geomorphological processes of a coastal zone. The other adopts a cyclic time structure to detect the activity patterns of peoples. It is argued that in order to fully make use of the spatio-temporal information, new spatio-temporal structures, visualisation and analytical tools are needed.

Harvey Miller then explores, in Chapter 16, the time geography of Torsten Hägerstrand who offered an alternative perspective that highlights the spatio-temporal conditions of human existence (Hägerstrand, 1970). While long recognised for its elegance, and even applied occasionally to real-world problems (e.g., Lenntorp, 1976), its power has been blunted by an unrealistic view of geographic space as homogeneous and difficulties in collecting and analysing space-time activity data. Harvey Miller discusses the growing potential for time geographic concepts to create a 'people-oriented' GIS. He discusses the use of GIS and related technologies to build more realistic time geographic constructs, collect more accurate space-time activity data cheaply, and explore complex information spaces implied by space-time activities. Miller also identifies research and development frontiers for a more complete and useful time geography in GIS. Combined with the potential for GIS in open planning processes, there is an opportunity for more sensitive and informed analysis in transportation and urban planning and policy.

Chapter 17, by Muetzelfeldt and Duckham, addresses the largely-neglected issue of how we can develop formal representations of spatio-temporal simulation models. Conventionally, such models are developed within a fixed spatial framework (usually raster-based), and implement the dynamic component in the form of programs written in a conventional programming language. This ignores the fact that developing such models is a design process, and should be supported by appropriate design tools.

Simile is a visual modelling environment that combines a System Dynamics and an object-based approach. System Dynamics is based on a conceptualisation of dynamic systems in terms of stocks and flows, variables and influences. The underlying mathematical form of such models is differential or difference equations, but the visual language enables such models to be specified in terms of network diagrams, which are easy to both construct and to communicate to others.

The object-based approach enables population of objects to be modelled, as well as the associations between them. This provides considerable flexibility in the specification of spatial frameworks, such that the distinction between (for example) raster and vector representations of space is part of the model specification, not hard-wired into the modelling environment. The modeller can equally well specify various other spatial frameworks, including novel ones that have not been used before, merely by specifying novel forms of association between objects.

The declarative modelling approach that Simile embodies has considerable significance for the practice of spatio-temporal modelling within the GI community. Rather than models being artefacts constructed by particular groups, lacking in transparency and locked into a particular spatial framework, they become statements of assumptions that are open to inspection and modification.

Finally, in Chapter 18, Subhrajit Guhathakurta is concerned with narratives in representation. Social theorists have long argued that all enquiries rely on 'pre-understanding' and 'pre-judices' since human consciousness is temporal in form. Phenomenological understanding occurs through the construction of a coherent plot or story in the mind through experiential learning. All argumentation is contextualised within this narrative, which is either communicated or inferred. Geographic models, like all other forms of analytical tools, are embedded within a narrative. However, temporal dimensions in geographic sciences have rarely been emphasised. Frequently, modellers have not consciously examined and conveyed this narrative before asserting an argument based on the empirical results. This problem is especially present in the case of empirical models that privilege associations discovered through statistical tests rather than uncovering the underlying story that generates these associations. The storyline in these cases is reduced to one or more hypotheses that are tested. Also, most empirical models are unduly restricted to measured data and have neglected the far richer and more complex body of information that exists in the experience of different individuals. These experiences can be easily described heuristically although precise measures may not be known. The critical element of understanding 'the story' is not necessarily the examination of associations but an analysis of interconnections. These interconnections are often too complex to comprehend without adequate methodological tools. One convenient approach for analysing the cumulative effect of multiple interconnections is available in the literature on dynamic, and often self-organising, systems.

Geographic models, based on dynamic models such as cellular automata or systems-based, provide some clues to capturing the narrative aspects of spatial representation. Some aspects of narratives are fairly self-evident in systems and cellular automata models. For example, systems models track the progress of an evolving process over time, such as the growth, decline and regeneration of species as determined by certain external as well as internal stimuli. This structure would be similar to a narrative construction of the same phenomenon (without the poetics). System models are also particularly adept at tracing different paths depending upon the sequence of events along the time-line. Therefore, both narrative and system model structures allow the examination of the temporal connections between events so that a coherent and unified experience is projected.

More importantly, acknowledging the narrative aspects of geographic models allows for a significant switch in our cognitive perception from the 'paradigmatic' to the 'narrative'. Bruner (1986) perceives the 'paradigmatic' realm to be the world of abstract and general theories that are empirically verified in the objective world. In contrast, he characterises the 'narrative' mode of thought as chronicles of particular events and experiences over time that gain credence through their lifelikeness. It is the quality of meaningfulness rather than factual accuracy that renders a narrative credible. Rendering meaning to a GIS model is as much related to an act of interpretation as is communicating a story because meaning does not pre-exist the interpretation of experience. Concepts such as 'explanation', 'validity', and 'verification' are redefined in the narrative forms of inquiry. The search is not for mathematical certainty but for results that are believable, meaningful, and verisimilar. This attribute of storytelling was poignantly stated by Parry and Doan (1994): 'The hearers of the story believed that it was true because it was meaningful, rather than it was meaningful because it was true'.

References

Bruner, J., 1986, *Actual Minds, Possible Worlds*, Cambridge, MA: Harvard University Press.

Hägerstrand, T., 1970, What about people in regional science? *Papers of the Regional Science Association*, **24**, 7–21.

Lenntorp, B., 1976, *Paths in Space-Time Environment: A Time Geography Study of Movement Possibilities of Individuals: Lund Studies in Geography 4*, Lund, Sweden: GWK Gleerup.

Parry, A. and R.E. Doan., 1994, *Story Re-visions*, New York, NY: Guilford Publications.

Raper, J.F., 2000, *Multidimensional Geographic Information Science*, London, Taylor and Francis.

14

Spatio-Temporal Ontology for Digital Geographies

Jonathan Raper

14.1 Introduction

In the last few years the geographic information (GI) technologies for creating, maintaining and visualising digital geographies have become mature. However, this progress in software has only served to highlight the impoverishment of our methodologies for generating the GI on which digital geographies can be built. Digital versions of paper maps have been shown to be inadequate on ontological, representational and communicative grounds (Raper *et al.*, 2002), suggesting that digital geographies must be constructed afresh.

This chapter will explore the ontological foundations of digital geographies, focusing in particular on their spatio-temporal formulation. This was a central theme of Raper (2000), who argued that geographic information systems (GIS) using digital versions of two-dimensional paper maps are stuck in 'flatland' (Abbott, 1884) where they cannot represent the multidimensional richness of the worlds in which we live.

14.2 Ontology and Digital Geographies

Ontology has recently emerged at the focus of a debate about how the geo-phenomena in experience and discourses can be best represented and communicated as digital geographies. There are a variety of threads to this emerging debate, including the socio-political reproduction of ontologies (Bowker, 2000), the ontology of natural kinds

Re-presenting GIS Edited by P. Fisher and D. Unwin

(Smith and Mark, 1998) and the influence of scale on the ontology of geo-phenomena (Freundshuh and Egenhofer, 1997; Mark *et al.*, 1999). This debate is also taking place in different forms and using different language in many different disciplinary settings: hence the use of the terminology of geo-phenomena and digital geographies to capture the generic aspects of the debate.

It is argued here (following Raper and Livingstone, 2001) that ontologies for geo-phenomena must be formulated on explicitly spatio-temporal grounds. Issues of scale, natural kinds and ontological reproduction are critically compromised by the limits of a static spatial framework. Hence, Massey (1999) argued that representation in geography should not be concerned with the 'spatialisation of the temporal...but with the representation of space-time' (Massey, 1999, p. 269). Such an objective requires a radical rethink of representation, which has become dominated by static frameworks as a consequence of the compromises required by paper maps. Whereas paper maps are fit-for-purpose as allocentric overviews of specific kinds of GI, they must not be allowed to condition the possibilities for representation.

Finally, note by way of introduction that this paper presupposes that ontology and representation are possible and desirable in the context of realism: the epistemological context and a methodological justification for this position can be found in Raper (2000).

14.3 The Spatio-temporal Nature of Geo-phenomena Identity

The conceptualisation of phenomena rests on the notion of identity, which is a foundation of ontology: in simple terms if phenomena exist then so should identity criteria for them. Specifying identity criteria involves making ontological commitments, for example, that all things can be divided into 'particulars' (individual things) or 'universals' (repeatable phenomena), which is a view held by 'metaphysical realists' (Loux, 1998). A commitment to a concept of 'universals' suggests that such repeatable phenomena could be defined by their 'properties' (such as shape), the 'kinds' of which they are members (e.g. taxonomic group) and the 'relations' (e.g. topological) that inter-connect them. Thus, in this account, the identity criteria for phenomena could be found in their properties, their kind or their relations. While there are other accounts of ontology, some of which reject realism (e.g. Heidegger), the notion of identity of phenomena is robust. The question for a digital geography is the appropriate theoretical framework by which identity is defined: in this chapter the question of the spatial and temporal framing of identity is addressed.

Raper (2000) argued that a four-dimensional space-time offered the most complete and versatile framework for the construction of geo-phenomena identity. This view is based on 'perdurantism', the ontological position that phenomena are spatially and temporally extended in a physical sense (Heller, 1990). In this framework geo-phenomena identity can be defined in terms of difference, i.e. the identity of a variety of geo-phenomena is defined in terms of their spatio-temporally extended inter-relations. The alternative view ('endurantism') that phenomena must exist separately and distinctly in all the different temporal parts of their 'lives' is more complex and more difficult to reconcile with an account of identity (Loux, 1998).

In this formulation, space and time must play a constitutive role in identity through their role as a framework for difference. If space-time is 'relational' then phenomena are different by virtue of their nature. Perdurantism requires a relational space-time since it posits a dynamic set of spatio-temporal relations. If space-time is 'absolute' then phenomena can be different by virtue of their spatio-temporal locality, which is the endurantist view.

The distinction between the endurantist, absolute view and the perdurantist, relational view can be illustrated by an example from GIS data modelling. In the endurantist, absolute approach geo-phenomena must be conceptualised as objects at-a-time e.g. a bounded traffic jam object on a highway identified at specified time intervals. In the perdurantist, relational approach geo-phenomena can be conceptualised as the closeness in space and time of a set of cars on the highway, making it possible to define a traffic jam as any specific set of inter-car relations.

In the perdurantist, relational view the dynamic set of spatio-temporal relations defining identity can be constituted by the interaction between causal processes and the contingent aspects of the environment. Hence, geo-phenomena identity can emerge from the continual interaction and feedback between causal processes such as physical forces and the actual contingent form of the environment local to the forces (Raper, 2000, Raper and Livingstone, 2001). Hence, a river in flood that avulses (breaks out of its channel) will form a new channel that will flow through the lowest-lying land in the flood plain. By definition this will be the land least often filled with sediment-laden floodwater in the past. The new flood channel will therefore gain its spatio-temporal identity from the relations between the current and previous flood avulsion paths.

Massey (1999) has argued that these causal processes should not be seen as immanent, i.e. unchanging, since that would imply time is closed. She instead proposes that time should be conceptualised in terms of 'open historicity' after Bergson's maxim that time is the 'continuous emergence of novelty'. By analogy she defines space as the 'sphere of open multiplicity' (Massey, 2001). The fusion of space and time in this way makes it possible to identify a multiplicity of spatio-temporal trajectories for geo-phenomena. However, the infinite future potential for human creativity is not constrained in the way that physical systems are by entropy in the Second Law of Thermodynamics.

Layered over these ontological considerations are cognitive ones. The identity of (geo-)phenomena can be said to emerge through the interaction of socially driven cognitive acts with the heterogeneous spatio-temporally extended structure of the world (Raper, 2000). Awareness of identity emerges from the interaction between human intentionality and the 'difference' that our senses, our viewpoint and our experience allow. Thus, human intentionality both constructs, and is constructed by, salient phenomena and processes in our worlds of experience. Consider, therefore, the expansion of geo-phenomena identified and named following the introduction of widespread aerial photography and remote sensing in the 1960s.

Thus, geo-phenomena identity is spatio-temporally constituted difference, driven by feedback between the causal and the contingent and highlighted by human intentionality. Since Massey (1999) has argued that 'the representation of space-time is itself an emergent product of the conceptualisation of the space-time entities themselves' (Massey, 1999, p. 269), it can be argued that representation should emerge from concepts of identity and not be imposed by software concepts.

The early history of GI saw representational concepts derived from computer graphics driving the creation of digital facsimiles of two-dimensional paper maps. More recently the development of object-oriented software architectures has allowed the development of much richer representational approaches in which geo-phenomena can be constructed according to rules, temporal sequence or inter-relations with other geo-phenomena. However, there is much work yet to do in order to develop a GI infrastructure that reflects the identity of geo-phenomena required by user communities.

14.4 Rethinking Representation from the Perspective of Identity

The impoverishment of our methodologies for generating GI is evident in the mismatch between concepts of identity and the forms of representation being used for geo-phenomena. The classic example of this problem is the representation of the coast, where the differing construction of identity between land and sea has led to incompatible ontologies across the shoreline. In this case, ontological work is needed to determine the spatio-temporal identity of coastal geo-phenomena. Densham and Raper (2001) described ontological fieldwork in which written evidence submitted to a British Houses of Parliament Select Committee was searched to produce a list of 1795 user-defined coastal concepts.

These concepts were clustered semantically to produce 35 'coastal objects' with common identity, whose spatio-temporal nature was explored. The modal change frequency was $\leq$3 months (see Figure 14.1), suggesting that a static ontology for a digital geography of the coast is, to say the least, problematic. Note also that a coastal GI collection will always be spatially and temporally heterogeneous, and could be temporally detailed and spatially sparse or spatially extensive and temporally integrating. An ontological framework for GI should reflect such heterogeneity.

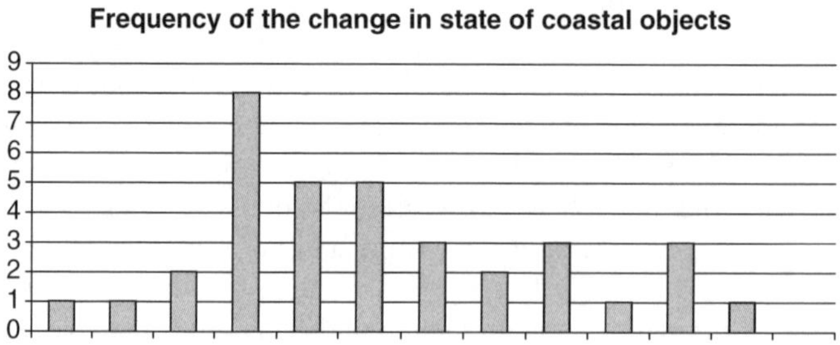

Figure 14.1 Frequency of changes of state (in one hour increments) for the 35 'coastal objects' in Densham and Raper (2001)

This impoverishment is also recognisable in the crisis in naming that has been exposed in the development of metadata schemes. Bowker (2000) has argued that in Biodiversity studies 'each (sub)discipline is an effective spokesperson . . . for the objects plus spatial and temporal units it produces', suggesting that 'these sciences deal on objects, spaces and times that cannot be readily normalised against each other' (Bowker, 2000, p. 755). Such a 'normalisation' is an example of an ontological 'grand challenge' suitable for a GI science research agenda.

A further solution to the impoverishment of GI that is orthogonal to representation and metadata is the issue of the spatial and temporal structures that are employed to evaluate identity. Raper (2000) suggested that there were three possible approaches to spatial and temporal structuring for geo-phenomena: first, the connections between space and time (whether hybrid or physically integrated); second, how space and time are discretised (whether continuous or discrete); and, third, the models of space and time (absolute or relative) that are employed. These three pairs of space-time structures can be combined to identify eight qualitative combinations of space-time properties and the associated geo-representations, which have been tabulated for each type in Table 14.1.

Table 14.1 *Space-time structures: for named models see references (from Raper, 2000)*

Space-time structure type	Properties	Geo-representations
Type 0	Integrated discrete absolute	4D GIS
Type 1	Integrated discrete relative	e.g. OOgeomorph model (Raper and Livingstone 1995)
Type 2	Hybrid discrete absolute	e.g. ESTDM model (Peuquet and Duan 1995)
Type 3	Hybrid discrete relative	Spatio-temporal autocorrelation
Type 4	Integrated continuous absolute	Field equations of physics
Type 5	Integrated continuous relative	Chaos theory
Type 6	Hybrid continuous absolute	4D process model
Type 7	Hybrid continuous relative	Catastrophe theory

These space-time structure types offer a range of representational options for the geographic information scientist: commercial GIS have only generally implemented Type 2. The alternative space-time structures outlined here are capable of supporting new concepts of identity, which in turn can be the foundation of new spatio-temporal ontologies. Many of these concepts have been explored in disciplines outside GI Science and lack digital implementations: this should not disqualify them for consideration as representational foundations for intractable research problems.

14.5 Conclusions

Thus, there is both the need and considerable scope for an ontological project that connects concepts of spatio-temporal identity with GI. There could be a historical perspective to this project to explore which socially driven cognitive acts actually did interact with what heterogeneous spatio-temporally extended structures of the world, to produce phenomena, as for example in the account given of the 19th century surveying of India in Keay (2000). There could be ontological fieldwork among the plethora of terminology and concepts that are currently quite uncritically encoded into homogeneous static representations of Type 2 in Table 14.1. There could also be exploration of the cognitive foundations of 'salience' and 'difference' in our discourses and hypotheses given different concepts of space and time and concepts of scale and detail. Only then will we tackle the impoverishment of our methodologies for generating geographic information.

References

Abbott, E.A., 1884, *Flatland: A Romance in many Dimensions*, (Original author: Square, A.)

Bowker, G.C., 2000, Mapping biodiversity, *International Journal of Geographical Information Science*, **14**(8), 739–754.

Densham, I. and Raper, J.F., 2001, First things first: an ontological approach to coastal mapping. In Wise, S., Brindley, P., Kim, Y-H. and Openshaw, C. (eds), *Proc. of the GIS Research UK 10th Annual Conference*, 3-5th April 2002, pp. 36–40.

Freundshuh, S.M. and Egenhofer, M.J., 1997, Human conceptions of spaces: implications for GIS, *Transactions of GIS*, **2**, 361–375.

Heller, M., 1990, *The Ontology of Physical Objects: Four Dimensional Hunks of Matter*, Cambridge, CUP.

Keay, J., 2000, *The Great Arc: The Dramatic Tale of how india was Mapped and Everest was Named*, New York, HarperCollins.

Loux, M.J., 1998, *Metaphysics*, Routledge, London.

Mark, D.M., Smith, B. and Tversky, B., 1999, Ontology and geographical objects: an empirical study of cognitive categorisation. In: *Spatial Information Theory*, Proc. COSIT '99, edited by Freksa, C. and Mark, D., Lecture Notes in Computer Science 1661, pp. 283–298

Massey, D., 1999, Space-time, 'science', and the relationship between physical geography and human geography, *Transactions of the Institute of British Geographers* NS **24**, 261–276.

Massey, D., 2001, Talking of space-time. *Transactions of the Institute of British Geographers,* NS **26**, 257–261.

Peuquet, D. and Duan, N., 1995, An event-based spatio-temporal data model (ESTDM) for temporal analysis of geographic data, *International Journal of Geographical Information Science*, **9**, 7–24.

Raper, J.F., 2000, *Multidimensional Geographic Information Science*, London, Taylor and Francis.

Raper, J.F. and Livingstone, D., 1995, Development of a geomorphological data model using object-oriented design. *International Journal of Geographical Information Systems*, **9**(4), 359–383.

Raper, J.F. and Livingstone, D.L., 2001, Let's get real: spatio-temporal identity and geographic entities, *Transactions of the Institute of British Geographers*, NS **26**(2), 237–242.

Raper, J.F., Dykes, J.D., Wood, J., Mountain, D., Krause, A. and Rhind, D.W., 2002, A framework for evaluating geographical information, *Journal of Information Science*, **28**(1), 39–50.

Smith, B. and Mark, D.M., 1998, Ontology and geographical kinds. In: *Proceedings, 8th International Symposium on Spatial Data Handling*, edited by Poiker, T and Chrisman N., 12–15 July, 1998, pp. 308–320.

15

Modeling and Visualizing Linear and Cyclic Changes

Tao Cheng

15.1 Introduction

Time and space are two important aspects of many real world phenomena and have significant implication for GIS research: 'For most purposes and situations space and time seem to be just given, standing right out there. Everyone possesses a wide background of spatial and temporal concepts: from very basic, intuitive and almost innate or automatically applied cognitive structures, to very sophisticated and specific concepts developed inside a particular scientific or professional community' (Nunes, 1991).

There are two major metaphors deeply embedded in human thought about time. One is linear time and the other is cyclic time. In the linear time metaphor, time is considered as a line, which represents the continuous novelty of the perceived world and that things always have an element of newness. In the cyclic metaphor, the perceived world is considered as a cycle, with constancy and continuity. There is change in the time cyclical model, but it forms part of a great continuum (Hazelton, 1992).

There are also two more metaphors about time. One is 'multi-strands of time' or 'multiple branches of time', which can be used to explain the generation of multiple scenarios, each with its own history and future. This is most useful when it is used in conjunction with 'What if?' considerations. The other is 'multi-dimensional time', which is used for dealing with concepts of multiple views of a single event or an object's history. Recently, this metaphor has been specialized as a means of discussing the

Re-presenting GIS Edited by P. Fisher and D. Unwin

differences between 'world time' and 'database time' in GIS (Hazelton, 1992; Snodgrass, 1992).

However, the 'multi-strands of time' model can be considered merely as an aggregation of separate linear ones, and also can be modified to include 'multi-dimensional time' by allowing strands to rejoin after separation. Therefore 'multi-strands of time' and 'multi-dimension of time' can be combined with the 'linear time' model (Hazelton, 1992).

Therefore, the main logic models of time are cyclic and linear time. The combination of cyclic and linear time may be realized through treating time as a parameter of dynamic modeling (Hazelton, 1992).

15.2 Spatio-Temporal Behaviors

Based on the changing characteristics of the spatial attributes of objects, such as location, boundary or shape, three types of spatio-temporal behaviors can be differentiated (Tryfona and Jensen, 1999; Hornsby and Engenhofer, 2000; Renolen, 2000):

- *Continuous* change: objects of this type are always considered to be in a changing state. For example, the flow of water, the moving plume of an oil spill or the occurrence of a storm are usually modeled as 'moving' objects with changing properties (e.g., intensity) and shape over time in environmental applications. In such cases, objects change in location as well as in shape in a continuous manner.
- *Discrete* change: objects of this type are always in static states but change instantaneously with events. This behavior involves objects located in space, whose characteristics, such as shape, as well as their position may change suddenly in time. The land that a person owns is a typical example.
- *Stepwise* change: objects of this type are sometimes static and sometimes change, for example, movement of people and transportation vehicles (car, plane, ship, etc.). In this type of application, objects change in spatial position, but not in shape. For example, a car moves on a road network; the location of the car is changing, but its shape remains unchanged.

The three types of spatial-temporal behaviors are illustrated in Figure 15.1. Panel (a) illustrates the continuous type of spatial-temporal behavior by a slant line. Panel (b) describes the discrete change by several horizontal and vertical lines. The horizontal line indicates the static states, and vertical lines represent the sudden changes between two static states. Panel (c) shows the stepwise change by horizontal and slant lines with the horizontal lines representing the static states and the slant lines representing states in motion.

The spatio-temporal process might be continuous, discrete or stepwise (Wang and Cheng, 2001), but it may proceed linearly or cyclically, or both. For example, most human and natural phenomena occur in both linear and cyclic modes (Table 15.1).

Different patterns (continuous, discrete or stepwise) occurring either linearly or cyclically urge us to re-think the spatio-temporal structures we may use. For example, the linear movement of coastal objects needs a linear time to match its continuous change, while the seasonal movement of migratory birds and human activities are

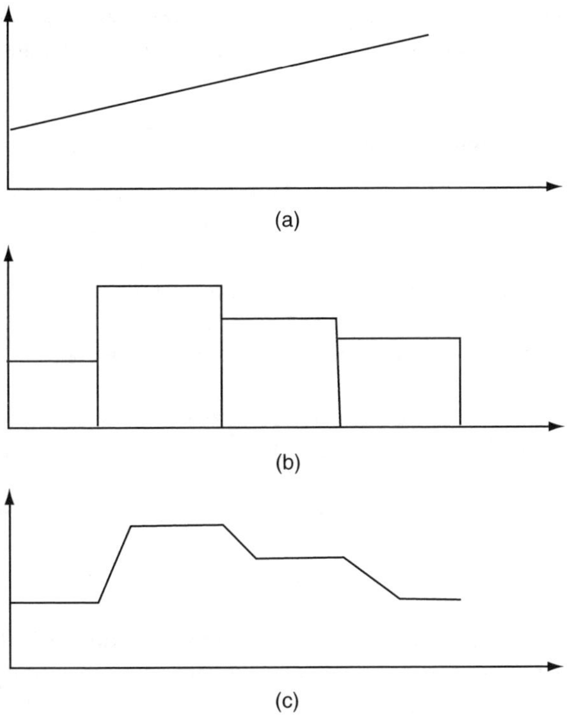

Figure 15.1 *Three types of spatio-temporal behaviors (Reproduced from Wang, D. and Cheng, T., 2001, a spatio-temporal data model for activity-based transport demand modeling, International Journal of Geographical Information Science,* ***15****, 561–586)*

Table 15.1 *Linear and cylic phenomena*

	Linear	Cyclic
Continuos	Storm Stock market	Storm Stock market
Discrete	Earthquake War	Earthquake War
Stepwise	Human daily activity Migration of bird	Human daily activity Migration of birds

conducted according to calendars of cyclic time. It is important for the representation of these latter kinds of phenomena to conform to their cyclic patterns.

15.3 Spatio-Temporal Structures

Since natural phenomena proceed linearly and cyclically, two kinds of time should be adopted and these lead to reasoning about time in the two ways. For example, Harrower *et al.* (2000) adopted two temporal legend styles to promote a linear and a cyclic understanding of time in a geovisualization tool to support earth science learning. Here we present two cases that make use of linear time and cyclic time respectively, for reasoning.

15.3.1 Reasoning in Linear Time

Combined with the linear model, time may be regarded as discrete, dense or continuous. If time is discrete it is isomorphic to the natural numbers; each point in time has a single successor. When time is dense, it is isomorphic to either the rational or the real numbers (between any two moments of time another moment exists), and when time is continuous, it is isomorphic to the real numbers. Although the natural phenomena are changing continuously, from the viewpoint of practice, the *discrete* model is preferred over the continuous model because measures are taken at specific points in time. Similarly, space may be regarded as discrete, dense or continuous (Snodgrass, 1992).

When time is regarded as linear and discrete, it is usually treated as another dimension similar and perpendicular to other spatial dimensions in a Cartesian coordinate system. Such presentation of time may be able to show states at particular moments, to display trends and evaluate time dependency.

Here we will study the dynamic process of the coastal zone in such a linear time structure. To monitor such phenomena they can be observed at a sequence of intervals, where at each interval they are sampled sparsely (Figure 15.2, Colour Plate 4). Afterwards, changes in objects are detected by comparing states at different intervals.

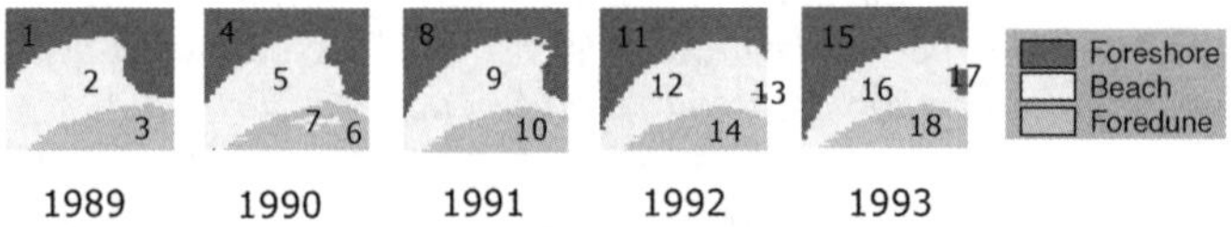

Figure 15.2 (Plate 4) *Classified regions (Reproduced from Cheng, T. and Molenaar, M, 1999, Diachronic analysis of fuzzy objects, GeoInformatica,* ***3****(4), 337–356; with kind permission of Springer Science and Business Media)*

Experts then compare the objects and the changes in the objects. For example, as shown in Figure 15.3, three states of objects were obtained from observation data. The historic lifeline can be constructed by linking the regions appearing in consecutive states. Region 1 is linked with Region 4, and 4 with 8, representing the lifeline of Object 1; Region 2 is linked with Region 5, representing the lifeline of Object 2; Region 3 with Region 6, represents the lifeline of Object 3; Region 3 with 7 and Region 7 with 10, represents the lifeline of Object 4. This procedure is usually done by domain experts who then use their knowledge of the environmental processes to infer the active processes. The temporal relationships between objects are also derived, e.g. Object 3 is split into two regions (7 and 6) in 1990, which means new object (Object 4) is created this year and Object 4 has merged into Object 3 in 1991. A method for the automatic analysis of the relationships of regions and identification of objects and their processes was proposed by Cheng and Molenaar (1999).

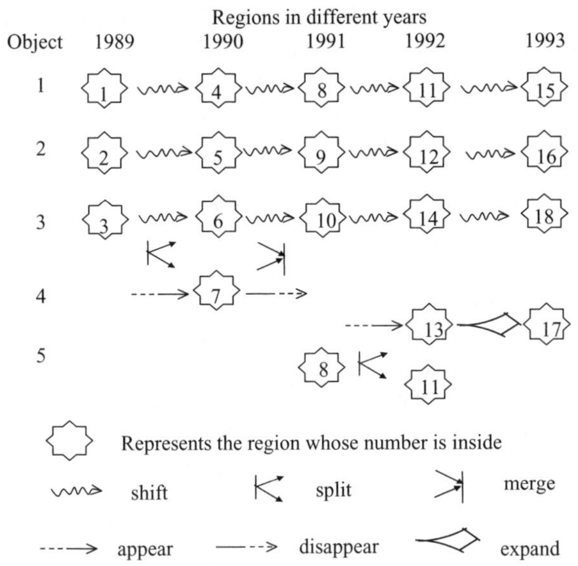

Figure 15.3 *Identified fuzzy objects and processes (Reproduced from Cheng, T. and Molenaar, M, 1999, Diachronic analysis of fuzzy objects, GeoInformatica, **3**(4), 337–356; with kind permission of Springer Science and Business Media)*

15.3.2 Reasoning in Cyclic Time

Linear time may not be appropriate for evaluation of time dependency (transient, periodic, renewable?), or where a cyclic time and a branch time are more appropriate (do they split, merge, spread, move?).

There are many natural and man-made events that occur in time cycles. In winter, migratory birds temporarily migrate from high latitudes to more tropical climes to avoid the harsh weather in the polar and mid-latitude regions; sea levels change with the tidal cycle; in most areas of the world, climate changes through four seasons every year; human beings arrange their activities according to calendar times, and they usually have schedules for each day, each month or each year. The economy of a country, or the whole world, may experience recessions every few decades or even years.

More than two decades ago, Moellering (1976) revealed through map animation the daily cycle of traffic accidents with peaks during rush hours periods and troughs between, as well as the temporal patterns of weekdays versus weekend. Moellering's innovation was to depict the location and time of accidents for a composite week in frames representing 15-minute intervals containing all accidents during that time on that day of the week for a three-year period. The result was a clear representation of an understanding of cyclic phenomena.

All the phenomena mentioned above have a common characteristic: they all occur in time cycles. It is important for the visualization of such patterns to conform to the cyclic nature of these phenomena (Hornsby and Egenhofer, 2002).

When an event changes over time, it is essential for the visualization of that event to include a variable that shows the time dimension. In other words, the time dimension needs to be presented together with the main phenomena that needs to be shown. Two types of representation have been adopted: legends in a separate display area (such as an analog or digital clock or side bar) or embedded into the map display as a variable on the map (Kraak *et al.*, 1997).

Legends in a separate display area can also be used. One example can be found in a geovisualization tool developed by Harrower *et al.* (2000). A four-square legend is adopted to represent cyclic time or cyclic understanding of time. We would like to say that the legend-based representation is still a snapshot-oriented representation. The dependency of time is viewed at particular points in time. Furthermore, the four-square legend is still quite different from the clocks people use in their daily life. According to Harrower *et al.* (2000), it was unclear whether the legend referred to the entire globe or only one portion of the globe. If the time is embedded into the map display, conventionally it is represented as a third dimension in the Cartesian coordinate system, orthogonal to the two spatial dimensions: x and y (Langran, 1992). This representation, however, assumes that events occur in linear time, not in cyclic time. As argued above, this linear time structure is not suitable to represent the cyclic patterns. Also, this view of time is very different from a human's general view of time: the clock. Therefore, we should think of spatio-temporal structures that could represent the time dimension in accordance with human being's general perception of time, i.e. similar to our clock time in our daily life.

In the polar coordinate system, the polar angle can be used to represent the time dimension, which can be a clock at daily, monthly or yearly resolution. It means the full range (360 °) can represent a cycle of 24-hours, 30-days, 12-months or any other time duration that is a time unit for an event to repeat. The status of the event (or a object) can be represented by the polar distance, which shows the distance between the current status and its reference status (or place) (such as the home of a person or an animal, or the average temperature of a month). In such a way, the changing pattern (e.g. the moving extension in space) can be clearly revealed.

Therefore, the spatio-temporal information of an object P is represented by three elements (r, θ, a), with $r \geq 0$ representing the distance of the current position of P to the central point O, angle θ in the anti-clock direction to the polar axis representing the time dimension, and a representing the attributes of the object. Based upon these three elements, the state of the object, i.e. the spatial, temporal and thematic information is represented. This supports the clock view of time.

In case we need a duration view of time, one more dimension of time is added to the three elements. Therefore, a four-tuple representation can be created as $P(r, \theta_1, \theta_2, a)$, which means during time (θ_1, θ_2) the object has the attribute a. Of course, attribute a can be a single or a set of thematic attribute(s) of P.

We may use home as the pole and all other activity destinations are represented in reference to this pole by polar distance, which indicates the actual spatial relations between activity destinations and home. As activity patterns can be considered as a series of stay (in activity destinations) and travel between (activity destinations), the key to representing activity patterns is thus to visualize accurately these two mobility states. Since a travel-between state describes the motion between two activity destinations, which have different distances from home, it is represented by a curve between two points: a starting point and an ending point. The starting point represents the location and the ending-time point of the previous stay in the activity pattern. The two coordinates of this starting point are respectively the polar distance r (indicating the spatial relation of the location of the previous stay to home) and the polar angle θ (indicating the clock time of the ending time of the previous stay). The ending point represents the location and the starting-time point of the next stay in the activity pattern. The two coordinates of this ending point are respectively the polar distance r (indicating the spatial relation of the location of the next stay to home) and the polar angle θ (indicating the clock time of the starting time of the next stay). The curve between the two points is not parallel to the cycles, because as travel-between proceeds the spatial relation of the person to home is constantly changing. The representation of stay is straightforward. It also requires two points to show its starting-time point and ending-time point. In this case, since the person stays in the same place, there is no change in the spatial relationship to home and the curve between the two points is thus a curve parallel to the cycles. The Polar Coordinate System (PCS; Figure 15.4) approach of representation can be applied to natural phenomena as well. To illustrate, we examine the temperature patterns in Beijing over several years.

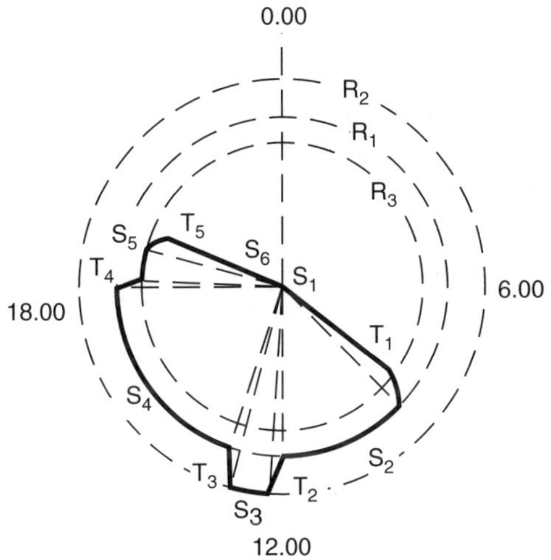

Figure 15.4 *Representing an activity pattern in PCS*

The National Climate Data Center of USA (NCDC, 2001) provides weather information on major cities in the world. The monthly maximum, minimum and average temperature information of Beijing was retrieved from this database for 1970, 1980 and 1990. To visualize these data using the Polar Coordinate System, we may define −20 degree Celsius (approximately the lowest temperature in Beijing) as the pole. The length on the polar axis represents temperature variations from this minimum. The polar angle represents the calendar months. Moving from the right (north) in the clockwise direction, the first 30 ° represents January, the second 30 ° represents February, and so on. Figure 15.5(a–c) shows the PCS representation of the monthly temperatures in Beijing

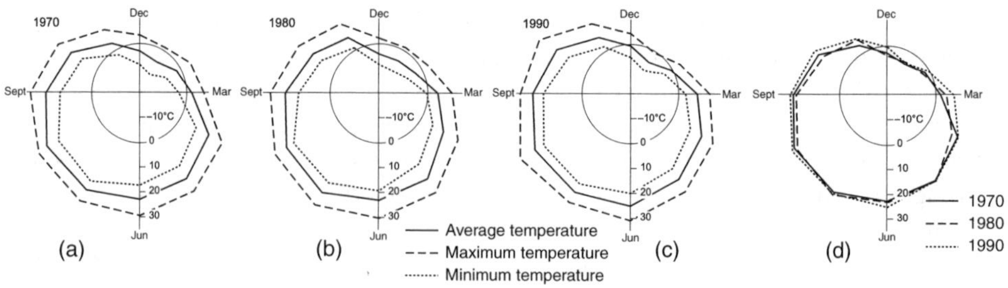

Figure 15.5 *Monthly temperature changes of Beijing, China*

for the three years. It is quite easy to see the dynamic pattern of temperature changes from month to month within a year. If we overlap the figures of the three years (Figure 15.5d), the cyclic pattern of the temperature changes are clearly illustrated. Repetition of the annual dynamics are quite obviously visible.

This subsection thus proposes the use of polar coordinate systems to visualize and analyze the cyclic spatio-temporal process/patterns. Polar angles are used to represent time, which can be at a daily, monthly or yearly resolution. Polar distances are applied to represent moving extensions in space, i.e., distances of natural or human objects to their reference places (such as the home of a person or an animal, or the average temperature of a month). The example of human activity patterns is used to illustrate and demonstrate how spatial and temporal patterns can be adequately represented in polar coordinate systems. The proposed approach has great advantages in identifying the characteristics of spatio-temporal patterns.

15.4 Discussion and Conclusions

This chapter proposes the use of a linear time structure to study the dynamic process of coastal zone and polar coordinate systems to analyze the cyclic spatio-temporal patterns. Different spatio-temporal structures are adapted for different situations to facilitate visualizing, analyzing and understanding of the spatio-temporal patterns. Based upon proper spatio-temporal structures, contemporary visualization opens opportunities to escape from iconic displays to more abstract representations in which space can be wrapped into non-spatial elements of display; i.e., not to represent ‘temporal data’ but to represent ‘processes occurring over time’.

Acknowledgments

The author wishes to thank The Hong Kong Polytechnic University for the Postdoctoral Research Fellowship (no. G-YW92).

References

Cheng, T. and Molenaar, M., 1999, Diachronic analysis of fuzzy objects, *GeoInformatica*, **3**(4), 337–356.

Harrower, M., MacEachren, A. and Griffin, A., 2000, Developing a geographic visualization tool to support earth science learning, *Cartography and Geographic Information Science*, **27**(4), 279–293.

Hazleton, N.W.J., 1992, Beyond the 2-D map: a new metaphor for multi-temporal 4-D GIS. In *Proceedings of GIS/LIS '92*, Vol. 2 (Falls Church, VA: ASPRS, ACSM), pp. 303–313.

Hornsby, K. and Egenhofer, M., 2000, Identity-based change: a foundation spatio-temporal knowledge representation. *International Journal of Geographical Information Systems*, **14**, 207–224.

Hornsby, K. and Egenhofer, M., 2002, Modeling moving objects over multiple granularities, Special issue on Spatial and Temporal Granularity, *Annals of Mathematics and Artificial Intelligence*, 177–194.

Kraak, M.J., Edsall, R. and MacEachren, A.M., 1997, Cartographic animation and legends for temporal maps: exploration and or interaction, In: *Proceedings 18th ICA/ACI International Cartographic Conference*, 23–27 June 1997, Stockholm, Sweden, 253–260.

Langran, G., 1992, *Time in Geographic Information Systems*, London: Taylor and Francis.

Moellering, H., 1976, The potential use of a computer animated film in the analysis of geographical patterns of traffic crashes, *Accident Analysis & Prevention*, **8**, 215–227.

NCDC, 2001, http://www.ncdc.noaa.gov/onlineprod/ghcnmcdwmonth/form.html accessed 24 January 2001.

Nunes, T.L., 1991, Geographic space as a set of concrete geographical entities, In: *Cognitive and Linguistic Aspects of Geographic Space*, edited by Mark, D. and Frank, A.U., Dordrecht: Kluwer, pp. 9–23.

Renolen, A., 2000, Modeling the real world: conceptual modeling in spatiotemporal information system design, *Transactions in GIS*, **4**, 23–42.

Snodgrass, R.T., 1992, Temporal database, In: *Theories and Methods of Spatio-Temporal Reasoning in Geographic Space*, edited by Frank, A.U., Campari, I. and Formentini, U., Berlin: Springer-Verlag, Lecture Notes in Computer Science, 639, pp. 22–64.

Tryfona, N. and Jensen, C., 1999, Conceptual data modeling for spatio-temporal applications. *Geoinformatica*, **3**, 245–268.

Wang, D. and Cheng, T., 2001, A spatio-temporal data model for activity-based transport demand modeling. *International Journal of Geographical Information Science*, **15**, 561–586.

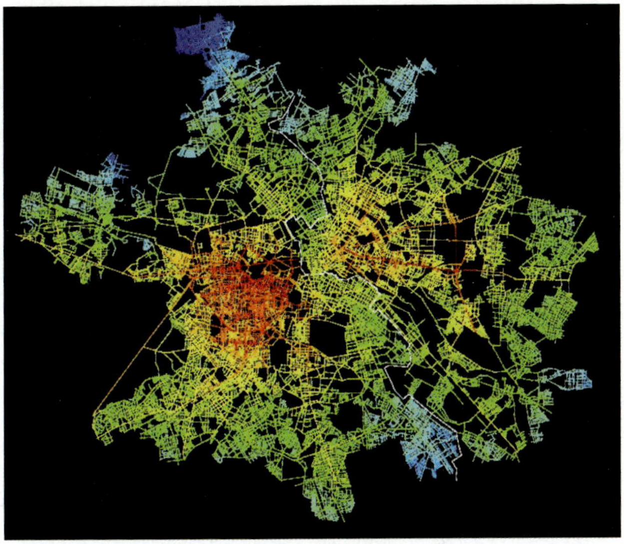

Berlin 1989 before unification

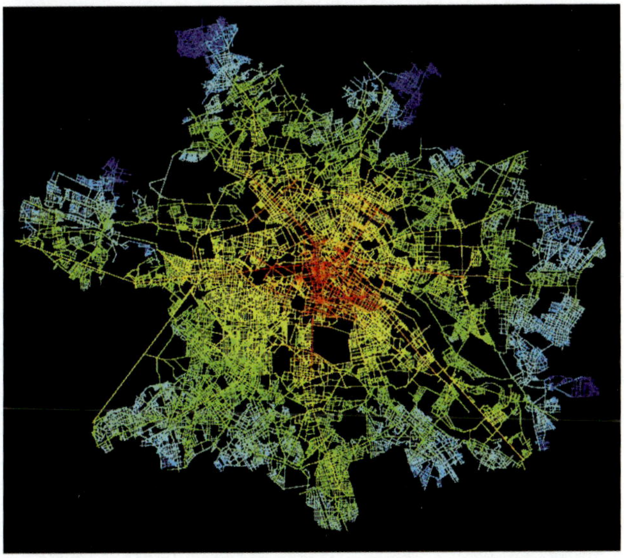

Berlin 1999 after unification

Plate 1 (Figure 11.3) *Accessibility in Berlin: before and after the wall is demolished. The colours indicate the street segment accessibility illustrating a massive increase in the central area after the wall is demolished. The scale for accessibility varies from high (red) through yellow and green to low (blue)*

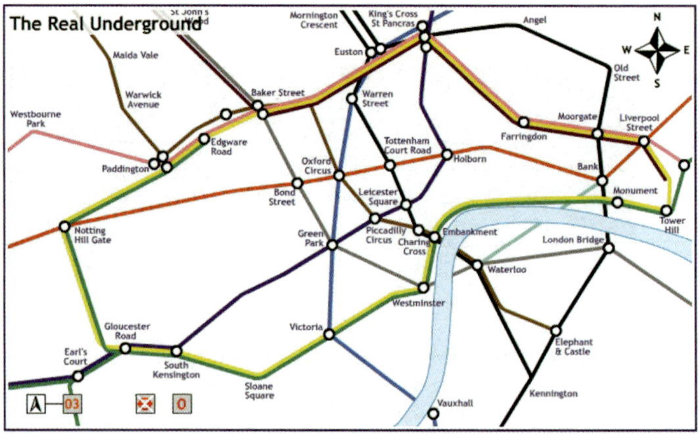

(a)

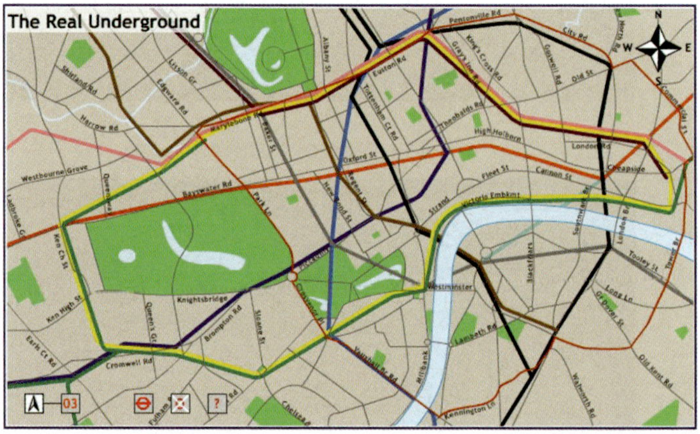

(b)

Plate 2 (Figure 11.6) *The way the London subway map connects to the real street pattern. (a) Morphing the topological subway map to the real geography. (b) Connecting the subway topology to the Euclidean pattern. A dynamic illustration of this morphing is at http://www.fourthway.co.uk/ (by permission of Sam Rich and Transport for London)*

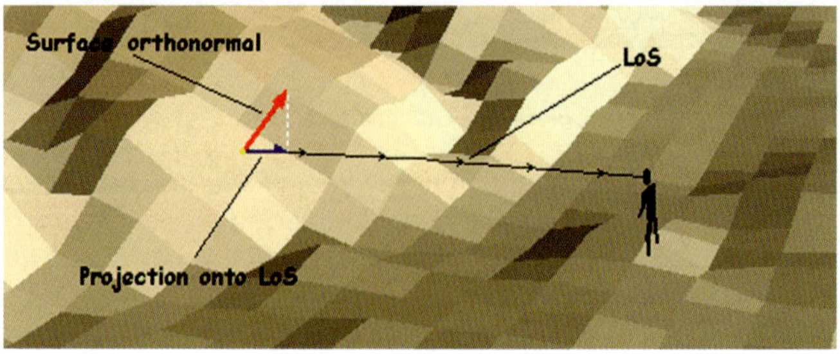

Plate 3 (Figure 12.3) *The vector nature of visual exposure*

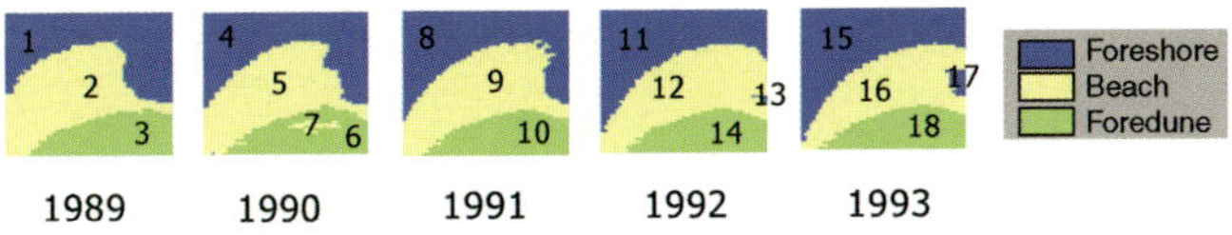

Plate 4 (Figure 15.2) *Classified regions (Reproduced from Cheng, T. and Molenaar, M, 1999, Diachronic analysis of fuzzy objects, GeoInformatica,* ***3****(4), 337–356; with kind permission of Springer Science and Business Media)*

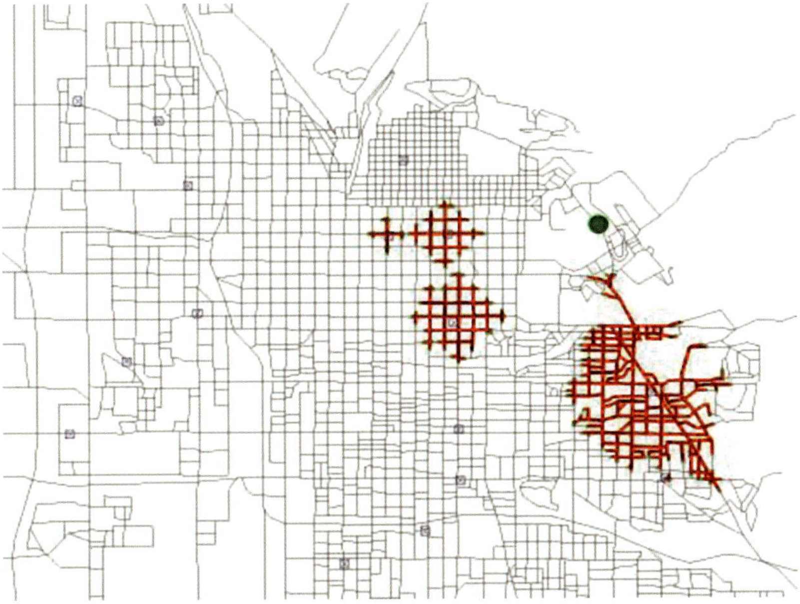

Plate 5 (Figure 16.5) *High accessibility locations within a network calculated using a potential network area (Reproduced from Miller, H.J. and Wu, Y.-H. (2000). GIS software for measuring space-time accessibility in transportation planning and analysis, GeoInformatica,* ***4****(2), 141–159. Figure 10, p. 157; with kind permission of Springer Science and Business Media)*

Plate 6 (Figure 17.4) *Various forms of spatial display provided by Simile*

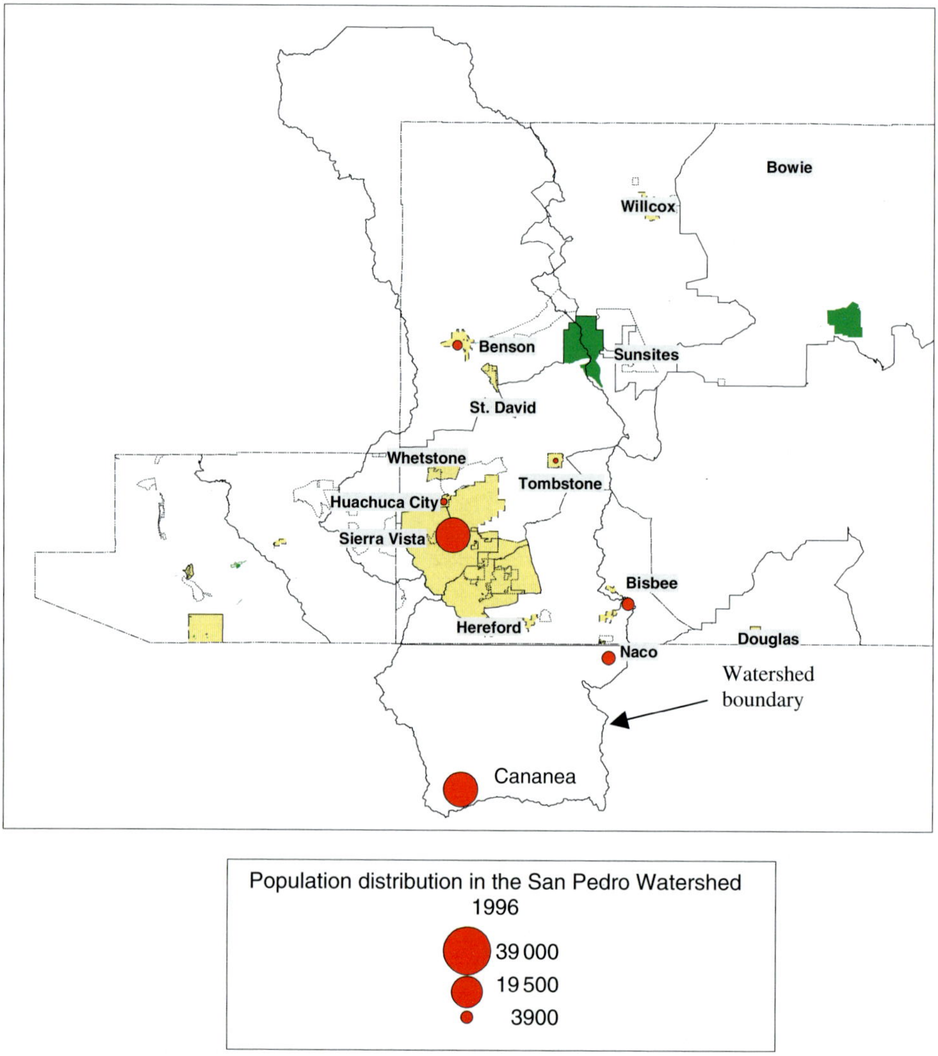

Plate 7 (Figure 18.1) *Communities in the San Pedro Watershed*

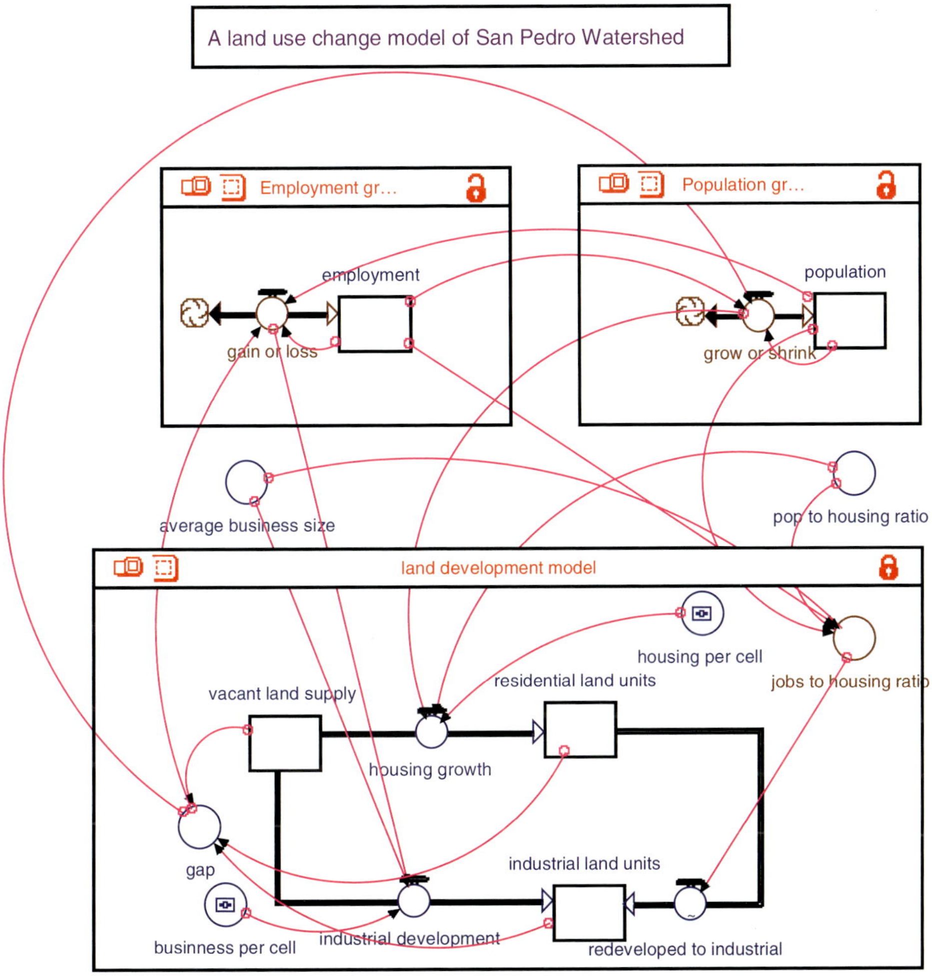

Plate 8 (Figure 18.2) *System-dynamic model of land development in Sierra Vista sub-watershed*

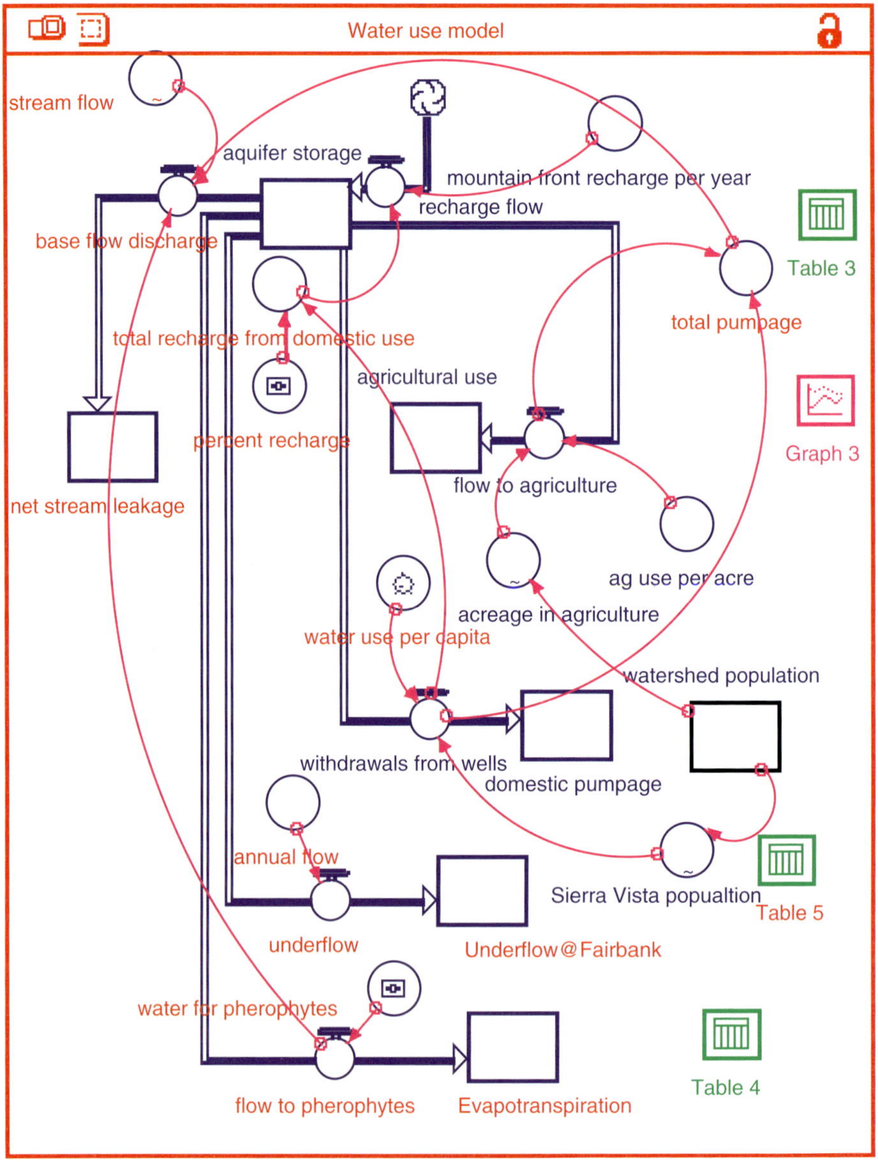

Plate 9 (Figure 18.4) *A system-dynamic model of hydrology in the Sierra Vista sub-watershed*

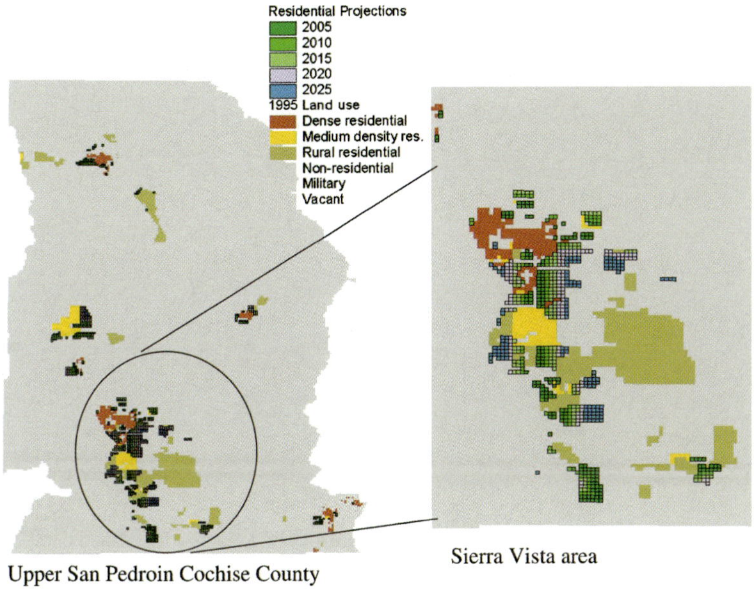

Plate 10 (Figure 18.5) *Projected residential development in the San Pedro Watershed region*

16

What about People in Geographic Information Science?

Harvey J. Miller

The title of this chapter is an homage to the classic paper by the late Torsten Hägerstrand, 'What about people in regional science?' (Hägerstrand, 1970).

16.1 Introduction

Our lives consist of activities in space and time. The basic activities that structure our lives, such as family, work, shopping, recreation and socializing, occur at a few geographic locations and for limited temporal durations. People have scarce time and resources to distribute among required (e.g. work, home) and desired (e.g. recreational, social) activities. Societies devote enormous amounts of energy and resources to overcoming spatial and temporal constraints.

Cities exist to compress a multitude of human lives into small geographic spaces to reduce the amount of time and energy required to access activities and resources. Transportation systems allow individuals to trade less time for more space when moving to activity and resource locations, as well as moving the resources themselves from supply to demand locations. Telecommunication systems allow humans to annihilate distance for some types of activities and interactions. Transportation, telecommunication and settlement systems grow and decline in response to human activities in space and time. They influence economic, social and knowledge networks, in turn shaping human activities and their locations in time and space.

Re-presenting GIS Edited by P. Fisher and D. Unwin

Geographic information systems (GIS) are convenient platforms for theoretical and applied transportation and urban analysis. The location-based organization of data and information from the cartographic roots of GIS is a good fit with the placed-based theories and models inherited from von Thunen's bid-rent theory of land-use. This includes methods such as travel demand models based on spatial interaction theory and market equilibrium models of urban spatial organization. These are *place-based* methods that represent transportation demand and urban form as a function of aggregate spatial units.

Place-based representations and methods were developed in an era when data were scarce, computational platforms weak and questions simpler (at least so we thought at the time). Despite growing theoretical and empirical evidence questioning their theoretical foundations (see Boyce *et al.*, 1994; Wegener, 1994), remarkably resilient methods such as the four-step travel demand model and urban equilibrium models still dominate GIS for transportation ('GIS-T' as it is commonly known) and urban GIS. Place-based methods ignore the basic spatio-temporal conditions of human existence and organization discussed above. Due to the drastic changes taking place in transportation, telecommunication and settlement systems, ignoring these spatio-temporal conditions is no longer tenable.

Important theoretical and policy questions require extending the place-based perspective in GIS-T and urban GIS to encompass a *people-based* perspective. A place-based perspective by itself is no longer viable in a world where transportation and telecommunication have altered dramatically the nature of space and time at the core of human existence. The world is shrinking in an absolute sense: transportation and communication costs have collapsed to an incredible degree over the last two centuries (see Janelle, 1969). The world is also *shriveling*: relative differences in transportation and telecommunications costs are increasing at most geographic scales (Tobler, 1999). The world is also *fragmenting*: people and activities are becoming disconnected from location (Couclelis and Getis, 2000). A place-based perspective is increasingly ill suited for answering questions of access, exclusion and evolution in a shrinking but shriveling and fragmenting world.

The time geographic perspective of Torsten Hägerstrand offers a people-oriented alternative to place-based tools in GIS-based transportation and urban analysis. This perspective views the person in space and time as the center of social and economic phenomena. Since they recognize constraints imposed by demographic, social, economic and cultural context (Kwan, 1998), time geographic methods are more sensitive measures of differences in accessibility and exclusion. The closely-related area of activity theory concerns the theory, measurement and analysis of how people organize activities in space and time, the relationship between these activity patterns and the evolution of transportation, communication and settlement systems, and how these evolving systems in turn influence the organization of activities in space and time. Space-time activity analysis also offers a more theoretically defensible view of networks and settlement systems as emergent from individual activities and shapers of these activities.

In recent years, time geography and activity theory has experienced a renaissance as, encouraged by developments in GIS, researchers have expanded their power and scope. The rapidly improving ability to collect space-time activity (STA) data through information technologies such as cellular/mobile phones, wireless personal digital assistants (PDA), global positioning system (GPS) receivers and radio-location methods is improving the quantity and quality of these data and reducing their cost. GIS allows more realistic and detailed depictions of accessibility and activities in space and time than imagined by the pioneers of time geography in the 1950s and 1960s. These developments can help expand GIS from its place-based representations to encompass the people-based perspective required by contemporary transportation and urban theory. However, much of the work on the 'new' time geography (including the previous work of the author) is *ad-hoc* and disconnected: there is no coherent framework for designing and developing GIS software tools or even thinking about the basic entities that should be represented within a people-oriented GIS.

This chapter is an attempt to review and assess time geography, activity theory and GIS. In it, I review the basic foundations of time geography and activity theory, improvements in geographic information technologies, and the state of the art in implementing time geographic and activity theory constructs within GIS. I also review formal representations of dynamic spatial objects in GIScience (GISci) and its relevance to time geography, with an eye towards developing a coherent framework for a 'people-oriented GIS'. Finally, I identify gaps in the research that must be addressed if time geographic and space-time activity techniques linked to GIS are to achieve breakthroughs in our understanding of human lives and urban environments.

A people-oriented GIS should complement rather than replace the traditional place-based GIS. Computational representations of geographic space can still serve as the basic framework with dynamic and mobile objects linked to the geo-spatial framework. This supports the traditional spatial applications of GIS such as inventory and mapping, but also supports advanced analysis of the dynamic and mobile objects within the geo-spatial frame. In some respects, the representational issues discussed in this chapter are not very different from the acknowledged 'object-centered' (vector) and 'space-centered' (raster) representational division in GIS (Goodchild, 2001). There are some recent attempts to bridge this gap and create integrated representations (see Cova and Goodchild, 2002; Yuan, 2001a). However, the objects of interest in this chapter are dynamic, mobile and *active* (they conduct activities that are relevant to analysis). As will be seen, this raises additional and complex representational issues.

The next section of this chapter discusses the relevance of a people-oriented GIS in transportation and urban analysis. Section 16.3 reviews the theoretical foundation including time geography, activity theory and GISci theories of dynamic spatial objects. Section 16.4 reviews current and potential tools for a people-based GIS. Section 16.5 concludes this chapter by summarizing the research and development frontiers.

16.2 Relevance

16.2.1 Activities and Accessibility

Accessibility to resources, opportunities and support networks such as employment, health care, education, shopping, recreation, friends and relatives is a central component of community livability (National Research Council 2001). Accessibility is an individual-level phenomenon with contextual effects related to demographic, social, economic and cultural factors. In most societies, life stage, social class, cultural identity, and even ethnicity, strongly influence the location of key *anchor points* in an individual's life such as home and work locations. The distance between affordable housing and employment opportunities can create severe constraints for some social groups (e.g. see Gober *et al.* (1993)). Scheduling constraints that compel presence at certain locations for fixed time intervals also vary by socio-economic factors and artifacts such as gender roles (Kwan, 1999). Socio-economic and demographic cohorts exhibit distinct space-time activity signatures that are remarkably stable over time, often persisting after many of the original members of the cohort have moved on and been replaced by new members (McNally, 1998).

Individuals also differ with respect to the transportation resources and information technologies (IT) available to overcome the constraints imposed by space-time anchor points. Because of the sparseness of the space-time network imposed by many public transportation systems in many parts of the world, such as the United States, individuals who are unwilling or unable to drive an automobile due to lack of resources, different abilities, or preference, are often at a disadvantage, shaped by socio-economic and demographic factors. A persistent digital gap exists between IT haves and have-nots and it is another dimension of differential accessibility among social groups as more resources and activities occur in *cyberspace*, the information space created by networked computers and IT (Shen, 2000).

16.2.2 Information Technology, Lives and Cities

The conditions that underlie our daily lives and influence the performance and development of our urban infrastructure are undergoing fundamental changes. These are due to the development and adoption of new IT and improvements in transportation and logistics systems that support the information economy and high consumption lifestyles. The increasing ability to manipulate and transmit data bits, combined with well-developed and managed systems for transporting atoms, are altering the fundamental relations between space, time, and human activities. Information technologies are not only influencing where people work, live, recreate and socialize, but also changing the very nature of the activities that occur in the home, office and automobile (Moss and Townsend, 2000).

There is a need to revise the theoretical shortcomings associated with traditional transportation and urban analyses in light of fundamental changes in individuals' abilities to interact across distance. The root of many of the changes occurring in the post-industrial city is an increasing disassociation between places and activities. The increasing power and scope of IT means that activities are becoming more person-based rather than place-based: activities are increasingly a function of the person in time and space rather than places. For example, with mobile computing and telecommunications, a person may work in an office, at home, in a coffee shop, or even in a public park. Place-based transportation or urban models generally only recognize work at the first location and not at the others. The increasing fragmentation of activity from space means that the assumption of strong structural correspondence between spatial and functional relationships at the basis of classical transportation and urban theory is increasingly untenable (Couclelis and Getis, 2000). Because of its place-based orientation, traditional transportation and urban theory is ill equipped to address many of the key questions regarding emerging lifestyles, urban form and differential access to activities and resources among social groups in the information age.

There is little detailed knowledge about the impacts of IT on lifestyles and life activities beyond some broad brushstrokes and generalities. The simplistic 'death of distance' argument in the popular press (e.g. Cairncross, 1997; Mitchell, 1995) does not hold up against evidence of the continuing draw of the city for home and work even for supposedly 'footloose' people and businesses such as high-level decision making and creative work (Graham and Marvin, 1996). Traditional urban theory views cities as land use configurations when in reality these are complex webs of individual activities, actions, reactions and interactions (Golledge and Stimson, 1997). An increasing number of these activities and actions occur in cyberspace rather than geographic space.

16.2.3 The Worldviews of GIS-T

Are current GIS suitable for the brave new worlds of transportation and urban analysis? Goodchild (2000) identifies three major worldviews required for current and potential GIS for transportation (GIS-T). Traditional GIS-T applications in transportation involve *map* representations for static inventory and display of transportation facilities and related geographic objects. One-dimensional networks derived from road center-lines and variable length, or 'dynamic' (which is not the appropriate term) segmentation data models, are examples of this perspective.

The emerging *navigation* perspective requires more demanding representations of geographic reality that can support routing applications, possibly in real time. This includes requirements for more complex topologies possibly across multiple modes (see Spear and Lakshmanan, 1998), dynamic attributes (e.g. congestion levels, travel speeds, temporary conditions such as obstructions), two-dimensional representations of transportation facilities and support for 'off-network' travel such as in parking lots or across unrecognized roads. All of these require dynamic attributes represented within the framework of static map geometry.

The *behavioral* perspective deals with the behavior of discrete objects within a dynamic geometry. A GIS must be able to represent space-time 'paths' (Hägerstrand 1970), 'trajectories' (Smyth, 2001) or 'lifelines' (Mark and Egenhofer, 1998; cited in Goodchild, 2000) as well as the information that emerges when these entities are aggregated. GIS tools should be able to maintain consistent linkages between individual objects and aggregate outcomes such as origin-destination flows in transportation links and between geographic locations as well as the evolution of transportation, telecommunication and settlement systems.

Commercial GIS software has made little progress beyond the traditional map worldview (Goodchild, 2000). Developing a GIS toolkit for answering increasingly important questions surrounding access and equity in a shrinking, shriveling and fragmenting world requires liberating GIS from its place-based representations to include people-based representations. Fortunately, there is a coherent body of theory to support the design of these systems as well as emerging technologies that allow detailed space-time data and information to be captured, handled and understood.

16.3 Theory

16.3.1 Time Geography

Hägerstrand's (1970) time geographic framework is a powerful and elegant perspective for analyzing constraints on individuals' participation in activities and opportunities. The time geographic framework recognizes that activity participation has both spatial and temporal dimensions. Activities occur at specific locations for limited time periods. Transportation resources allow the individuals to trade time for space, to travel and participate in activities at dispersed locations. Travel is anchored by certain activities that are relatively fixed in space and time. For example, a person's work often cannot be easily rescheduled or moved in space, at least in the short-run. The space-time framework dictates the *necessary*, but not *sufficient*, conditions for most human interaction.

At the heart of time geography is the notion that all activities and events that make up an individual's existence have both spatial and temporal dimensions. The basic conceptual tool in the framework is the *space-time path*, which traces the movement of an individual in space and time. In addition to tracing movement in geographic space from location to location, it also traces simultaneous movement in time. Figure 16.1 illustrates a space-time path. Note that the path is vertical when the individual is stationary in space (but always moving in time) and that a shallower slope indicates that the person is moving faster (i.e. they are trading less time for more space).

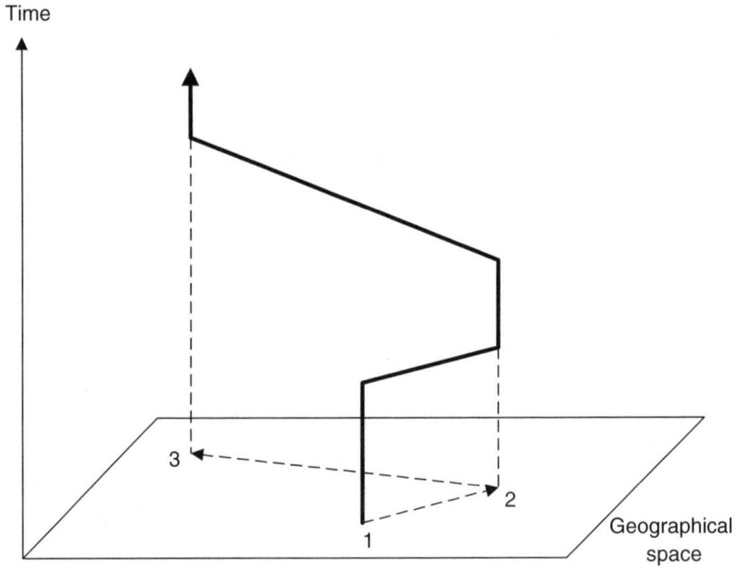

Figure 16.1 *The space-time path*

The following types of constraint dictate the locations that the space-time path can occupy:

1. *Capability constraints* limit the activities of individuals through their own physical capabilities and/or the resources they can command. For example, individuals with private automobiles can generally travel faster through space than individuals who walk or rely on public transportation.

2. *Coupling constraints* define where, when, and for how long, an individual has to join with others to produce, transact or consume. Coupling constraints define space-time bundles of individuals existing in a particular space and time. For example, having to be at work for certain time periods is a coupling constraint. *Space-time bundles* or groupings of individual space-time paths within a limited domain of space and time (see below) are evidence of these constraints.
3. *Authority* or 'steering' *constraints* impose certain conditions of access in particular space-time domains. For example, a private shopping mall can impose more constraints than a traditional city center on individuals' space-time autonomy since private space can be more effectively restricted from occupancy during certain hours and days and for some purposes. Gated suburban communities can prevent certain 'undesirable' individuals from occupying their space, particularly at certain time periods (e.g., from dusk to dawn).

A *space-time prism* (STP) is an extension of the space-time path that measures accessibility to events in space and time. Figure 16.2 illustrates a simple STP. In this, the person must be at a given location (say work) until time t_1 and then must return again at time t_2. If we measure or assume an average travel velocity, we can delimit the *potential path space* (PPS) showing all locations in space and time that the person can occupy. If he or she wants to visit an activity location, its space-time path must intersect the potential path space. Projecting the PPS to the two-dimensional geographic plane

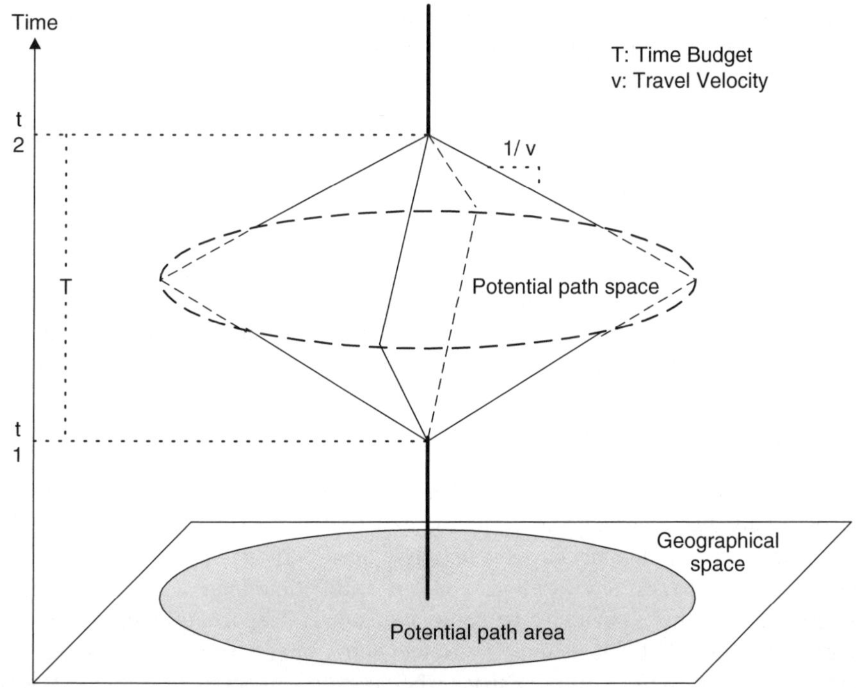

Figure 16.2 *The space-time prism (Reproduced from Wu, Y.-H. and Miller, H.J. (2002). Computational tools for measuring space-time accessibility within transportation networks with dynamic flow, Journal of Transportation and Statistics,* ***4****, 1–14)*

delimits the *potential path area* (PPA). These are the set of geographic locations that the person can occupy. This is a simple example: the STP, PPS and PPA can be more complex with non-coincident fixed locations and different travel metrics (see Burns, 1979).

Although the fundamental level of analysis in time geography is the individual and its path through space and time, time geography provides conceptual linkages between the individual and the broader socio-economic system. A space-time *project* consists of the specific tasks required to complete any goal-directed behavior. The tasks associated with a project usually have a logical order, e.g. task A must be completed before task B and so on. This concept can be applied at a variety of scales including the individual, family, society, and the state, as well as for national and transnational organizations. At aggregate levels, the requirement for tasks to be sequenced logically and coordinated with other individuals leads to the formation of activity *bundles* or the convergence of two or more space-time paths, or the convergence of a single space-time path with one or more physically tangible resources such as equipment, materials or buildings. Activity bundles tend to form at *stations*, which are fixed locations with limited temporal durations that support activity bundling, usually conceptualized as *tubes* in space and time. Examples include offices, retail outlets and schools. The activity bundle and station concepts provide a direct interface between the external actions of the individual with the observable workings of the socio-economic system (Pred, 1981).

Applying the project concept at broad levels such as a city, region or nation also leads to the concept of an *activity system*. This is a synoptic view of space-time activities as a market where a finite supply of time must be allocated among competing activities. Time demand results from the interplay among the population of the system and the multitude of interrelated projects at the individual and organizational levels (Golledge and Stimson, 1997).

The emphasis on time in time geography provides a natural fit with emerging perspectives that view time as the scarce commodity of the information economy and accelerated modern lifestyles (see Gleick, 1999; Goldhaber, 1997). This can provide an effective link between the geographic space of traditional transportation and urban theory with the cyberspace of the information age. Locations in cyberspace can be treated as logical locations in information space or related to geographic locations or different geographic scales of interaction. There has been some preliminary conceptual work formulating these linkages (e.g., Adams, 1995; Batty and Miller, 2000; Kwan, 2000a).

16.3.2 Activity Theory

Activity theory focuses on people rather than places as the source of travel and location demands. Individuals participate in periodic activities that have varying levels of necessity and urgency. The resources that satisfy these activities are sparsely distributed in space and time, i.e., at few locations and for limited time intervals. This can include requirements that other individuals be contemporaneous in space, time or both (such as work or socializing). Individuals must distribute their limited time among these activities, using transportation to trade time for space when traveling to activity locations. They can also substitute *in-situ* activities that do not consume transportation services, for example, using IT. Aggregate-level outcomes such as transportation system performance, urban development, lifestyle decisions and long-term mobility both condition and are conditioned by these individual-level activity sequences (Ben-Akiva and Bowman 1998; Bhat and Koppelman, 2000; Thill and Thomas, 1987). Wang and Cheng, (2001) provide an elegant summary of the basic components in activity theory (based an Axhausen, 1994) and Table 16.1 summarizes these entities.

Table 16.1 *Basic components of activity theory*

Entity	Definition
Activity	The main purpose carried out at a location, including any waiting time before or afterwards. Activities can be classified into different types depending on purpose
Activity frequency	The number of times the activity occurs during a given time period
Activity destination	The location where an activity occurs
Trip	Movement between two activity destinations
Transport mode(s)	Methods of conveyance used to perform a trip
Activity program	Set of activities to be performed within a given time period
Activity schedule	The planned ordering of activities in space and time within a given time period
Activity pattern	The activities in space and time actually conducted within a given time period
Activity space	A composite of the locations where an individual conducts routine activities
Physical environment	Spatial configuration of activity destinations and transportation services between these destinations
Institutional environment	Set of formal rules that regulate the individual's activities in space and time (e.g., store hours, working hours)

A major application of activity theory is empirical measurement and analysis of *space-time activity* (STA) data, or records of where and when individuals conducted activities over a daily, weekly or monthly cycle. Empirical measurement of STA behavior dates back to Chapin (1965), the landmark study in Halifax, Nova Scotia in the 1970's (Goodchild and Janelle, 1984) and continued in transportation studies in the early 1990s in cities such as Boston, Salt Lake City, Portland and Dallas-Forth Worth, USA (Greaves and Stopher, 1998). These efforts need to be continued and expanded to track changes in the impacts of IT on daily lives and urban form.

16.3.3 Representing Dynamic Spatial Objects

Time geography and activity theory provide a powerful perspective from which to view the interactions between people, transportation, telecommunication, socio-economic and settlement systems. Although these two perspectives are highly complementary, they have different conceptualizations of the person and his or her activities in time and space. A true people-oriented GIS should encompass both theories through a unified representational theory and tools that can support both and exploit their commonalities and complementarities. Representational principles from GISci can allow rigorous development of time geographic and activity theory and the functional requirements for GIS software to support these theories and models.

Spatial objects can exhibit three major types of change that often occur in concert but nevertheless can be meaningfully separated (Yuan, 2001b). One class is *motion* or change in the position or geometric form of the object over time. The second class encompasses changes in the temporal identity of the object; we refer to this as *life* (Frank, 2001). A third is change in the semantics or non-spatial attributes of the object; referred to as the object's *state*. The next three subsections review GISci and related literature that addresses these types of change in spatial objects. The discussion suggests linkages between this literature and time geography/activity theory that should be explored in greater detail (and rigor) in continuing research and software development.

Motion. As Galton (1995, 1997) points out, any theory of motion should incorporate theories of time, space, position and objects. With respect to time, essential properties include *duration* and *direction*: the former consists of *intervals* or 'chunks' of time bounded by *instants*. Direction is determined by an ordering of intervals and instants. Another issue is temporal *granularity* or the minimum resolution for measuring time. This results in two natural idealizations of time, namely, *dense* (or continuous) and *discrete* time, where the former allows infinitely fine subdivisions while the latter consists of finite intervals. The theory of space is analogous to the theory of time, although spatial ordering is more complex and can be absolute (e.g. contiguity) or relative (e.g. north). The basic entities of space are the fundamental entities in physical reality, e.g., points, lengths, areas and volumes. Traditional space theories such as point-set topology are too primitive in the sense that they allow arbitrary spatial objects that have no meaning in physical reality and therefore are irrelevant to motion.

With respect to motion, important properties of objects include whether they are conceptualized as rigid or non-rigid, unified wholes, or comprised of parts, individuals or collectives and are concrete or abstract. *Rigid* objects maintain a constant shape and size while *non-rigid* objects, in addition to position, can change size and shape. In practice, it can be difficult to separate motion from changes in size or shape. Some objects' motions are best represented as a unified *whole*. Other spatial objects can be best described in terms of the movements of its *parts*, such as an object that maintains its overall position although each part is moving (e.g. a spinning phonograph or compact disc). Motion of a *collective* is sometimes best described by the motions of its constitute individuals, but some collective motion, such as the flow of water in a river, cannot be reduced to the movements of discrete individuals. Finally, some spatial objects have physical existence (*concrete*) while others are *abstract* in the sense that they are uniform but non-essential properties such as political entities or ownership (these are sometimes referred to as 'shadows'; see Frank, 2001).

The *position* of an object is the region of space it occupies during a given unit of time. This can be exact or qualitative. The concept of position allows a definition of *motion* as a mapping from time to position: for each time unit a position of the object can be specified. If space is discrete, direct motion can only occur between neighboring positions. If time is discrete, then the discrete time unit places an upper limit on apparent motion. If both space and time are continuous, than motion is continuous and we need to specify its exact position at each moment in time (subject to the temporal granularity).

Table 16.2 *Conceptualizing motion for time geographic entities*

	Movement properties				
Entity	Time	Space	Objects	Position	Motion
Space-time path	Dense	Dense	Rigid Whole Individual Abstract	Exact	Continuous
Space-time prism	Dense	Dense	Non-rigid Whole Individual Abstract	Exact	Continuous
Activity bundle	Dense	Dense	Rigid Parts Collective Abstract	Exact	Continuous
Trip	Discrete	Discrete	Rigid Parts/Whole Individual/Collective Abstract	Inexact	Discrete
Activity pattern	Discrete	Discrete	Rigid Parts Collective Abstract	Inexact	Discrete
Activity space	Discrete	Dense	Non-rigid Whole Individual Abstract	Inexact	Discrete

However, exact positions can be aggregated into *locations* or the sum of all positions occupied by the object over a time interval (Galton, 1997).

Table 16.2 is a tentative conceptualization of motion properties for some time geographic and activity theory entities. Several broad observations become evident. First, there is a discontinuity between time geography and activity theory with respect to the apparent (represented) motion of entities: time geography conceptualizes its entities as moving through continuous space and time where (theoretically) the position of the entity is known exactly at all times. In contrast, activity theory treats time and space as discrete: motion is an interaction between two activity locations and, beyond the demand for transportation services it generates, its exact spatio-temporal geometry is irrelevant. However, these movement conceptualizations will be merged as new IT facilitates real-time geo-location in activity analysis and a corresponding closer integration of time geography into activity theory (see Section 16.4).

Another observation from Table 16.2 concerns the diversity of moving objects. Some objects such as the space-time path are rigid, comprised of individualistic entities that are abstract in the sense that they have no physical existence in the real world. While the individual has a physical existence, his or her path is abstract. In contrast, the space-time prism is non-rigid: its boundaries and form can change. Space-time bundles are best understood as collectives of rigid parts. The trip can be treated as a rigid, whole, individual and abstract object, except when two or more individuals share a trip (e.g. pooling cars), in which case it may be best represented as a collective of parts.

Table 16.2 provides an initial foray into a theory of motion that can support the spatial objects of interest in time geography and activity theory. If we are to have detailed functional requirements for a people-oriented GIS, a unified and rigorous theory of motion for this domain is required. Nevertheless, due to the requirements to support diverse spatial objects and motions, a people-oriented GIS will have some difficult design challenges.

Table 16.3 *Major life events (based on Frank 2001; Medak 2001)*

Life events	Definition
Create	New object comes into existence
Destroy	Object's existence is terminated (although information on the object may be maintained)
Kill	Similar to *Destroy* but allows object to be *Reincarnated* at a later time
Reincarnate	Re-active an object that was previously *Killed*
Evolve	Combination of *Create* and *Destroy*: one object is destroyed but a new object is created that has information about its ancestor
Identify	Retain the identity of an object that is merged with other objects
Spawn	New objects are created from an existing object (and the existing object continues its identity)
Aggregate	Two or more objects merge into a collective object (but retain their identities)
Disaggregate	Inverse of *Aggregate*
Fusion	Two or more objects merge into a collective object (and lose their identities)
Fission	A single object is broken into parts that become new objects

Life. Life refers to essential changes in the identity of a spatial object over time. Several types of life changes (reviewed in Frank, 2001; Medak, 2001) can occur to spatial objects. Table 16.3 summarizes these generic life events.

A *lifestyle* is a coherent set of life events that are appropriate for a particular domain. For example, in time geography it is logical for space-time paths to be created, destroyed, identified, aggregated (e.g., space-time bundles) and disaggregated. A space-time path may also spawn a new space-time path (i.e. the individual represented by the path has a child). It makes less sense for space-time paths to be killed, reincarnated, evolved, fused or fissioned. Therefore, the life events *Create*, *Destroy*, *Identify*, *Spawn*, *Aggregate* and *Disaggregate* form a lifestyle for the space-time path (Frank, 2001). A formal theory of lifestyles for time geography and activity theory would help identify the types of dynamic spatial operations that a people-oriented GIS needs to support. This theory is still an open research question.

States. In time geography and activity theory, the state of an object comprises the relevant socio-economic, demographic and cultural attributes, and the activity (or activities) conducted at a given moment in time. Activities can span the gamut of human experience, including production, education, shopping, socializing, community activities, recreation, entertainment, church, and political behavior, as well as the use of transportation and telecommunication services to participate in these activities (Golledge and Stimson, 1997). Consequently, systematic classification of activities is a critical decision in STA research design.

Unfortunately, there are almost an unlimited number of activity classification systems available to the researcher. Activity classification systems are often 'one-off' schemes developed for particular research projects and standardization is required for comparisons across studies (Golledge and Stimson, 1997). Some national and international classification systems have been developed (see United Nations 2000). Nevertheless, detailed comparisons across different studies can be difficult since there is no universal agreement with respect to the fundamental categories of human activities. Nor is it clear that this is even possible. Even if a universal classification system is not possible, there is still a need to translate between the classification systems developed for different cultural, geographic and temporal settings.

A related challenge is determining the linkages and interrelationships among activities, particularly with respect to space-time projects. At the individual level, some tasks (e.g. purchasing gasoline) may be required to support other activities (e.g. traveling to a grocery store to buy food) in order to support a broader project (e.g. hosting a dinner party that evening). At the organizational and higher levels, these interactions become even more complicated and there is still little understanding of how project-related tasks lead to the formation and dissolution of activity bundles at some stations in space and time. Without an understanding of these activity linkages, it will be difficult to understand the emergence of aggregation spatio-temporal systems (such as cities) from individual behavior and how these aggregate systems in turn constrain individual behavior.

16.4 Tools

As discussed above, IT is changing lifestyles in ways that are poorly understood. However, IT can also facilitate the collection of more accurate, comprehensive and detailed STA data, including data on IT-mediated interactions. GIS can allow the capture, representation, analysis and exploration of massive STA databases, potentially leading to unexpected new knowledge about the interactions between people, technologies and urban infrastructures. This section reviews current and emerging GIS tools that can support time geography and activity theory, including existing attempts to enhance these concepts and theories using GIS.

16.4.1 Collecting Space-Time Activity Data

There are four traditional methods for collecting space-time activity (STA) data. *Recall* methods require subjects to recall and report activities during some previous time period. *Stylized* recall methods require subjects to report 'normal' activities that occur during some typical time period. *Diary* methods require subjects to record activities in a diary, either in a free-format manner or at pre-determined time periods. 'Beeper studies' complement this approach by prompting subjects via a pager at selected time intervals to record their current activities. *Prospective* methods are typically game-based and employed in conjunction with other methods to investigate the effect of potential changes in the activity environment.

These traditional methods for recording STA data all have substantial problems. The recall method relies on the subjects' abilities to remember activities and their locations at a later time period. Stylized recall methods suffer from definitional problems with respect to 'typical' activities during a 'normal' time period. These definitions can be vague, fluid and variable among individuals and over time (Golledge and Zhou, 2001). Previous research suggests that the best data are obtained from activity diaries (Ettema *et al.*, 1996;

Pas and Harvey, 1996). Nevertheless, this method has significant problems. Free-format diaries offer little guidance to individuals with respect to specifying activities and locations and therefore can have high degrees of recording error. These data can also be difficult to code. Individuals are sometimes unwilling to report certain activities and often under-report short trips and the number of stops during a multi-purpose trip (Brog *et al.*, 1982; Golledge and Zhou, 2001; Purvis, 1990).

New IT can greatly enhance the collection of activity data (Greaves and Stopher, 1998). Global positioning systems (GPS) combined with recording devices such as personal digital assistants (PDA), in-vehicle navigation systems, and cellular/mobile telephones can allow for more accurate and detailed recording of activities in space and time (Murakami and Wagner, 1999). Although currently limited by clumsy keypads and pen interfaces, continuing advances in voice recognition software and natural language processing will allow voice interfaces to be integrated into in-vehicle navigation systems, cell/mobile phones and PDAs, as well as the activity diary software that could be designed for these platforms. This will greatly facilitate diary methods for collecting activity data by reducing the burden on subjects through easier, more natural data entry, perhaps even reducing under-reporting and related errors. GPS receivers can also collect network travel time information during the travel event, allowing calibration with aggregate travel time data (see Guo and Poling, 1995). Even without an activity-recording device, the detailed location and time information available from a vehicle-mounted GPS receiver can facilitate the subjects' memory of the activity purpose after the event using recall methods (Stopher and Wilmot, 2000).

Although a potential improvement over traditional methods, there are some problems with GPS-based activity recording that must be resolved if they are to be effective in collecting STA data. An obvious difficulty is the reliance of the receivers on line-of-sight communication with the GPS satellite constellation. This can be problematic in city centers where tall buildings can block line-of-sight communication. It also negates tracing activity patterns within architectural structures such as shopping malls. The present state of the technology is limited to motorized vehicle-based travel due to the size and weight considerations; however, these problems will be resolved over time as technological improvements allow smaller GPS receivers even down to the size of a microchip. A subtler problem is automating GPS receiver data collection. If these are vehicle mounted they can automatically activate when the engine starts, but personal devices need to be activated manually, which can lead to under-reporting problems similar to activity diaries (Golledge and Zhou, 2001).

The rise of *location-based services* (LBS) through wireless communication networks offers another vehicle for collecting STA data. LBS provide specific, targeted information to individuals based on their geographic location, typically though wireless communication networks and devices such as PDAs, cell phones and in-vehicle navigation systems (Benson, 2001). LBS are widely expected to be the 'killer application' for wireless Internet devices: some predict worldwide deployment levels reaching one billion devices by 2010 (Bennahum, 2001; Smyth, 2001). LBS technology can allow for analysis of individuals trajectories in space and time combined with users' information access patterns (Smyth, 2001).

LBS technologies require a high degree of positional accuracy as well as complete coverage across geographic space to be effective. GPS can play a central role, although this will need to be complemented by other technologies for the reasons discussed above. Inertial navigation systems such as gyroscopes and accelerometers can complement GPS technology for in-vehicle LBS. Personal devices can exploit the wireless communication network through high precision radiolocation methods that use the angles of arrival, absolute arrival times or relative differences in arrival times of signals at the base station to calculate the user's location (see Zagami *et al.*, 1998). Hybrid systems that use more than one method provide the best accuracy, particularly in challenging environments for signal propagation such as urban areas (Reed *et al.*, 1998).

LBS offer several advantages for collecting STA data. Non-response biases may be lower since these technologies will be more ubiquitous and accepted than special-purpose, 'unusual' data collection efforts. Changes in space-time activity behavior induced by the data collection effort may also be lower. Finally, LBS can lower the per-unit cost of collecting STA data since new technologies and special data collection efforts are not necessary: STA data are required by LBS and therefore will be a necessary by-product of these services (Smyth, 2001).

We should also note that LBS can benefit from the time geographic and activity analysis available through a people-based GIS. One possible benefit is supporting *space-time queries*. Queries such as 'Which locations can I reach in 15 minutes?', 'Who can attend this event?' or 'Can I meet my friends at the pub this evening?' are in fact queries against space-time prisms. Another area is in *adaptive mobile computing*. Adaptive mobile computing refers a mobile computing and communication system adjusting itself in response to current and anticipated user events (Kanter, 2001). Demand for LBS will be high and also highly uneven across space and time. Strategies must be developed to ensure real-time or near-real-time responsiveness in a situation when many users want information at the same time (Miller, 2001).

There are, of course, important privacy and ethical issues surrounding the tracking and recording of individuals' activities in space and time. The author shares these concerns but contends that these data can be used in an ethical and respectful manner using standard or perhaps expanded human subjects review protocols in place at most universities and research institutions. The application or modification of these protocols to the primary STA data collection or the use of secondary (LBS-derived) data are not the subjects of this current chapter but are highly worthwhile research topics. Regardless of academic debates, the private sector will be using LBS to market products and services more effectively; and it would be nice if we can find ethical ways to use these data to also make our cities more livable and sustainable.

16.4.2 Extracting Activities and Projects

As noted in Section 16.3, there is no universally accepted standard classification for human activities and linkages among activities to form space-time projects at the individual, organizational and settlement system levels are not well understood.

STA data collection methods often involve textual descriptions in natural language: recall, diary and game methods all use subjects' textual descriptions of activities to some degree. Voice-based interfaces in PDAs, cellular phones and in-vehicle navigation systems mean that natural language narratives will become an even more important and potentially rich source of activity data.

Kuhn (2001) develops a method for designing GIS software to support human activities in geographic space. The method extracts the ontology of activities in geographic space based on natural language textual descriptions. The method is based on a type of 'activity theory' that represents human activities and the objects to which activities are directed as the basic units of analysis (see Engeström and Miettinen, 1999). The method exploits the inseparability of semantics and objects and the hierarchical nature of many activities and objects. For example, activity hierarchies range from goal-directed actions to lower-level actions that satisfy higher-level goals. The method consists of the following generic steps:

1. select a natural language text describing activities in a domain;
2. extract actions from the verbs found in the text;
3. identify the object classes that afford these actions from the nouns in the text;
4. order actions according to relations among the verbs;
5. produce an *action hierarchy* that comprises a hierarchical theory of the domain.

Kuhn (2001) uses the German traffic code as a case study. While this is an 'easy' case study since legal codes tend to be complete and consistent, the method appears promising for developing ontologies of human activities and projects in geographic space. The method could be applied to data modeling and designing analytic and exploratory time geographic software. In addition to their use in software engineering, STA ontologies could guide research design. Extracting and formalizing STA ontologies across different socio-economic and geographic settings can support evaluation of classification systems as well as derive mappings between different systems. Formal STA ontologies are also a potential source of direct scientific knowledge.

16.4.3 Data Models

Similar to any GIS-based analysis and modeling, time geography requires careful database and system design. Due to the requirements for representing time and mobile objects, both of which are not handled well by the static, place-based perspective of most GIS software, the design challenges are particularly onerous in this domain, but tools and data models are emerging at the research frontiers in GISci.

Spatio-temporal Data Modeling. Traditional vector and raster GIS data models are inadequate for representing the mobile spatial entities of interest in time geography and activity theory. Vector GIS is limited since it uses location as a basis for organizing data. Each time an entity's location changes, the locational basis and the associated properties (attributes, topology) of the corresponding object in the database must be updated. While this may not be a major difficulty for point objects, entities that have complex geometry or non-rigid boundaries cannot be represented well. Raster GIS offers slightly more flexibility in this regard. In this case, each location in space (subject to a finite spatial resolution) has a unique attribute. By varying these attribute values sequentially, we can simulate the movement of the mobile entity over time. However, this approach is awkward since the entire raster field must be updated in each time step to simulate the movement of just a few entities (Bian, 2000).

There are a number of spatio-temporal data models that can be adapted to represent the dynamic spatial entities of increasing interest in transportation (Miller and Shaw, 2001).

A *space-time object model* proposed by Yearsley and Worboys (1995) integrates abstract spatial data types with a geometric layer to construct a higher-level topological data model. A geometric object can belong to several higher-level spatio-temporal objects. Each geometric object is linked to both real time (when the event occurred in the real world) as well as 'database time' (when the database records the event). *OOgeomorph* is an object-oriented approach that represents dynamic spatial processes as spatio-temporal aggregations of point objects (Raper and Livingstone, 1995).

The *three-domain model* treats time as a temporal object instead of an attribute. In contrast to the location-centric emphasis of most spatio-temporal data models, spatial, temporal and semantic domains have equal emphasis in this design. An event list represents time while a spatial graph maintains the history of spatial object changes such as birth, death, merging and splitting. Semantic objects have unique identifiers and therefore can maintain identity across these changes. Domain links associate objects across the location-centered, semantic-centered, and time-centered perspectives (Yuan 1996, 2001b).

The *event-based spatio-temporal data model* (ESTDM) maintains spatio-temporal data as a sequence of temporal events associated with a spatial object (Peuquet and Duan 1995). A base layer maintains the initial spatial configuration of an attribute. An event list maintains time stamps of when a change occurs and points to the set of locations and features that changed at that point in time. The ESTDM does not maintain object identity beyond its spatial location and therefore cannot handle processes such as merging and splitting (Yuan, 2001a). Yuan (2001a) integrates aspects of the space-time object, OOgeomorph and ESTDM models to formulate a conceptual framework for dynamic geographic processes that display properties of both fields and objects.

Data Models for Mobile Entities. Object-orientation (OO) is a natural strategy for representing the behavior of mobile entities over space and time (Bian, 2000). Objects can easily represent several critical attribute dimensions of mobile entities. One dimension is physical attributes: these are the non-spatial properties of the entity. A second dimension is geometry, including size and shape. The third dimension is motion attributes, including direction, speed and acceleration. As Galton (1995, 1997) argued, this latter dimension requires a mapping of time to positions. Westervelt and Hopkins (1999) and Bian (2000) use OO for modeling the behavior of mobile entities through continuous space, specifically, predator-prey relationships among land animals and fish growth in aquatic environments respectively.

OO can also provide necessary linkages between mobile entities and placed-based data on transportation systems and related land-uses. The *multidimensional multi-modal location referencing system* (MDLRS) conceptual data model being developed through the US National Cooperative Highway Research Program (NCHRP) supports relationships between mobile entities and fixed geographic entities. The MDLRS data model extends the functionality of linear location referencing system (LRS) data models that allow the determination of an unknown location within a transportation network based on reference from a known point (Vonderohe *et al.*, 1997). The MDLRS supports locational referencing of entities in four dimensions (the three dimensions of geographic space and time) relative to a transportation network and related geographic entities. This allows transportation analysts to reference data collected through GPS receivers and other position-aware technologies (Koncz and Adams, 2002).

The MDLRS foundation is the three-domain strategy since this allows representation of dynamics that is not possible in other spatio-temporal data models (Yuan, 1997). *Transportation Features* are atomic real world or virtual entities within the transportation system. *Spatial Objects* maintain the spatial properties (geometric, topological) of a Transportation Feature while *Time Objects* maintain the temporal properties of a Transportation Feature or its behavior, either in real world or database time. *Event Objects* represent occurrences that generate changes in the attributes of the Transportation Feature while *Experience Objects* record those changes as a history of the Transportation Feature. This supports a full range of temporal referencing, storage strategies and topological relations for analyzing change (see Koncz and Adams, 2002). Figure 16.3 illustrates the basic MDLRS object model using unified modeling language (UML) notation (see Booch *et al.*, 1999).

The MDLRS supports the movement navigation of entities within the transportation system through a *Conveyance* object and the temporal attributes of the Transportation

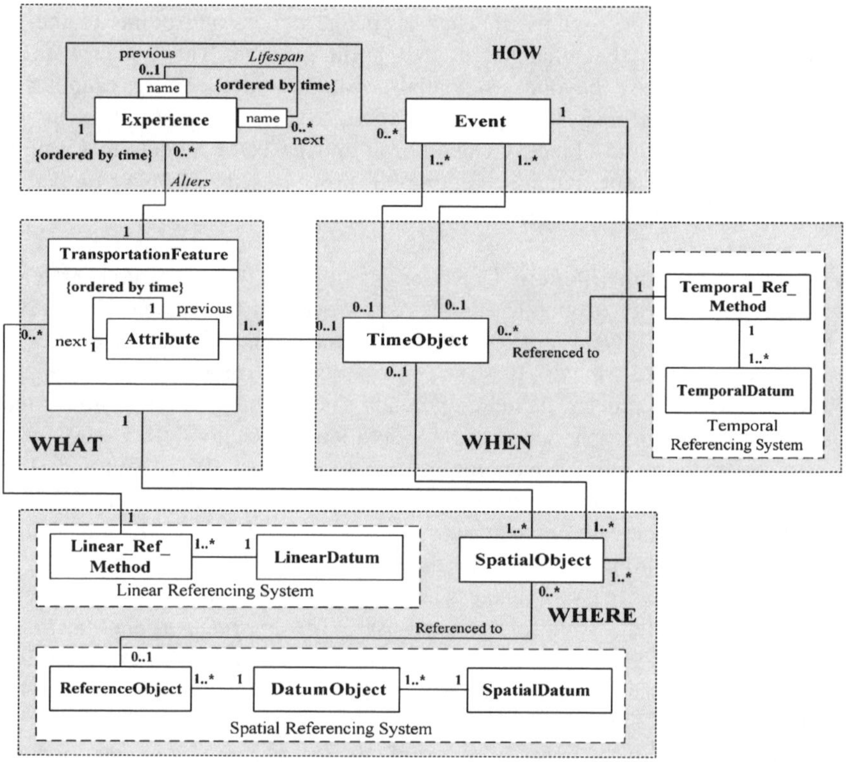

Figure 16.3 *The multidimensional location referencing system (MDLRS) object model (Reproduced from Koncz, N. and Adams, T. (2002) A data model for multi-dimensional transportation applications, International Journal of Geographical Information Science,* ***16****, 551–569. Figure 2, p. 555)*

Feature and Event objects. The *Conveyance* object represents anything that moves within a spatial or temporal reference frame and contains the navigation methods *Track* (descriptive) or *Route* (prescriptive). Positions are expressed as a function of time along a *Traversal*, where this consists of positions referenced to the transportation network.

The MDLRS is a potential breakthrough in referencing and linking data on mobile entities, transportation infrastructure and geographic data. However, in its basic form it does not support well some of the concepts required by time geography and activity theory. Events are tied to transportation features and the mobile entity in the MDLRS does not directly support relevant attributes such as socio-economic and demographic factors and activity plans (although it does allow differences between planned and executed routes). The MDLRS cannot easily support stations or activity locations where space-time paths bundle with each other and with resources. These weaknesses are understandable, since the MDLRS function requirements center on referencing the transportation infrastructure and the vehicles that operate within it. Extending this system to support time geography and activity theory is a worthwhile effort.

Data Models for Activity Analysis. Due to the increasing prominence of activity-based approaches to transportation and urban analysis, as well as increasing abilities to collect STA data, there have been several recent attempts to develop conceptual and logical data models to support activity analysis. Despite the complex, many-to-many relationships inherent in these data, including the relationships between individuals, households, travel, activities and their spatial and temporal dimensions, one of the major challenges in developing activity-based data models is to eliminate redundancy as much as possible (Shaw and Wang, 2000). Another consideration is representing complex temporal dynamics where an individual is sometimes moving along a continuous space-time trajectory but at other times is stationary in space (Wang and Cheng, 2001). As mentioned in Section 16.3, this is required to link the continuous spatio-temporal representations of time geography with the discrete spatio-temporal representations of activity theory.

Shaw and Wang (2001) develop a relational data model for handling disaggregate STA data. The central entity of their model is the 'trip' or a movement from one location to another. Other data such as the trip location (spatial), trip timing (temporal), the trip maker (a person within a household of other trip makers) and relevant trip attributes are linked to the trip. Trip locations are represented as paths through a network maintained using a variable-length segmentation model. While effective at maintaining data on multi-stop/multi-purpose trips, this data model does not contain the support for activity data required by time geography and activity theory. In these theories, it is activities and projects in space and time that drive travel and telecommunication demand.

Wang and Cheng (2001) formulate a STA data model that encompasses activities and projects to a greater degree. Figure 16.4 illustrates their conceptual data model using entity-relation notation (see Elmasri and Navathe, 1994). For clarity, the entity's attributes are suppressed; see Wang and Cheng (2001) for the complete depiction. The *Household* entity allows capturing of interactions among individuals and their activity patterns as required in time geography and activity theory. Each *Person* in the Household has a planned *Activity Program* that can be realized as an *Activity Pattern*. An Activity Pattern links the Person to a *Location* in one of two ways, namely, either by staying at (*Stay_At*) a

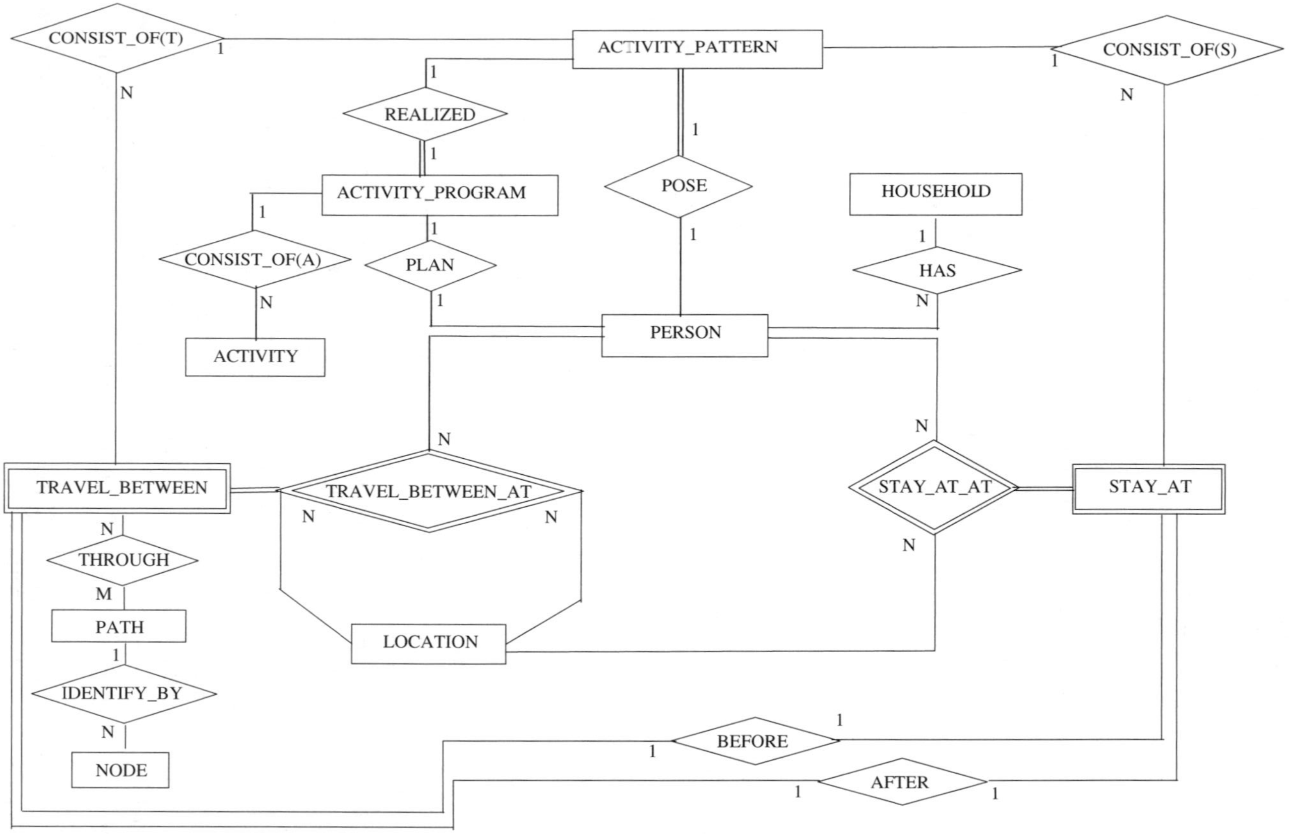

Figure 16.4 *Extended entity-relationship diagram for an activity-based data model (Reproduced from Wang, D. and Cheng, T. (2001) A spatio-temporal data model for activity-based transport demand modeling, International Journal of Geographical Information Science, 15, 561–585. Figure 5, p. 573)*

location while performing an activity or by traveling between (*Travel_Between*) locations to perform planned activities. Their data model also enforces space-time constraints among activity and travel locations and timings.

16.4.4 Exploring and Visualizing Space-Time Activity Data

A difficulty with analysis of STA data is the combinatorial explosion of the information space. Decisions such as the number of activities within a time period, sequencing, timing, interaction mode and route choice are interlinked, implying an information space that is exponential with respect to choice dimensions (Ben-Akiva and Bowman, 1998). Consequently, traditional methods for activity analysis require substantial reduction of the information space.

Econometric and statistical approaches require *a priori* specification and testing multidimensional utility functions or shallow first-order summaries of inter-activity linkages from data (see O'Kelly and Miller, 1984). Utility maximizing approaches include behavioral models that also require *a priori* specification of utility structures, meaning that few alternatives within the universe of plausible structures can be explored (see Kitamura, 1984). Rule-based reasoning systems construct activity and travel schedules based on decision heuristics derived from cognitive science (see Garling *et al.*, 1994; Hayes-Roth and Hayes-Roth, 1979; Vause, 1997). Simulation methods derive plausible choice sets and simulate individual choices from those sets (Ben-Akiva and Bowman, 1998). All of these techniques can only explore a very small subset of the complex and vast information space of space-time activities in geographic and cyber-space.

New IT for data storage, integration and analysis can break the combinatorial barrier that has prevented full exploration and discovery of the spatio-temporal patterns in activity data. Data warehousing techniques are available for integrated and efficient storage of digital geographic data (Bedard *et al.*, 2001). However, existing conceptual database design and storage/access techniques for geographic data warehousing must be modified to handle the temporal dimension of STA data.

Data mining and exploratory visualization techniques for digital geographic data are also emerging (see Miller and Han, 2001). There are so far only a few techniques available that can address STA data; most techniques are oriented towards analyzing flow data within network structures (see Marble *et al.*, 1997). Huisman and Forer (1998), Kwan (2000b) and van der Knaap (1997) develop cartographic visualization techniques for exploring STA patterns. Arentze *et al.* (2000) apply decision tree induction methods to STA data. Joh *et al.* (2001) adapt multidimensional sequencing methods from genome research to measure similarities among activity patterns.

16.4.5 Time Geographic Analysis

An unrealistic assumption of the STP is that travel velocities are uniform and continuous across time and space. In most settlement systems, travel is restricted to transportation networks. Travel velocities within these networks vary by location and time based on the capacity of the infrastructure, and the movement of other individuals through the system. If STPs are to be useful as a technique and not just a conceptual device, the assumption of a uniform velocity across time and space must be relaxed.

Miller (1991) relaxes the uniform velocity across space assumption by developing an algorithm for constructing a network *potential path tree* (PPT). This network analog to the PPA demarcates all nodes in a transportation that a person can reach given fixed anchor locations, a time budget and travel times within the network. A problem with this approach is that it focuses on nodes and can leave unresolved gaps in the network. Miller (1999) adopts a network based market area delimitation technique developed by Okabe and Kitamura (1996) to construct the *potential network area* (PNA). This shows all locations within a network that a person can occupy. Miller (1999) and Miller and Wu (2000) show that space-time constraints can be integrated into traditional accessibility measures and calculated for locations within a network using the PNA. Figure 16.5 (Colour Plate 5) illustrates a PNA-based accessibility measurement. O'Sullivan *et al.* (2000) develop GIS methods for calculating space-time access to public transportation; this illustrates that STP products can be extended to multi-modal travel. Wu and Miller (2001) relax the uniform velocity across time assumption by developing time-dependent STP measures and developing computational tools linked to a dynamic network flow model.

Figure 16.5 (Plate 5) *High accessibility locations within a network calculated using a potential network area (Reproduced from Miller, H.J. and Wu, Y.-H. (2000). GIS software for measuring space-time accessibility in transportation planning and analysis, GeoInformatica,* ***4****(2), 141–159. Figure 10, p. 157; with kind permission of Springer Science and Business Media)*

The time geographic tools discussed above still only recognize physical movement and travel. They do not recognize the ability of some individuals to use IT as a substitute or complement for transportation. They are also still loosely place-based: although accessibility is attributed to individuals, these individuals are identified through the locations of fixed geographic anchor points such as home and work.

Some progress is being made in disconnecting the STP from geographic space. Adams (2000) develops graphical representations of space-time paths in both physical and virtual space. *Extensibility diagrams* are extensions of the space-time path that encompass communication at a range of geographic scales from local to global. Extensibility diagrams can illustrate general characteristics of the relationships between IT and transportation in activity participation. Individuals can be compared with respect to the frequency, duration, time and geographic scale of travel, incoming communication and outgoing communication. However, these visual tools are only viable for very small datasets, even though geographic space is restricted to only a crude ordinal scale (local, regional, national, etc). Tools with higher spatial resolution are still needed along with the ability to support synoptic summaries, spatio-temporal aggregation, drill-down analysis and other exploratory and data mining techniques.

Another weakness of time geography is that it ignores the fact that people often have imperfect information and uncertainty about transportation system performance and the outcomes from travel (Hall, 1983). Although Hägerstrand (1970) argued that we should ignore preference and choice and instead focus on constraints, lack of information can be as strong a constraint as lack of time. In addition, we must consider imperfect information if we are to extend the theory of accessibility from its transportation context to encompass cyberspace. Since IT is about information search and retrieval, it is difficult to imagine an integrated theory of accessibility that assumes omnipotent beings.

Some initial but limited efforts have been made. Hall (1983) analyzes the impact of uncertainty about transportation system performance (i.e. travel velocity) on the potential path space. He also analyzes the impact of random coupling constraints and simple random search for activity locations. Kwan and Hong (1998) integrate cognitive constraints (e.g., preferences or lack of information) into a STP through an effective but ad-hoc overlay procedure. An extended research effort is required that re-examines time geography from its foundation, reformulates it as an analytical theory (similar to Burns (1979)), and develops computational tools that recognize imperfect information.

16.5 Conclusion: Research and Development Frontiers

Place-based representations and methods were developed in an era when data were scarce, computational platforms weak and questions apparently simpler. Urban and transportation theory and policy for a shrinking, shriveling and fragmenting world requires a people-based perspective. This perspective focuses on individuals in space and time and their interactions using transportation and telecommunication infrastructure and services. The increasing availability of digital data on people and objects in space and time and abilities to store, process and understand these data can make it possible. The deployment of location-based services (LBS) means that the private and public sectors will be collecting and using space-time activity (STA) data to sell and promote their products and programs. Researchers should also use these data and tools to make our transportation, telecommunication and settlement systems more livable and sustainable.

There has been a great deal of research in domains such as time geography, activity theory and GISci to support an extension of the place-based perspective in GIS to a people-based perspective. Research and development efforts along the following frontiers will provide better support for a people-based perspective in GIS:

1. a rigorous, formal representational theory of the dynamic spatial objects of interest in time geography and activity theory;
2. new data collection protocols and methods that exploit advances in IT, location-aware technologies and LBS, including exploiting detailed but noisy spatio-temporal referencing data as well as extracting activities and projects from incomplete and perhaps inconsistent vocal descriptions and queries from LBS;
3. new database designs that can support activities and project planning by mobile entities, their socio-economic and demographic characteristics and their movements within a detailed, georeferenced representation of the transportation infrastructure;
4. efficient geospatial data warehousing techniques for handling massive, noisy STA data;
5. spatio-temporal data mining and exploratory visualization techniques that can handle the massive, noisy data STA data;
6. enhanced versions of the space-time path, prism and other constructs that can be disconnected from geographic space and referenced within cyberspace (information space loosely connected to geographic space);
7. a time geography that recognizes imperfect information, information search and learning.

Acknowledgments

Thanks to Mike Goodchild and Dave Unwin for helpful comments on earlier versions of this chapter.

References

Adams, P.C., 1995, A reconsideration of personal boundaries in space-time, *Annals of the Association of American Geographers*, **85**, 267–285.

Adams, P.C., 2000, Application of a CAD-based accessibility model, In: *Information, Place and Cyberspace: Issues in Accessibility*, edited by Janelle, D.G. and Hodge, D.C., Berlin: Springer, 217–239.

Arentze, T.A., Hofman, F., van Mourik, H., Timmermans, H.J.P. and Wets, G. (2000) Using decision tree induction systems for modeling space-time behavior, *Geographical Analysis*, **32**, 330–350.

Axhausen, K.W., 1994, Travel diaries: an annotated catalogue. Working paper Imperial College, London.

Batty, M. and Miller, H.J., 2000, Representing and visualizing physical, virtual and hybrid information spaces, In: *Information, Place and Cyberspace: Issues in Accessibility*, edited by Janelle, D.G. and Hodge, D.C., Berlin: Springer, 133–146.

Bedard, Y., Merrett, T. and Han, J., 2001, Fundamentals of geospatial data warehousing for geographic knowledge discovery, In: *Geographic Data Mining and Knowledge Discovery*, edited by Miller, H.J. and Han, J., London: Taylor and Francis, 53–73.

Ben-Akiva, M.E. and Bowman, M.E., 1998, Activity based travel demand model systems, In: *Equilibrium and Advanced Transportation Modeling*, edited by Marcotte, P. and Nguyen, S., Boston: Kluwer Academic, 27–46.

Bennahum, D. S., 2001, Be here now, *Wired*, **9**(11), 159–163.
Benson, J., 2001, LBS technology delivers information where and when it's needed, *Business Geographics*, **9**(2), 20–22.
Bhat, C. and Koppelman, F.S., 2000, Activity-based travel demand analysis: History, results and future directions, In: *Proceedings, 79th annual meeting of the Transportation Research Board*, Washington, DC, January 9–13, CD-ROM.
Bian, L., 2000, Object-oriented representation for modelling mobile objects in an aquatic environment, *International Journal of Geographical Information Science*, **14**, 603–623.
Booch, G., Rumbaugh, J. and Jacobson, I., 1999, *The Unified Modeling Language Users Guide*, Reading, MA: Addison-Wesley.
Boyce, D.E., Zhang, Y.-F. and Lupa, M.R., 1994, Introducing feedback into four-step travel forecasting procedure versus equilibrium solution of combined model, *Transportation Research Record*, 1443, 65–74.
Brog, W., Erl, E., Meyburg, A.H. and Wermuth, M.J., 1982, Problems of non-reported trips in surveys of nonhome activity patterns, *Transportations Research Record*, **891**, 1–5.
Burns, L.D., 1979, *Transportation, Temporal and Spatial Components of Accessibility*, Lexington, MA: Lexington Books.
Cairncross, F., 1997, *The Death of Distance*, Boston; Harvard Business School Press.
Chapin, F.S., 1965, *Urban Land Use Planning*, 2nd edition, Urbana, IL: University of Illinois Press.
Couclelis, H. and Getis, A., 2000, Conceptualizing and measuring accessibility within physical and virtual spaces, In: *Information, Place and Cyberspace: Issues in Accessibility*, edited by Janelle, D.G. and Hodge, D.C., Berlin: Springer, 15–20.
Cova, T. and Goodchild, M.F., 2002, Extending geographic representation to include fields of spatial objects, *International Journal of Geographical Information Science*, **16**, 509–532.
Elmasri, R. and Navathe S.B., 1994, *Fundamentals of Database Systems*, 2nd edition, Redwood City, CA: Benjamin/Cummings.
Engeström, Y. and Miettinen, R., 1999, Introduction, In: *Perspectives on Activity Theory*, edited by Engeström, Y., Miettinen, R. and Punnamäki, R.-L., Cambridge: Cambridge University Press, 1–16.
Ettema, D., Timmermans, H. and van Veghel, L., 1996, *Effects of Data Collection Methods in Travel and Activity Research*, Eindhoven: European Institute of Retailing and Services Studies.
Frank, A., 2001, Socio-economic units: Their life and motion, In: *Life and Motion of Socio-economic Units*, edited by Frank, A., Raper, J. and Cheylan, J.-P., London: Taylor and Francis, 21–34.
Galton, A., 1995, Towards a qualitative theory of movement, In: *Spatial Information Theory: Proceedings of the European Conference on Spatial Information Theory (COSIT'95)*, edited by Frank, A. and Kuhn, W., Berlin: Springer-Verlag, Lecture Notes in Computer Science, 377–396.
Galton, A., 1997, Space, time and movement, In: *Spatio and Temporal Reasoning*, edited by Stock, O., Dordrecht, U.K.: Kluwer Academic, 321–352.
Garling, T., Kwan, M.-P. and Golledge, R.G., 1994, Computational-process modeling of household activity scheduling, *Transportation Research B*, **26**, 355–364.
Gleick, J., 1999, *Faster: The Acceleration of Just About Everything*, New York Pantheon Books.
Gober, P., McHugh, K.E. and Lecerc, D., 1993, Job-rich but housing-poor: The dilemma of a western amenity town, *Professional Geographer*, **45**, 12–20.
Goldhaber, M.H., 1997, Attention shoppers! *Wired*, 5.12, available at www.wired.com.
Golledge, R.G. and Stimson, R.J., 1997, *Spatial Behavior: A Geographic Perspective*, New York: Guilford Press.
Golledge, R.G. and Zhou, J., 2001, *GPS Tracking of Daily Activities*, final report, UCTC Grant # DTRS99-G-0009, Department of Geography and Research Unit on Spatial Cognition and Choice, University of California at Santa Barbara.
Goodchild, M.F., 2000, GIS and transportation: Status and challenges, *GeoInformatica*, **4**, 127–139.
Goodchild, M.F., 2001, A geographer looks at spatial information theory, paper presented at the Conference on Spatial Information Theory (COSIT), Morro Bay, CA USA, 19–23 September.
Goodchild, M.F. and Janelle, D.G., 1984, The city around the clock: Space-time patterns of urban ecological structure, *Environment and Planning A*, **16**, 807–820.

Graham, S. and Marvin, S., 1996, *Telecommunications and the City: Electronic Spaces, Urban Places*, New York: Routledge.
Greaves, S. and Stopher, P., 1998, A synthesis of GIS and activity-based travel-forecasting, *Geographical Systems*, **5**, 59–89.
Guo, B. and Poling, A.D., 1995, Geographic information systems/global positioning systems design for network travel time study, *Transportation Research Record*, **1497**, 135–144.
Hägerstrand, T., 1970, What about people in regional science? *Papers of the Regional Science Association*, **24**, 7–21.
Hall, R.W., 1983, Travel outcome and performance: The effect of uncertainty on accessibility, *Transportation Research B*, **17B**, 275–290.
Hayes-Roth, B. and Hayes-Roth, F., 1979, A cognitive model of planning, *Cognitive Science*, **3**, 275–310.
Huisman, O. and Forer, P., 1998, Computational agents and urban life spaces: A preliminary realisation of the time-geography of student lifestyles, In: *Proceedings, GeoComputation 98*, www.geocomputation.org.
Janelle, D.G., 1969, Spatial organization: A model and concept, *Annals of the Association of American Geographers*, **59**, 348–364.
Joh, C.-H., Arentze, T.A. and Timmermans, H.J.P., 2001, Multidimensional sequence alignment methods for activity-travel pattern analysis: A comparison of dynamic programming and genetic algorithms, *Geographical Analysis*, **33**, 247–270.
Kanter, T.G., 2001, *Adaptive Personal Mobile Communication: Service Architecture and Protocols*, Ph.D. dissertation, Department of Microelectronics and Information Technology, Royal Institute of Technology, Stockholm, Sweden.
Kitamura, R., 1984, Incorporating trip chaining into analysis of destination choice, *Transportation Research B*, **18B**, 67–81.
Koncz, N. and Adams, T., 2002, A data model for multi-dimensional transportation applications, *International Journal of Geographical Information Science*, **16**, 551–569.
Kuhn, W., 2001, Ontologies in support of activities in geographical space, *International Journal of Geographical Information Science*, **15**, 613–631.
Kwan, M.-P., 1998, Space-time and integral measures of accessibility: A comparative analysis using a point-based framework, *Geographical Analysis*, **30**, 191–216.
Kwan, M.-P., 1999, Gender and individual access to urban opportunities: A study using space-time measures, *Professional Geographer*, **51**, 210–227.
Kwan, M-P., 2000a, Human extensibility and individual hybrid-accessibility in space-time: A multiscale representation using GIS, In: *Information, Place and Cyberspace: Issues in Accessibility*, edited by Janelle, D.G. and Hodge, D.C., Berlin: Springer, 241–256.
Kwan, M-P., 2000b, Interactive geovisualization of activity-travel patterns using three-dimensional geographical information systems: A methodological exploration with a large dataset, *Transportation Research C: Emerging Technologies*, **8**, 185–203.
Kwan, M.-P. and Hong, X.-D., 1998, Network-based constraints-oriented choice set formation using GIS, *Geographical Systems*, **5**, 139–162.
Marble, D.F., Gou, Z., Liu, L. and Saunders, J., 1997, Recent advances in the exploratory analysis of interregional flows in space and time, In: *Innovations in GIS 4*, edited by Kemp, Z., London: Taylor and Francis.
Mark, D.M. and Egenhofer, M.J., 1998, Geospatial lifelines, In: *Integrating Spatial and Temporal Databases*, edited by Günther, O., Sellis, T. and Theodoulidis, B., Dagstuhl Seminar Report 228.
McNally, M.G., 1998, Activity-based forecasting models integrating GIS, *Geographical Systems*, **5**, 163–187.
Medak, D., 2001, Lifestyles, In: *Life and Motion of Socio-economic Units*, edited by Frank, A., Raper, J. and Cheylan, J.-P., London: Taylor and Francis,139–153.
Miller, H.J., 1991, Modeling accessibility using space-time prism concepts within geographical information systems, *International Journal of Geographical Information Systems*, **5**, 287–301.
Miller, H.J., 1999, Measuring space-time accessibility benefits within transportation networks: Basic theory and computational methods, *Geographical Analysis*, **31**, 187–212.

Miller, H.J., 2001, Location-based services and space-time activity analysis, position paper, Center for Spatially Integrated Social Science Specialist Meeting on Location-based Services, Santa Barbara, CA USA, December 15–17.

Miller, H.J. and Han, J. (eds), 2001, *Geographic Data Mining and Knowledge Discovery*, London: Taylor and Francis.

Miller, H.J. and Shaw, S.-L., 2001, *Geographic Information Systems for Transportation: Principles and Applications*, Oxford University Press.

Miller, H.J. and Wu, Y.-H., 2000, GIS software for measuring space-time accessibility in transportation planning and analysis, *GeoInformatica*, **4**, 141–159.

Mitchell, W.J., 1995, *City of Bits*, Cambridge, MA: MIT Press.

Moss, M.L. and Townsend, A.M., 2000, How telecommunications systems are transforming urban spaces, In: *Cities in the Telecommunications Age: The Fracturing of Geographies*, edited by Wheeler, J.O., Aoyama, Y. and Warf, B., New York: Routledge, 31–41.

Murakami, E. and Wagner, D.P., 1999, Can using global positioning system (GPS) improve trip reporting? *Transportation Research C: Emerging Technologies*, **7**, 149–165.

National Research Council, 2001, *Community and Livability: Data Needs for Improved Decision Making – Transportation and Quality of Life*, Washington, DC: National Academy Press.

Okabe, A. and Kitamura, M., 1996, A computational method for market area analysis on a network, *Geographical Analysis*, **28**, 330–349.

O'Kelly, M.E. and Miller, E.J., 1984, Characteristics of multistop multipurpose travel: An empirical study of trip length, *Transportation Research Record*, **976**, 33–39.

O'Sullivan, D., Morrison, A. and Shearer, J., 2000, Using desktop GIS for the investigation of accessibility by public transport: An isochrone approach, *International Journal of Geographical Information Science*, **14**, 85–104.

Pas, E. and Harvey, A.S., 1996, Time-use research and travel demand analysis and modelling. In: *Understanding Travel Behavior in an Era of Change*, edited by Ryuichi, K. and Lee-Gosselin, M., New York: Pergamon Press.

Peuquet, D. and Duan, N, 1995, An event-based spatiotemporal data model (ESTDM) for temporal analysis of geographical data, *International Journal of Geographical Information Science*, **9**, 7–24.

Pred, A., 1981, Of paths and projects: Individual behavior and its societal context, In: *Behavioral Problems in Geography Revisited*, edited by Cox, K.R. and Golledge, R.G., New York: Methuen, 231–255.

Purvis, C.L., 1990, Survey of travel surveys II, *Transportation Research Record*, **1271**, 23–32.

Raper, J. and Livingstone, D., 1995, Development of a geomorphological spatial model using object-oriented design, *International Journal of Geographical Information Systems*, **9**, 359–383.

Reed, J.H., Krizman, K.J., Woerner, B.D. and Rappaport, T.S., 1998, An overview of the challenges and progress in meeting E-911 requirement for location service, *IEEE Communications Magazine*, **36**(4), 30–37.

Shaw, S.L. and Wang, D., 2000, Handling disaggregate spatiotemporal travel data in GIS, *GeoInformatica*, **4**, 161–178.

Shen, Q., 2000, Transportation, telecommunications and the changing geography of opportunity, *Urban Geography*, 20, 334–355; reprinted In: *Information, Place and Cyberspace: Issues in Accessibility*, edited by Janelle, D.G. and Hodge, D.C., Berlin: Springer, 47–72.

Smyth, C.S., 2001, Mining mobile trajectories, In: *Geographic Data Mining and Knowledge Discovery*, edited by Miller, H.J. and Han, J., London: Taylor and Francis, 337–361.

Spear, B.D. and Lakshmanan, T.R., 1998, The role of GIS in transportation planning and analysis, *Geographical Systems*, **5**, 45–58.

Stopher, P.R. and Wilmot, C.G., 2000, Some new approaches to designing household travel surveys: Time-use diaries and GPS, In: *Proceedings, 79th annual meeting of the Transportation Research Board*, Washington, DC, January 9–13, CD-ROM.

Thill, J-C. and Thomas, I., 1987, Toward conceptualizing trip-chaining behavior: A review, *Geographical Analysis*, **19**, 1–17.

Tobler, W.R., 1999, Lecture at the 1999 ESRI User conference, San Diego, CA.

United Nations, 2000, *United Nations Statistics Division – Time-use Surveys*, webpage available at http://www.un.org/Depts/unsd/timeuse/tusindex.htm.

van der Knaap, W.G.M., 1997, Analysis of time-space activity patterns in tourist recreation complexes: A GIS-oriented methodology, In: *Activity- based Approaches to Travel Analysis*, edited by Ettema, D.F. and H.J.P. Timmermans, Oxford: Elsevier Science, 283–311.

Vause, M., 1997, A rule-based model of activity scheduling behavior, In: *Activity-based Approaches to Travel Analysis*, edited by Ettema, D.F. and H.J.P. Timmermans, Oxford: Elsevier Science, 73–88.

Vonderohe, A., Chou, C.L., Sun, F. and Adams, T.M., 1997, A generic data model for linear referencing systems, *Research Results Digest 218, National Cooperative Highway Research Program,* Transportation Research Board, Washington, DC.

Wang, D. and Cheng, T., 2001, A spatio-temporal data model for activity-based transport demand modeling, *International Journal of Geographical Information Science*, **15**, 561–585.

Wegener, M., 1994, Operational urban models: State of the art, *Journal of the American Planning Association,* **60** (Winter), 17–29.

Westervelt, J.D. and Hopkins, L.D., 1999, Modeling mobile individuals in dynamic landscapes, *International Journal of Geographical Information Science*, **13**, 191–208.

Wu, Y.-H. and Miller, H.J., 2001, Computational tools for measuring space-time accessibility within transportation networks with dynamic flow, *Journal of Transportation and Statistics*, **4**, 1–14.

Yearsley, C. M. and Worboys, M.F., 1995, A deductive model of planar spatio-temporal objects, In: *Innovations in GIS 2*, edited by Fisher, P., London: Taylor and Francis, 43–51.

Yuan, M., 1996, Modeling semantic, temporal and spatial information in geographic information systems, In: *Geographic Information Research: Bridging the Atlantic*, edited by Craglia, M. and Couclelis, H., London: Taylor and Francis, 334–347.

Yuan, M., 1997, Use of knowledge acquisition to build wildfire representation in geographic information systems, *International Journal of Geographical Information Systems*, **11**, 723–745.

Yuan, 2001a, Representing complex geographic phenomena in GIS, *Cartography and Geographic Information Systems*, **28**, 83–96.

Yuan, M., 2001b, Representing geographic information to support queries about life and motion of socio-economic units, In: *Life and Motion of Socio-economic Units*, edited by Frank, Raper, A., J. and Cheylan, J.-P., London: Taylor and Francis, GISDATA, **8**, 217–234.

Zagami, J.M., Parl, S.A., Bussgang, J.J. and Melillo, K.D., 1998, Providing universal location services using a wireless E911 location network, *IEEE Communications Magazine*, **36**(4) 66–71.

17

Dynamic Spatial Modelling in the Simile Visual Modelling Environment

Robert Muetzelfeldt and Matt Duckham

17.1 Introduction

This chapter addresses two issues. The first concerns the integration of dynamic and spatial modelling. There are numerous tools available for doing one or the other, but very few for supporting both. The second issue concerns the flexibility of available tools for representing space. Most spatial modelling systems and GIS tend to constrain the user to a limited number of spatial representation frameworks. Simile (Muetzelfeldt and Taylor, 2001; Muetzelfeldt and Massheder, 2003; see www.simulistics.com) is a visual modelling environment, developed originally for the dynamic modelling of ecological and environmental systems, which supports a wide range of ways of representing space. Therefore, it addresses both of the above issues in a single environment.

Dynamic simulation models are important in environmental research and decision-making, and a wide variety of approaches are used (Rizzoli and Young, 1997). The great majority of models are implemented as computer programs in a conventional programming language. Most models are non-spatial; those that are spatial tend to be based on a regular grid, using the 2D array as the basic data structure.

There are a number of software packages available for modelling the dynamics of natural systems (for example, Stella (www.hps-inc.com), ModelMaker (www.modelkinetix.com), Powersim (www.powersim.com), and Vensim (www.vensim.com)). Typically, these are based on the System Dynamics modelling paradigm (Forrester, 1971), in which a system is conceptualised in terms of storages and flows. These packages greatly facilitate the process of designing and running models, since the user builds a model by

Re-presenting GIS Edited by P. Fisher and D. Unwin

drawing a compartment-flow diagram, then adds appropriate numeric values and equations. However, these packages themselves have limited or non-existent capabilities for spatial modelling. Conversely, GIS have evolved into very powerful tools for static, purely spatial modelling. Incorporating dynamics into GIS is generally difficult and typically demands *ad hoc* solutions, such as linking to external software through shared data files or embedding program modules within the GIS. Related to the literature on spatio-temporal GIS, a range of research projects have tackled the problems of integrated dynamic and spatial modelling capabilities within GIS. Chomicki and Revesz (1999) and Saunders *et al.* (1999), for example, have developed functional representations of dynamic spatial phenomena suitable for use within GIS. Raper and Livingstone (1995, 1996) and Claramunt and Thériault (1995, 1996) have addressed the difficulties surrounding the inclusion of process-based models of dynamic spatial information within GIS more generally. The Spatial Modelling Environment (iee.umces.edu/SME3) enables the equations for a single-patch model developed in Stella to be incorporated into a GIS. Applications for dynamic spatial modelling are also common in the literature, for example, dynamic resource management for shepherds and their sheep (Cheylan and Lardon, 1993) and dynamic spatial predator-prey relationships in both ecological and military application domains (Westervelt and Hopkins, 1999). Despite this relatively rich literature on general theory and specific potential applications, conventional GIS are not well suited to dynamic spatial modelling, and the topic has remained largely a research activity.

Simile, like the other modelling packages mentioned above, supports the construction of System Dynamics models in a visual (diagramming) environment. However, in contrast to these other packages, Simile is also object-based. Objects in Simile can be used to represent geographic phenomena. This does not mean that objects in Simile are necessarily objects in the sense of the object versus field debate (see Couclelis, 1992). Objects in Simile (or indeed in any object-based spatial information system – c.f. Laurini and Pariente, 1996) can be considered to represent, say, elements of a spatial field (e.g. cells in a raster of temperature, polygons in an irregular tessellation of ecotopes) or discrete entities located in space (e.g. nodes and edges in a road vector).

In fact, Simile has no particular spatial representations built into it, rather these are specified by the user in Simile's modelling language and this means that modellers have considerable flexibility in just how space is represented. They are not restricted to some pre-defined spatial framework. One model can include both field and object views, polygonal, rectangular and hexagonal areal units, 3D units (e.g. cubes), and point and linear features, all referenced to a common co-ordinate system. Together with appropriate visualisation tools, this flexibility enables a very wide range of dynamic spatial models to be developed.

Flexibility of representation is important not only to ensure the spatial modelling system achieves a degree of generality, but also in the context of work on the social theory of GIS. A core critique of GIS in the literature is that the dominant spatial frameworks embedded in GIS restrict the types of spatial information that can be admitted (Sheppard, 1995) and limit the 'ways of knowing' that can be achieved using these systems (Pickles, 1999). GIS-based research initiatives, such as the Varenius project, have begun to consider critically the causes and effects of such restrictions (see Goodchild *et al.*, 1999; Sheppard *et al.*, 1999). By providing users with systems that do not have particular in-built spatial representations, these systems should allow users to model a much wider variety of spatial concepts. However, developing more flexible GIS will never permit the user's imagination to have a completely free rein: using any computational system inevitably constrains as well as supports certain ways of understanding our geographic environment. However, we will argue that the language Simile provides for spatial representation allows considerably more scope for users' own spatial modelling and concepts than conventional GIS.

In this chapter we will first overview Simile's capabilities as a modelling environment, and give an example of its use for simple (non-spatial) dynamic modelling. We will then consider a range of spatial configurations, and show how they can be represented in Simile. Finally, we will discuss the potential for the approaches that Simile supports in modelling space and time.

17.2 Simile

Simile has a number of features:

Visual modelling: Simile supports a two-phase approach to model construction. The first involves the drawing of diagrams that show the main features of the model and the second involves fleshing-out the model-diagram elements with quantitative information on the relevant values and equations.

System Dynamics: Simile allows models to be formulated in System Dynamics terms, as compartments (stocks, levels) whose values are governed by flows in and out. This can be considered as a visual language for representing differential-equation models, with a compartment representing a state variable, and the rate-of-change being the net sum of inflows minus outflows. Although this is the primary method of representing dynamics within Simile models, it is not elaborated here because it is not directly relevant to issues of spatial representation.

Disaggregation: Simile allows the modeller to express many forms of disaggregation: e.g. age/size/sex/species classes. This is achieved by defining how one class behaves, then specifying that there are many such classes.

Object-based modelling: Simile allows a population of objects to be modelled. As with disaggregation, model designers state how one member behaves, then specify that there are many such members. In this case, the designer can add in symbols denoting the rules for creating new members of the population, and for killing off existing members. Individual members of the population can interact with others.

Spatial modelling: It follows that spatial modelling in the system is simply a special form of disaggregation. One spatial unit (grid square, hexagon, polygon, etc.) is modelled, then many such units are specified. Each can be given spatial attributes, such as area or location, and the proximity of one unit to another can be represented.

Modular modelling: Simile allows any model to be inserted as a submodel into another. Having done this, the modeller can then manually make the links between variables in the two components (in the case where the submodel was not designed to plug into the main model); or links can be made automatically, giving a 'plug-and-play' capability. Conversely, any submodel can be extracted and run as a stand-alone model, greatly facilitating testing of the submodels of a complex model.

Efficient computation: Models can be run as compiled C++ programs. In many cases, these will run as fast as a hand-coded C++ program, enabling Simile to cope with complex models (100s equations; 1000s object instances). While larger or institutional spatial databases are likely to contain millions of object instances, the complexity of modelling, rather than the efficiency of computation, means that dynamic spatial modelling tasks are often more modest in size.

Customisable output displays and input tools: Simile users can design and implement their own input/output procedures independently. In particular, they can develop displays for model output that are specific to disciplinary norms or other requirements. Once developed, these can be shared with others in the relevant research community.

Declarative representation of model structure: A Simile model is saved in an open format as a text file (in Prolog syntax). This means that others can develop tools for processing Simile models in novel ways. For example, one group may develop a new way of reporting on model structure, while another may wish to undertake automatic comparison of the structure of two similar models. It also opens the way for the efficient transmission of models across the Internet (as XML files), and for the sharing of models between different modelling environments.

Further information may be obtained from the website at www.simulistics.com.

In common with other visual modelling environments, Simile models are built in two phases. First, the modeller produces a diagram, using a set of 11 icons to represent the model structure. Then, the modeller quantifies the model by entering numeric values and equations for the various components of the model.

The diagramming icons are chosen from the Toolbar shown as Figure 17.1. Four of these, the compartment, flow, variable and influence, are concerned with conventional System Dynamics modelling, and will be considered first. The remaining seven icons are concerned with submodels and objects, and will be considered subsequently.

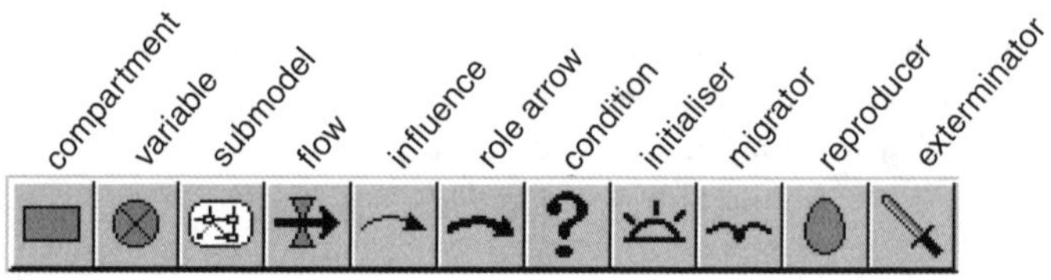

Figure 17.1 *Simile's toolbar icons*

System Dynamics (SD) is a common dynamic modelling paradigm within the ecological and environmental research communities. A SD model consists of compartments (stocks, levels, storages) connected by flows, with subsidiary variables for representing parameters and intermediate variables, and influence arrows to show which compartments and variables are used in the calculation of flows and other variables. Essentially, SD is a cosmetic language for defining differential- or difference-equation models: differential equations if the equations are taken to define continuous change; difference equations if the time step is taken to be unity.

Figure 17.2 shows a simple SD model in Simile. The model represents the interaction between crop growth and soil water. The crop is represented in terms of biomass, with

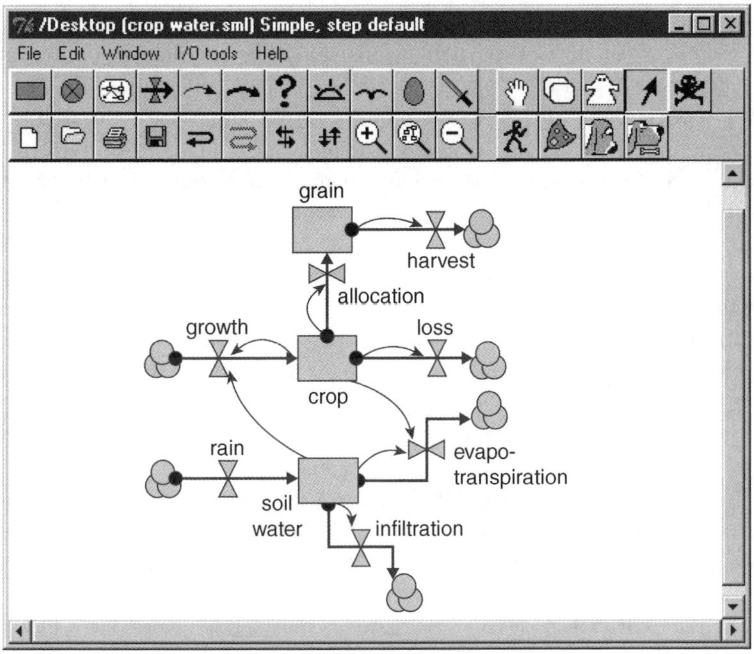

Figure 17.2 *A simple System Dynamics model*

one compartment for the vegetative component and one for the grain produced. A single compartment represents soil water. The dynamic behaviour of the crop is governed by flows corresponding to the biological processes of growth, allocation to grain, and losses. The dynamic behaviour of the soil water is governed by flows for rainfall, evaporation and infiltration of water into the soil. Influence arrows show the controlling effect of compartments on the flows: in biological systems, growth processes tend to be influenced by the amount of growing material; while loss processes tend to be influenced by the amount in the donor compartment. Finally, soil water influences crop growth, and crop biomass influences soil water transpiration.

The model is completed by entering a value or an equation into a dialogue window. For compartments, the value represents the initial value at the start of the simulation. For flows and variables, depending on whether that model component has influence arrows pointing at it, a constant value or equation is entered. Once every component has been instantiated with a value or an equation, the model is ready for running. The user specifies various run settings, such as time step for integration and run duration, and required displays for selected model variables. 'Running the model' then involves the computer simulating the dynamic behaviour of the modelled system, by iteratively calculating all the flows for each compartment, then updating the compartment contents over each time step.

This model is clearly non-spatial. We could make it into a very crude spatial model by duplicating the compartment-flow structures in a larger model diagram, to represent, for example, different fields, each with its own crop and water components. Each could then be modified to represent different soil and crop characteristics. Obviously, this is not worth considering as a way of doing serious spatial modelling. Other modelling environments enable model components to be declared as arrays: this gives some potential to model spatial disaggregation, but is cumbersome and non-intuitive.

What is required is the ability to specify the dynamics of a class of objects, such as a spatial unit, *once*, then specify that we have many instances of object belonging to that class. Using its submodel construct, this is exactly what Simile enables the modeller to do.

The `submodel` is the key to Simile's ability to handle a wide variety of model-design requirements. Essentially, the submodel is a container for some collection of model elements, including System Dynamics elements and other submodels. It is constructed by selecting the submodel tool from the Toolbar, then dragging an envelope in the model-diagram window. This may enclose existing model components, or can be in a blank area of the screen, ready to receive model components later.

The submodel has a wide range of roles in Simile. At the simplest level, rather like the 'sector' in Stella, it can be used to divide a complex model visually into different sections. Used in this way, it has no implications for the model's mathematical structure. The submodel can also be used for modular modelling, since a submodel can be saved to file and loaded from file independently of the model it is in (rather like the 'co-model' in PowerSim or the 'submodel' in ModelMaker).

However, the real power of the submodel comes when we specify that there are multiple instances of a submodel. This is roughly analogous to the notion of 'class' and 'object' in object-oriented software engineering: the submodel represents the class, and the multiple instances represent multiple objects belonging to the class. In general, the submodel is used whenever we realise, during the model-design process, that we have multiple things of the same type, each instance following the same rules for its behaviour. So, for the example above in Figure 17.2, a multiple-instance submodel might be used for multiple crop types (e.g. wheat, barley), or multiple soil layers for modelling soil water dynamics.

17.3 Spatial Modelling in Simile

How, despite having no spatial modelling constructs built into it, can Simile do spatial modelling? The answer is simple. Spatial units are objects just like any other and so can be given attributes that we happen to interpret as spatial or that are appropriate for spatial units (e.g. elevation), but, as far as Simile is concerned, these are just like any other attribute an object can have. It does not read anything into the fact that a variable is called 'area' or 'x' (and indeed we could, if we wished to be obscure, call the variables holding x and y coordinates 'fred' and 'susan'). This section, and the following examples, contains a minimal description of spatial modelling in Simile. Clearly, real applications would involve many more model variables for representing fixed and dynamic attributes within spatial units than the simplified examples discussed.

Figure 17.3 shows the model diagram for a model containing a single 'multiple-instance submodel'. Note the multiple border on the Patch submodel, which, by analogy with a deck of cards, denotes multiplicity. This is interpreted as representing multiple spatial units ('Patch'), each having the attributes of x and y coordinates, area and elevation.

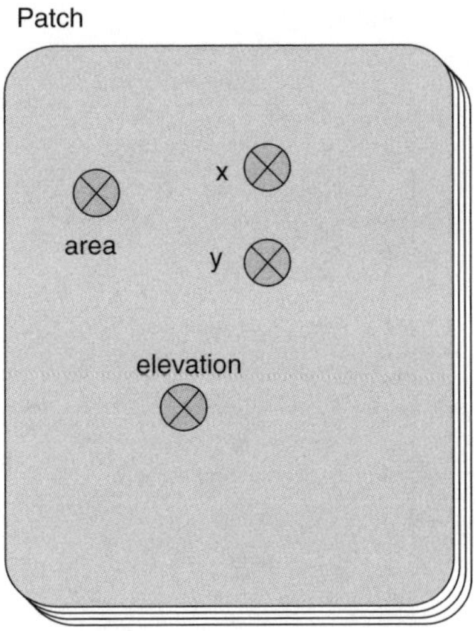

Figure 17.3 *A simple (non-dynamic) spatial model with no spatial interaction*

Simile provides various tools for displaying even such simple models in a spatial manner. Figure 17.4 (Colour Plate 6) illustrates some that are possible, including standard raster- and vector-type display, as well as a display representing the spatial units as points in space, or even interpreting them as trees scattered over a landscape, in which case 'elevation' is being interpreted as tree height. The only additional requirement is to provide the boundary coordinates for the vector-based display, either within each instance or in an external file used only by the display tool.

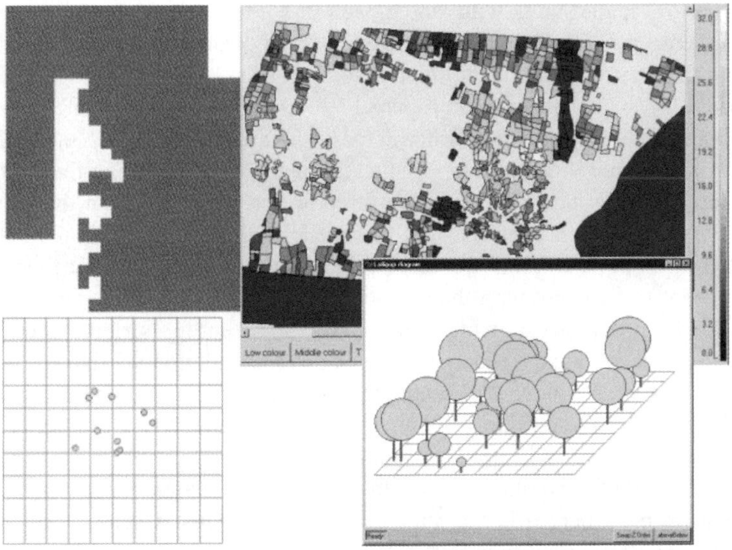

Figure 17.4 (Plate 6) *Various forms of spatial display provided by Simile*

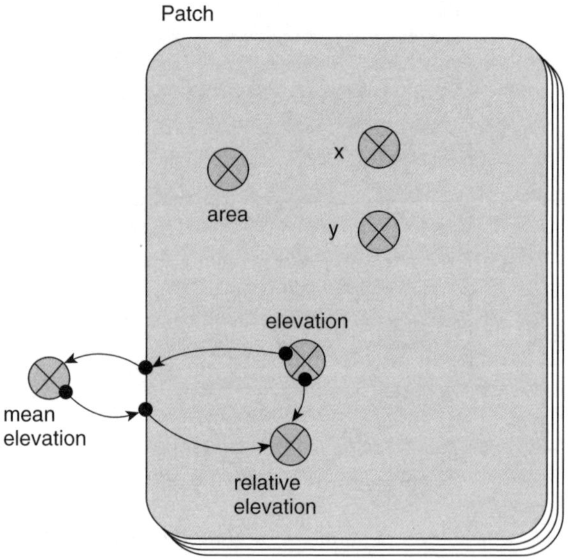

Figure 17.5 *Calculation of aggregate statistics, and attributes related to the aggregate*

Given a model like this, it is straightforward to calculate aggregate statistics, and to relate the value of each spatial unit to the aggregate value. Figure 17.5 illustrates this, with a calculation of mean elevation outside the Patch submodel, and the subsequent calculation of the relative elevation of each spatial unit with respect to the mean.

Many dynamic models require some form of interaction between spatial units, typically, but not necessarily, neighbouring ones. This raises the question: how do we define the set of spatial units with which each unit can interact? In the simple analysis presented here, we restrict ourselves to the case where the 'interaction' involves calculating the elevation of a patch relative to a set of neighbouring patches, as opposed to the set of all patches presented in Figure 17.5. This could, for example, form the basis for a model in which the flow of water across a landscape was being modelled.

In the first example (Figure 17.6), we tackle the problem by taking the values for x, y and elevation outside the Patch, forming three arrays, which contain the values for all patches. We then bring these back into the Patch submodel, first calculating the set of distances from each patch to all the others, then using this to select just those elevations which we want to compare with the elevation of the current patch.

Thus, we have a pretty flexible notation for representing various forms of spatial relationship, since we have freedom to do whatever calculation we like for working out 'distances' (it does not have to be simple Euclidean distance), and for using this in deciding which elevation values to use. However, there are two limitations with this solution, one computational, the other conceptual.

The computational limitation is that the number of values being passed around is proportional to n^2, where n is the number of patches. This is because each patch has access to information on all the other patches. If we have 10 000 patches, then we have 100 million values to process! The conceptual limitation is that the model diagram does

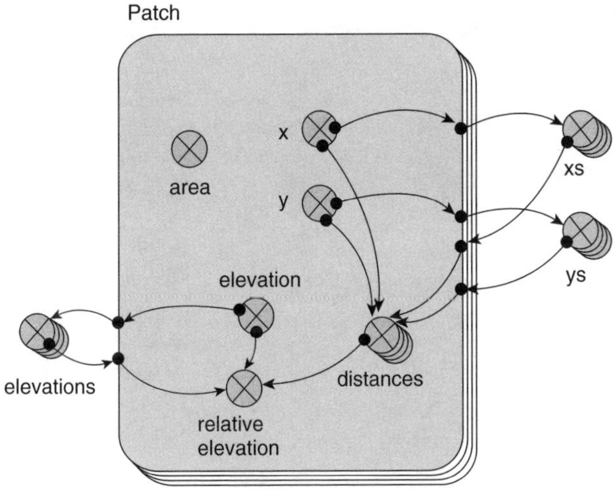

Figure 17.6 *Interaction between patches, using arrays*

not really communicate what is going on. What we are really saying is that only some of the patches can interact with a particular patch, but that is not obvious from the model diagram. Yet the role of the diagram is to communicate the main assumptions made in the model design down to a certain level of detail. It would be so much nicer if we could show diagrammatically that *some* form of relationship between patches was being modelled, even if we do not show the detail of just how we define that relationship. This is analogous to a Simile influence arrow showing that a variable has some form of influence on another, even though we have to look at the equation to see just how that relationship has been represented mathematically.

Simile allows the user to express the concept of a relationship ('association') between objects, of the same type or of different types. UML, the Unified Modelling Language (Stevens and Pooley, 2000), uses a similar construct in a UML class diagram. In the present context, we wish to express an association between objects of the same type – patch to patch – and we can denote this as a 'neighbour' association.

Figure 17.7 shows a re-expression of the previous model in which a neighbour association between the patches is expressed. This is done by introducing a new Simile submodel, drawing two thick 'role' arrows onto it from the Patch submodel, and adding in a condition that determines when the neighbour association is deemed to hold between two patches. The elevation value is now passed through the association submodel, so that each patch now only sees the elevation values of the patches that are its neighbour.

A key point here is the condition symbol inside the Neighbour submodel. Since this contains an expression that determines when the 'neighbour' association holds between two patches, it is the place where a view on the nature of the spatial arrangement of, and relationship between, patches can be captured. In this case, it can be seen that this is based solely on their respective coordinates. One cannot see *how* the coordinates are used to determine the association. It could be that the 4 nearest or 8 nearest on a grid-square basis, or the 6 nearest on a hexagonal basis, are used, but this is a detail to be found by looking at the expression inside the condition symbol. The point here is that the model

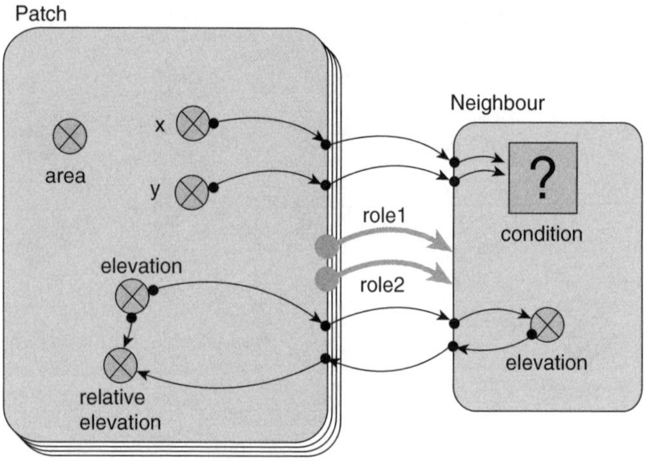

Figure 17.7 Interaction between patches, using an association submodel

diagram shows that some form of association exists, and that we have considerable flexibility in expressing the nature of this association.

In the next example (Figure 17.8), we show how to handle the case where we cannot infer the existence of an association from attributes of the patches alone. If there is a polygonal set of patches without boundary coordinate information; or a network between points on a landscape, it is necessary to enter the information on each association link explicitly. In Simile, this can be done by having a separate submodel ('Neighbour data') with as many instances as there are associations links, and then picking up the identifiers for the two participants in each association from a file. This allows considerable freedom of expression: for example, links between spatial patches that are far apart can be incorporated, as can weights on each link that have nothing to do with the Euclidean distance between the linked patches.

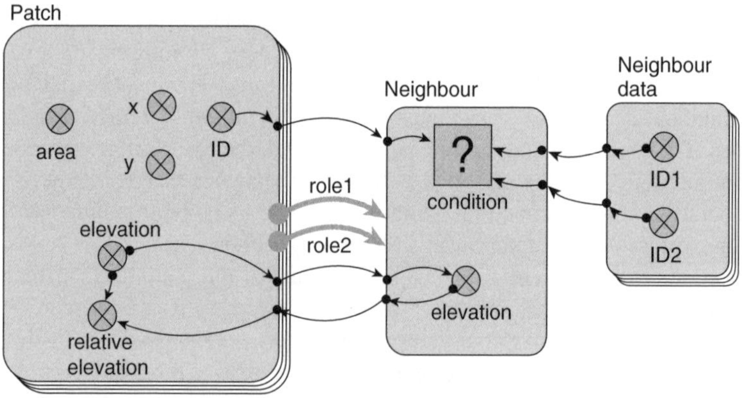

Figure 17.8 Data-specified associations between patches

17.4 Discussion

These examples indicate some of the key features of Simile with respect to dynamic spatial modelling. In this section we examine the implications of using Simile for dynamic spatial modelling, by looking at how Simile might affect the issues highlighted at the beginning of the chapter: dynamism and flexibility.

Dynamism is an important component of many geographic applications, yet the capability for dynamic spatial modelling is absent from conventional GIS. Traditional GIS-based applications are often characterised by relatively fine granularity spatial information, but relatively coarse granularity temporal information. In map production, for example, spatial surveys will commonly resolve geographic phenomena to within metres or finer, while individual spatial surveys are usually only conducted infrequently, often with a number of years separating consecutive surveys (see, for example, Harley, 1976). The lack of dynamism is often tolerated in these applications because of their coarse temporal granularity, with each spatial survey treated as a *snapshot* of the spatial situation at a particular time (see Peuquet, 1999). However, as Chrisman (1998) points out, the idea of a snapshot can be misleading since even superficially instantaneous observations, such as aerial photographs and satellite imagery, often form time series, while individual ground surveys may take years to complete. Consequently, even traditional GIS-based applications might benefit from some dynamic modelling capabilities.

At the other extreme, many environmental data are collected at relatively coarse spatial and fine temporal resolutions, typically using automated data-logging techniques. For example, based on information at four different locations, but collected with a frequency of up to one sample every 5–20 minutes over a 10 month period, Durand *et al.* (1999) have examined the dynamic spatial variations in overland water flow. Similarly, Tranter *et al.* (1997) use fine temporal and coarse spatial granularities to study the chemical composition of sub-glacial water, with samples taken from 21 sites several times a day for a period of weeks in a number of consecutive years. These data provide the basis for developing detailed, process-based models of environmental phenomena, but lack spatial detail.

Many real-world problems require models with both fine spatial and fine temporal resolution for their solution. These include problems in areas as diverse as atmospheric pollution, ground water hydrology, the spread of disease, and transport studies. Modelling is required because there is a need to predict the future and to explore alternative options. Fine spatial resolution is required to avoid errors of aggregation, and fine temporal resolution is required because many key processes happen over short periods of time or show rapid responses to changing conditions. GIS can provide the site data, boundary conditions and validation data for such models. Dynamic models, developed from fine-resolution temporal data, can capture our understanding of process.

In addition to the potential applications of dynamic spatial modelling, the discussion in Section 17.3 highlights the flexible nature of the spatial frameworks that can be constructed within Simile. This flexibility enables Simile to support the existing dominant spatial models and data structures, such as object and field models, and vector and raster data structures. As indicated in Section 17.1, and if required, Simile might also be used to mix different spatial frameworks. Such capabilities already exist to some extent in many conventional GIS. However, the flexibility to adopt dramatically different spatial frameworks is a feature not found in conventional GIS. For example, whether object- or field-based, conventional GIS usually adopt a Euclidean (normally two-dimensional) space for storage and analysis of spatial information. Yet many common models of geographic information do not fit particularly well into Euclidean space. Many geographic distance relationships, such as travel time and nearness, do not fulfil the properties of a metric space (see Worboys, 1996), and consequently are non-Euclidean. Fractal geometry provides a better approximation of the shape of many natural features than Euclidean geometry, and the distortions introduced into spatial data by approximating such features using Euclidean approaches have been observed by Duckham and Drummond (2000). Simile offers the flexibility to adopt non-metric, fractal, or any of a wealth of other possible spatial frameworks without resorting to Euclidean approximations. Furthermore, there is nothing to prevent the spatial frameworks themselves participating in dynamic relationships. It is conceivable that Simile could be used to model spaces that reconfigure, adapt or change over time, for example morphing between field-based and object-based representations of geographic phenomena.

The flexible and explicit spatial relationships and processes expressible in Simile are in stark contrast to the implicit relationships and processes assumed to be 'self evident' within typical GIS (see Sheppard, 1995). This flexibility inevitably comes at a cost. The effort involved in constructing novel spatial frameworks within Simile, both technical and conceptual, will always be greater than needed simply to adopt conventional spatial models embedded in existing GIS. However, Simile's submodels construct can help, by providing a mechanism for developing off-the-shelf modules for common spatial frameworks, such as raster data structures. Furthermore, the social theory literature reviewed in Section 17.1 provides a clear indication that more flexible spatial models do have an important role to play in the future development of GIS.

To some extent, any standard object-oriented design language, such as UML (Stevens and Pooley, 2000), might be used in a similar fashion to specify spatial configurations. For example, the spatial configuration shown in Figure 17.7 could be represented in terms of a UML class diagram showing a single class Patch, with a reflexive Neighbour association. However, unlike UML, Simile allows the definition of the conditions under which the association holds. In Simile, the declarative representation of spatial configuration enables us to distinguish between, for example, a spatial configuration where cells in a raster have 4 immediate neighbouring cells (rook's-case) and a configuration where cells have 8 immediate neighbours (queen's-case).

Another apparent limitation of Simile's representational language for defining spatial configurations is that labels have no inherent domain-specific meaning. Thus, Simile has no way of knowing what 'plant biomass' or 'soil water content' means. In the present context, this means that our use of labels such as 'area', 'x' or 'neighbour' is arbitrary: we may expect humans to use these labels as part of the process of 'reading' a Simile model diagram, but we cannot expect Simile to interpret them spatially. However, a community of spatial modellers is likely to reach agreement on a set of terms and the meaning to be attached to these terms. In that case, it is possible to envisage the development of software tools, external to Simile, that are capable of processing Simile models from the point of view of the particular spatial configuration or configurations used in the various models. Potentially, this could permit the development of, for example, a tool that checked whether a model implementation of a field-based view was consistent with such a view.

17.5 Conclusions

In this chapter, we have demonstrated that Simile is capable of integrating dynamic and spatial modelling in a single package. Moreover, it does this in a way that greatly increases the ease with which non-programmers can construct models: modelling is treated as it should be, as a design activity rather than as a programming activity.

A key feature of Simile's approach is that spatial modelling is achieved without any pre-programmed ('hard-wired') spatial modelling constructs. This is deliberate: in designing Simile's modelling language the intention was to provide a minimum set of generic constructs, while at the same time supporting the construction of a wide range of model types, and providing a language that would be intuitive both for designing models and for 'reading' other people's models. It is this generic nature of Simile's model-specification language that enables it to be used to represent such a broad range of spatial configurations.

Finally, Simile is a modelling environment that supports *dynamic* spatial modelling. It not only has an expressive language capable of representing a wide variety of spatial configurations, but it can also represent the dynamics of the modelled system, both in terms of continuous processes and in terms of the creation and destruction of objects (spatial or non-spatial). Thus, it has a role to play in showing what is possible as the temporal dimension assumes greater importance in geographic information science.

References

Cheylan, J.-P. and Lardon, S., 1993, Towards a conceptual data model for the analysis of spatiotemporal processes: the example of the search for optimal strategies. In: *Spatial Information Theory*, edited by Frank, A. and Capari, I., Volume 716 in *Lecture Notes in Computer Science*, Berlin: Springer-Verlag, pp. 158–176.

Chomicki, J. and Revesz, P., 1999, Constraint-based interoperability of spatiotemporal databases. *Geoinformatica*, **3**, 211–243.

Chrisman, N., 1998, Beyond the snapshot: changing the approach to change error and process. In: *Spatial and Temporal Reasoning in Geographic Information Systems*, edited by Egenhofer, M. and Golledge, R., Oxford: Oxford University Press, pp. 85–93.

Claramunt, C. and Thériault, M., 1995, Managing time in GIS: an event-oriented approach. In: *Recent Advances in Temporal Databases*, edited by Clifford, J. and Campari, I., Berlin: Springer-Verlag, pp. 158–176.

Claramunt, C. and Thériault, M., 1996, Toward semantics for modelling spatio-temporal processes within GIS. In: *7th International Symposium on Spatial Data Handling*, edited by Kraak, M. and Molenaar, M., Volume 1, Delft: TU Delft, pp. 2.27–2.43.

Couclelis, H., 1992, People manipulate objects (but cultivate fields): beyond the raster-vector debate in GIS. In: *Theories and Methods of Spatio-Temporal Reasoning in Geographic Space*, edited by Frank, A.U., Campari, I. and Formentini, U., Volume 639 in *Lecture Notes in Computer Science*, Berlin: Springer-Verlag, pp. 65–77..

Duckham, M. and Drummond, J., 2000, Assessment of error in digital vector data using fractal geometry. *International Journal of Geographical Information Science*, **14**, 67–84.

Durand, P., Cayot-Cros, S., Gascuel-Odoux, C. and Heddadj, D., 1999, Solute concentrations of overland flow water in a cultivated field: spatial variations and intra- and inter-storm trends, *Hydrological Processes*, **13**, 1465–1477.

Forrester, J.W., 1971, *World Dynamics*, Cambridge, MA: Wright-Allen.

Goodchild, M.F., Egenhofer, M.J., Kemp, K., Mark, M. and Sheppard, E., 1999, Introduction to the Varenius project, *International Journal of Geographical Information Science*, **13**, 731–745.

Harley, J., 1976, *Ordnance Survey Maps: A Descriptive Manual*, Southampton: Ordnance Survey.

Laurini, R. and Pariente, D., 1996, Towards a field-oriented language: first specifications. In: *Geographic Objects with Indeterminate Boundaries*, edited by Burrough, P.A. and Frank, A.U., London: Taylor and Francis, pp. 225–235.

Muetzelfeldt, R.I. and Taylor, J., 2001, Developing forest models in the Simile visual modelling environment. Paper presented at the IUFRO 4.11 Conference on Forest Biometry, Modelling and Information Science. University of Greenwich, 25–29 June, 2001, (available at http://www.ierm. ed.ac.uk/simile/documents/iufro3.pdf).

Muetzelfeldt, R.I. and Massheder, J., 2003, The Simile visual modelling environment, *European Journal of Agronomy*, **18**, 345–358

Peuquet, D., 1999, Time and GIS and geographical databases. In: *Geographical Information Systems*, edited by Longley, P., Goodchild, M., Maguire, D. and Rhind, D., Volume 1, New York: John Wiley and Sons, pp. 91–103.

Pickles, J., 1999, Arguments, debates and dialogues: the GIS-social theory debate and the concern for alternatives. In: *Geographical Information Systems*, edited by Longley, P., Goodchild, M., Maguire, D. and Rhind, D., Volume 1, New York: John Wiley and Sons Ltd, pp. 49–60.

Raper, J. and Livingstone, D., 1995, Development of a geomorphological spatial model using object oriented design. *International Journal of Geographical Information Systems*, **9**, 359–384.

Raper, J. and Livingstone, D., 1996, Spatio-temporal interpolation in four dimensional costal process models, In: *Proceedings Third International Conference/Workshop on Integrating GIS and Environmental Modeling*, http://www.ncgia.ucsb.edu/conf/SANTA FE CD-ROM.

Rizzoli, A.E. and Young, W.J., 1997, Delivering environmental decision support systems: software tools and techniques, *Environmental Modelling and Software*, **12**, 237–249

Saunders, S., Chen, J., Drummer, T. and Crow, T., 1999, Modeling temperature gradients across edges over time in a managed landscape, *Forest Ecology and Management*, **117**, 17–31.

Sheppard, E., 1995, GIS and society: towards a research agenda. *Cartography and Geographic Information Systems*, **22**, 5–16.

Sheppard, E., Couclelis, Graham, H., S., Harrington, J.W. and Onsrud, H., 1999, Geographies of the information society. *International Journal of Geographical Information Science*, **13**, 797–823.

Stevens, P and Pooley, R., 2000, *Using UML: Software Engineering with Objects and Components*, Boston, MA: Addison-Wesley.

Tranter, M., Sharp, M., Brown, G., Willis, I., Hubbard, B., Nielsen, M., Smart, C., Gordon, S., Tulley, M. and Lamb, H., 1997, Variability in the chemical composition of in situ subglacial meltwaters. *Hydrological Processes*, **11**, 59–77.

Westervelt, J. and Hopkins, L., 1999, Modeling mobile individuals in dynamic landscapes. *International Journal of Geographical Information Science*, **13**, 191–208.

Worboys, M.F., 1996, Metrics and topologies for geographic space. In: *Proceedings 7th International Symposium on Spatial Data Handling*, edited by Kraak, M.J. and Molenaar, M., Volume 1, Delft: TU Delft, pp. 7A.1–7A.10.

18

Telling Stories with Models: Reflecting on Land Use and Ecological Trends in the San Pedro Watershed

Subhrajit Guhathakurta

18.1 Introduction

Essentially, telling stories is similar to painting pictures. Both are forms of artistic expression intrinsically coloured by individual perspectives. Such artistic expressions include subjective response to stimuli received from the environment. In contrast, models – that is, mathematical models – are rooted in the scientific method framed within the logico-positivistic epistemology. Quite clearly Ludwig Wittgenstein had a very different conception of a picture when he stated: 'A model is a picture of reality' (1921, p. 8). Wittgenstein's model eschews subjective responses that could colour the purity of objective analysis of the underlying data. The picture of such models is supposedly outside the realm of individual human emotions and particularistic perceptions. As a result, stories derived from such pictures would be limited to specific objectively verifiable facts and processes perhaps disembodied from the more meaningful complex relationships experienced in everyday life. In fact, these would not be stories but scientific observations.

In the eight decades since Wittgenstein's *Tractatus Logico-Philosophicus*, paradigmatic changes in epistemology have progressed towards a stage 'beyond objectivism and relativism', a stage that Bernstein (1983, 1991) describes as the incorporation of the positivistic epistemology within the interpretive and hermeneutic traditions. It is now widely accepted that all enquiries rely on 'pre-understanding' and 'pre-judices' since

Re-presenting GIS Edited by P. Fisher and D. Unwin

human consciousness is temporal in form. Phenomenological understanding occurs through the construction of a coherent plot or story in the mind through experiential learning. All argumentation is contextualized within this narrative, which is either communicated or inferred. Geographic models, like all other forms of analytical tools, are embedded within a narrative. In this chapter, I argue first that a narrative form of modelling provides a powerful and persuasive means for advancing complex arguments. Second, I demonstrate how the systems dynamic approach offers a broad intellectual framework for interdisciplinary model building. Third, I discuss the potential for developing loosely integrated models that incorporate various domains of knowledge. Finally, I offer one such model that uses elements of both system-dynamic and discrete choice approaches to construct a story for the San Pedro Watershed in Southern Arizona.

18.2 The Power of Narratives

Narratives and stories have long been considered important in securing and endorsing the premises needed to make decisions under conditions of uncertainty and complexity (Rein, 1976; Simon, 1976; Krieger, 1981; Neustadt and May, 1986;). Stories define the issues and provide a means to make subtle but powerful arguments. They shape the course of a discourse by reshaping the initial positions of the actors concerned. The fact that every great religion and culture has expressed their central tenets through one or more narratives is testimony to the power of stories in preserving and solidifying these tenets. Policy narratives, despite being narrower in scope and less grand in aspirations, have similar objectives – to convey visions of order amidst disorder, and to persuade others to adopt a course of action. In this respect, policy narratives are effectively functioning as policy arguments. The arguments produced for the purpose of policy are not judged solely by their technical merit but in significant part by their power to persuade and congeal public opinion.

The craft of persuasion is rhetoric. The use of rhetoric in planning and policy analysis distinguishes these disciplines from academic social science on the one hand and from problem solving methodologies like operations research or spatial analysis on the other (Majone, 1989). Fact and values are so intertwined in planning and policy oriented disciplines that factual arguments that are not persuasive seldom play a significant role in public debate. However, rhetoric in such cases does not dispense with science and analysis. The use of logical deduction and rational argumentation are essential tools in the rhetoric of planning. Rhetoric in these disciplines acknowledges the fallibility of scientific analysis given that scientific results are always accepted through convention and the fulfillment of methodological and professional norms. As Forester (1993) poignantly points out: 'We forget too easily that science is a cultural form of argument, not a valueless, passionless use of magical techniques'.

The use Geographic Information Systems and related modelling tools are also embedded within a narrative. Frequently, the narrative aspects of such tools have not been consciously examined or conveyed before asserting an argument based on the empirical results. This problem is present especially in the case of empirical models that privilege associations discovered through statistical tests rather than uncovering the underlying story that generates these associations. Also, most empirical models are restricted unduly to measured data and have neglected the far richer and more complex body of information that exists in the experience of different individuals. These experiences can be easily described heuristically although precise measures may not be known. The critical element of understanding 'the story' is not necessarily the examination of associations but an analysis of interconnections. One convenient approach for analysing the cumulative effect of multiple interconnections is available in the literature on dynamic systems.

18.2.1 Narrative Aspects of Dynamic Systems

All simulation models trace the progression of a narrative in a selective manner. The narrative in this case is understood as a sequence of connected events evolving in time. The progression of a narrative is selective because the events are chosen and structured by individuals specifically to suggest a coherent plot. A narrative is therefore inter-subjective as well as communicative since the plot renders meaning to specific experiences or logical deductions. It is also a powerful means of communicating an argument.

System-dynamic models are particularly adept at articulating the narrative aspects of our mental models. For example, systems models track the progress of an evolving process over time, such as the growth, decline and regeneration of species as determined by certain external as well as internal stimuli. This structure would be similar to a narrative construction of the same phenomenon (without the poetics). System models are also particularly adept at tracing different paths depending upon the sequence of events along the time-line. The narrative equivalent of this is captured in movies such as 'Sliding Doors' (directed by Peter Howitt, UK, 1998) and 'Blind Chance' (directed by Krzysztof, Kieslowski, Poland, 1981). Essentially, these movies follow two (or more) separate narratives depending upon the outcome of specific events, such as being able to catch the train or not. The movies depict separate scenarios based on chance but in a systems model we can derive an understanding of multiple scenarios based on both chance and deliberate action. However, both narrative and system model structures allow the examination of the temporal connections between events in a manner such that a coherent and unified experience is projected.

More importantly, acknowledging the narrative aspects of general systems models allows for a significant switch in our cognitive perception from the 'paradigmatic' to the 'narrative'. Bruner (1986) perceives the 'paradigmatic' realm to be the world of abstract and general theories that are verified empirically in the objective world. In contrast, he characterizes the 'narrative' mode of thought as chronicles of particular events and experiences over time that gain credence through their lifelikeness. It is the quality of meaningfulness rather than factual accuracy that renders a narrative credible. Rendering meaning to a system model is as much related to an act of interpretation as is communicating a story because meaning does not pre-exist the interpretation of experience. Concepts such as 'explanation', 'validity', and 'verification' are redefined in the narrative forms of inquiry. The search is not for mathematical certainty but for results that are believable, meaningful, and verisimilar. This attribute of storytelling was described poignantly by Parry and Doan (1994): 'The hearers of the story believed that it was true because it was meaningful, rather than it was meaningful because it was true'.

18.2.2 Integrating Narratives Through Multiple Models

Scholars in the applied social sciences have long realized that an interdisciplinary perspective is critical in understanding collective human and ecosystem behaviour. However, models of ecosystem dynamics and urban processes have developed in separate knowledge domains. Given that human and natural systems interact upon and affect each other's behaviour, an integrated framework is crucial for modelling urban and ecological processes. Most urban models are still limited in their ability to address environmental issues. These models have focused primarily on economic and spatial interaction among jobs, housing, and transportation. The economic framework has limitations in incorporating ecological dynamics since price signals play a marginal role in environmental processes. Similarly, until recently ecological modellers have concentrated on modelling species behaviour in non-urban landscapes primarily by accounting for the flows of energy and matter through various natural systems. Also, urban and environmental modellers have distinctly different concept of spatial processes. However, recent advances in the literature on agent-based models, system-dynamic processes, and complexity theory offer important insights about integrating social and ecological knowledge domains.

Agent-based processes examine the dynamic interaction between the choices made by various entities such as institutions, governments, businesses, and households. Ecosystem modellers have been using agent-based models to simulate population growth and decline as well as changes in environmental resource endowments. Typically, the components entered in an agent-based model interact with each other in the form of feedback processes. Such feedback processes can be negative or positive. Negative feedback from one component in the model leads to a response in other components that counteract the original change. Positive feedback does the opposite, evoking a response from other components of the model that strengthens the original change. The interplay between the negative and positive feedback processes lead to the dynamic characteristics of the system being modelled.

The system-dynamic approach is ideally suited for modelling agent-based processes. It is also well adept at capturing emergent processes that exhibit complexity. Complexity is often manifested from simple rules applied to local phenomena such that aggregate patterns are clearly distinct from local behaviour. Complexity studies now contend that detailed micro level studies and their dynamic properties are essential to understanding macro behaviour. This is in contrast to the reductionist perspective that assumes that simpler, local level characteristics can be disaggregated from macro processes. The systems approach provides an elegant means of observing complexity that is emerging from simple rules of expected behaviour.

Although system-dynamic approaches have been conceived as a 'grand' approach that attempts to tie together multiple domains of knowledge, it cannot be expected to integrate different theoretical and epistemological domains that have framed disciplinary advances. System-dynamics is also limited in constructing theories since its principle purpose is to clarify, test and unify *a priori* theoretical insights or 'mental models'. It is, therefore, a tool to refine and develop existing theories and extract insights about these theories as they play out in the real world. In addition, systems models lose their simplicity and elegance when spatial aspects of a system are included. The amount of computation increases exponentially with increasing resolution of spatial categories. However, spatially disaggregated dynamic-system models are being developed and tested. Some current and ongoing projects of such spatial models include the Spatial Modeling Environment being developed at the University of Maryland (Costanza *et al.*, 1995; Voinov *et al.*, 1999) and UrbanSim, a project housed in the University of Washington (Waddel, 2000).

While there are several models available to address various aspects of the environment and economy, any one model would be too limited to capture this larger system. What is needed is a computing interface in which various models can be linked at spatial, temporal, and functional frames. The use of multiple models that interface with one another has several advantages. First, this approach requires fewer compromises and preserves, to a large extent, the integrity of the submodels. Second, it allows greater examination of the model substructures and hence facilitates more vigorous discussions about the different epistemologies guiding model development. Third, such models are not 'owned' by any knowledge domain and are likely be truly interdisciplinary. Fourth, the use of multiple models requires a more conscious examination of the embedded narratives and allows the construction of a coherent plot. Thus, the overarching narrative that weaves the model together serves as the glue for integrating different approaches. The San Pedro Watershed model described below is one such attempt to link different modelling approaches for the purpose of constructing a coherent story.

18.3 The San Pedro Watershed Model

18.3.1 Preamble

The San Pedro Watershed is the drainage basin for the San Pedro River, which flows from its headwaters near Cananea, Mexico, northward approximately 140 miles until it meets the Gila River near Winkelman, Arizona. The Arizona Department of Water Resources (ADWR) has delineated a number of smaller sub-watersheds within the San Pedro Watershed region, the largest and the most critical of these being the Sierra Vista sub-watershed. This sub-watershed encompasses 1200 square miles, more than half of which lies within Sonora, Mexico. The Sierra Vista sub-watershed supported approximately 100 000 persons in 1995 and includes two of the largest places in the entire watershed – Cananea, Sonora and Sierra Vista, Arizona. The location of principal population centres in the San Pedro Watershed is shown in Figure 18.1 (Colour Plate 7).

The area within the Sierra Vista sub-watershed has been rapidly urbanizing since the mid to late 1980s. About one-third of the population of the Sierra Vista sub-watershed is located on the Mexican side of the international border, mostly in the urban centres of Cananea and Naco. The remaining two-thirds is concentrated in eight communities in Cochise County, Arizona. At the 2000 census, these Arizona communities ranged in size from 1504 in Tombstone to over 37 775 in Sierra Vista. A significant and growing

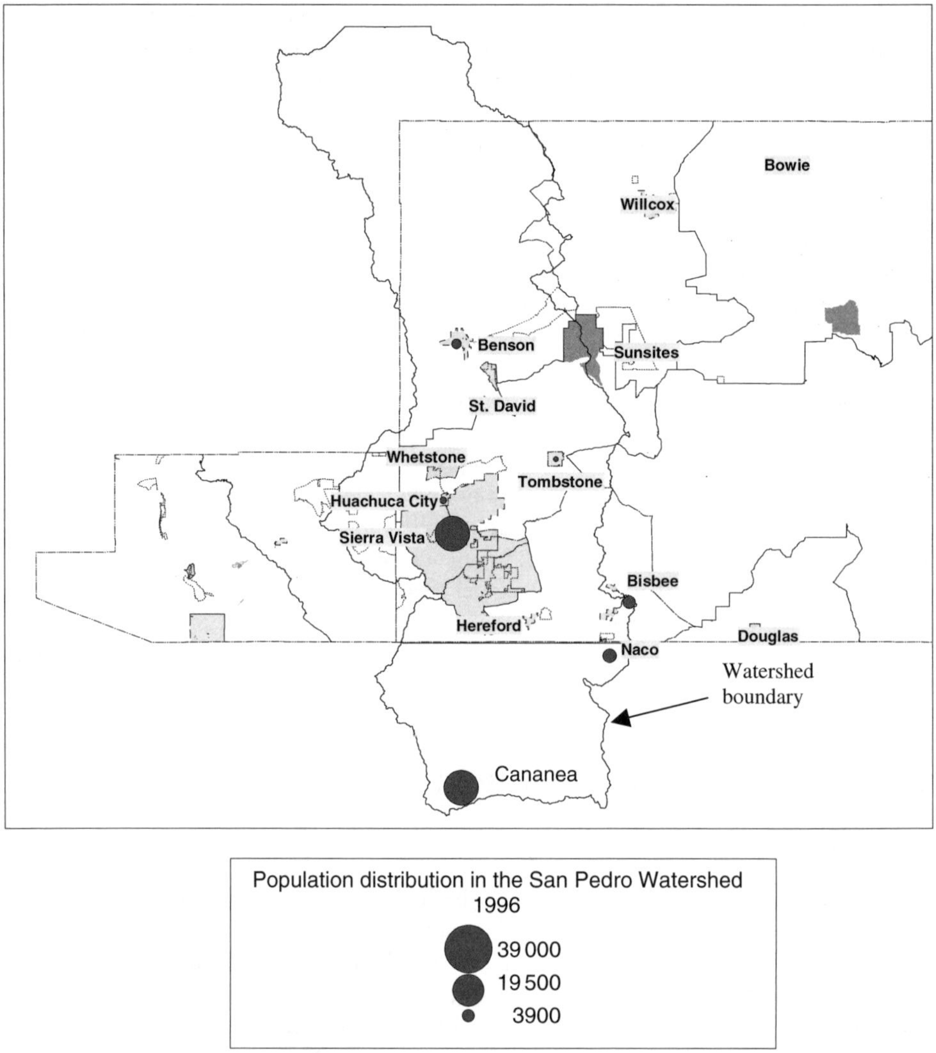

Figure 18.1 (Plate 7) *Communities in the San Pedro Watershed*

number of persons also live in unincorporated areas of Cochise county. Trends in population growth, shown in Table 18.1, suggest that the population around Sierra Vista is growing more rapidly than in any other part of the watershed. Between 1980 and 1990, Sierra Vista grew at about 2.8% per annum. In the same period, the population of communities such as Benson, Bisbee, and Tombstone actually declined. More recently, from 1990 to 2000, most of the communities in the watershed increased in population. This growth was led by Sierra Vista and Tombstone, with compound annual growth rates of 1.4% and 2.1% respectively. Population projections conducted by the Arizona

Table 18.1 *Population trends in the San Pedro Watershed*

	1980	1990	2000	Compounded annual change 1980–1990 (%)	Compounded annual change 1990–2000 (%)
In Mexico					
Cananea	25 327	26 931	NA	0.6	NA
Naco	4441	4645	NA	0.5	NA
In United States					
Benson	4190	3824	4711	−0.9	2.1
Bisbee	7154	6288	6090	−1.3	−0.3
Huachuca City	1661	12782	1751	0.7	−0.2
Sierra Vista	24 937	32 983	37 775	2.8	1.4
Tombstone	1632	1220	1504	−2.9	2.1

Source: US Census of Population and Housing, 1980, 1990, and 2000.

Department of Economic Security suggest continued robust growth in Sierra Vista with stable and slow growth in the other Cochise county communities in the watershed.

To appreciate the human footprint on the Sierra Vista sub-watershed, the population distribution in the communities mentioned earlier has to be seen in the context of the entire region. The most reliable data that provide this detailed picture are available in the decennial censuses. For the purpose of this study, the 1980 and 1990 census information was obtained at the level of the block group (approximately 0.06 square miles for each block group). These data at the block group level were then abstracted to a finer level of rectilinear grid-cells measuring 285 metres square. At this resolution, trends in housing construction in the Sierra Vista sub-watershed indicate a continuous low-density build-up between 1980 and 1990, which has been targeting natural and underdeveloped areas. Some of the most significant of these changes are happening just outside the Sierra Vista city limits.

Preliminary estimates suggest that the pace of housing construction in Sierra Vista has in fact accelerated in the 1990s. Another interesting characteristic of the housing growth is the increasing number of homes that are vacant on account of being second homes or vacation homes of families that have primary residences outside this area. As a result of this and other demographic trends persons, per housing unit has declined from an average of 2.42 in 1980 to 2.08 in 1990.

The decline in persons per housing unit together with rapid housing growth has serious implications for the use of land, water and other natural resources. Moreover, the housing units being built are overwhelmingly single-family tract homes. These homes have also targeted unincorporated areas with few municipal services and regulations. For example, the City of Sierra Vista approved between 178 and 385 single-family detached units each year between 1992 and 1996. In contrast, only 64 multi-family units were approved during this four-year period. Such significant increases in single-family tract homes are also reflected in all other communities in the US portion of the upper San Pedro Watershed. Another important trend in the housing market is the increasing proportion of rentals. This indicates an increasing number of transient and seasonal populations in the upper San Pedro watershed in Cochise County.

The San Pedro River and the Sierra Vista urban agglomeration is the setting for this story, which is essentially about the use and distribution of water resources. The San Pedro River is Arizona's last un-dammed, free-flowing, river and one of the nation's hot spots for biodiversity. According to the Nature Conservancy, more than 80 species of mammals, 40 species of reptiles and amphibians, 100 species of butterflies, 20 species of bats, and 100 species of breeding birds rely on the river. Another 250 species of migratory birds are part-year residents. The river was recently named as the first of the 'Last Great Places on Earth', a programme of listing important global habitats by The Nature Conservancy. Along the banks of the river is a riparian ecosystem consisting of Cottonwood and Willow and several other non-native species of plants. This ecosystem is under considerable stress from several fronts. The burgeoning urban area in and around Sierra Vista is pumping out the aquifer and changing the hydrology of the region. The dwindling surface water runoff to the river is beginning to threaten its very existence. The riparian area along the river is also impacted by development. Humans are introducing non-native species of plants that are overwhelming the fragile native Cottonwood and Willow trees.

18.3.2 Methodological Issues

Given the size of study area, land use changes in the San Pedro watershed were estimated from spectral imagery. These images were classified with the help of spectral signatures and supervised comparisons to available ortho-photos and other maps. For the analysis of socioeconomic impacts, classified images from 1986 and 1995 were used. These years were chosen because they were about ten years apart and provided the best match with decennial census data, albeit with an approximately five year lag. Given that land use changes follow from socioeconomic changes, the lag from the census data is pertinent in this analysis. However, the period of lag (five or six years) is dictated by the specific years for which spectral imagery was available.

The level of detail for the classification of land use was limited by the information that could be gleaned from the satellite images. Six types of land uses were identified. These were dense residential, medium density residential, high density residential, non-residential, military, and vacant. The resolution of the classified image was 28 m. However, this resolution was scaled up to 285 m to make the individual grid-cells comparable in size to an average block group. The total number of such cells numbered 47 687 in the upper San Pedro Watershed basin. These cells were used as the unit of analysis in the land use change model.

The data from the 1980 and 1990 census were used to derive population and housing characteristics at the level of the block group. The information was then abstracted to the grid-cell layer to match the data on land use changes. Given that the irregular boundaries of block groups do not correspond to the rectilinear grid-cell layer, an algorithm was developed to assign population to grid cells that overlapped several block groups in proportion to the area of overlap. The spatial analysis for this and all subsequent tasks was accomplished using ArcView GIS.

Besides the census information, several other data layers were compiled to examine the impact of socioeconomic drivers on land use. These included land ownership information, distances from the nearest highway, slope, and the characteristics of land use in the adjacent cells. Additional information was also obtained from the Assessor of Cochise County about the detailed characteristics of the built structures in the area. This additional information allowed comparison with the land use classifications obtained from the satellite imagery with parcel level data. The modelling framework used in this analysis is explained below.

18.3.3 The Modelling Framework

The model developed for this study is composed of three parts – two are system-dynamic, and the third is analytical. This is one of the few attempts to stitch together different modelling strategies so that the best attributes of each strategy are preserved. The first component of the land use model is a systems-dynamic model that seeks to simulate growth in population, employment, and land required for residential and non-residential uses. The model has interconnected feedback loops that control population and employment growth according to land availability for each period. The objective of this component of the modelling exercise is to determine the aggregate amount of land for residential and non-residential uses that will be required for accommodating the population growth in the future. The second component is the spatial component, which extracts analytically the factors determining changes in land use by land units. In this stage, the required number of land units derived from the systems-dynamic model is allocated spatially to those parcels that are most likely to convert to residential uses. The third component adds the other critical aspect to this modelling exercise – the use of water. In this component, water usage, aquifer levels, and base-flows to the river are simulated according to population and employment parameters.

The Dynamic Land Utilization Component. The first component of the land use change model is developed using Stella software package available from High Performance Systems. The initial parameters of the model were generated to simulate the growth in the region during the past 15 years. Once the initial model was performing robustly, several ‘What if?’ scenarios were generated by changing parameters such as density of development and rates of population and/or economic growth. These tests were performed to check how well the systems-dynamic model performs under the selected parameters.

The land utilization component is comprised of three sub-models: the population growth sub-model, the employment growth sub-model, and the land unit change sub-model. The relationships among the three sub-models, as described in Figure 18.2 (Colour Plate 8), are based on a dynamic link between employment growth and population growth, which then generates demand for residential and non-residential land. As land becomes dearer (measured through the ‘gap’ between available unused land and used land), the price signals translate into lowering the growth of employment and population. The model is also able to generate ‘What if?’ scenarios with different sets of density parameters.

The Spatial Discrete Choice Component. While the simulated results from the first model show the demand for new land units, the discrete choice component is designed to translate that demand spatially to those cells that have the highest probability for

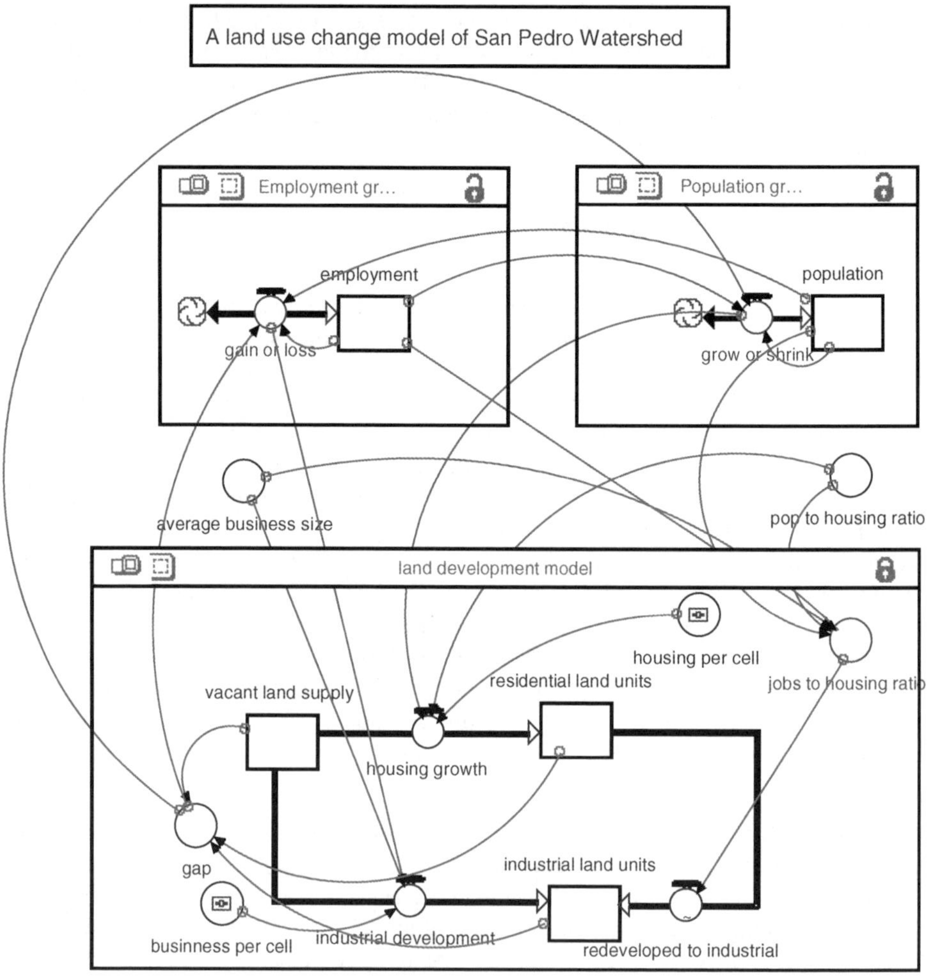

Figure 18.2 (Plate 8) *System-dynamic model of land development in Sierra Vista sub-watershed*

residential land use change. The schematic diagram of this interface is provided in Figure 18.3.

Given the discrete nature of the process of land use change, empirical modelling of land use is generally undertaken with the help of a class of models known as logit and probit. A range of techniques are associated with this general category of models that are tailored to the specific nature of the discrete variable being modelled as well as to the characteristics of the cases used in the models. Logit models have been used extensively in the past for modelling land use change (see Landis, 1994, 1995; Landis and Zhang, 1998a, 1998b). Most recently, Landis used the multinomial logit framework to analyse and project land use changes in the nine-county San Francisco Bay region. The modelling technique used in this study is a variation on the multinomial logit framework. Given that

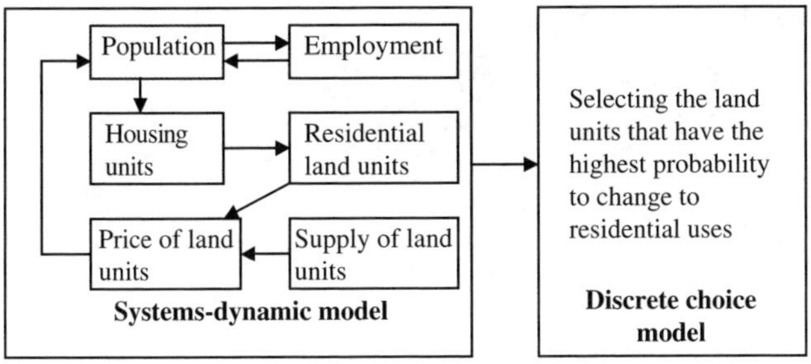

Figure 18.3 *A schematic showing the integration of system-dynamic and discrete choice components of the land-use model*

the data highlight mostly residential developments, the different intensity of such development can be described within an ordinal framework. The generalized version of the model is provided below:

$$Log\left(\frac{\gamma_j}{1-\gamma_j}\right) = \theta_j - [\beta_1 x_1 + \beta_2 x_2 + \beta_3 x_3 \cdots + \beta_k x_k]$$

where γ_j = the cumulative probability for the jth category, θ_j is the threshold for the jth category, $\beta_1 \ldots \beta_2$ are the regression coefficients, and $x_1 \ldots x_k$ are the (k) explanatory variables.

The interpretation of the model is based upon several assumptions. First, the model should suggest a latent continuous variable, which is made discrete by the use of *j* ordinal categories. In this case, the latent variable can be perceived to be the intensity of development. Second, the constants in the model are only determined by the category's probability of being predicted (without the contribution of the independent variables). Third, the prediction part of the model depends only on the predictors. The second and third assumptions guarantee that the results will be a set of parallel planes (or lines), one for each category of the outcome variable.

The categories selected for this analysis are those cells that were vacant in the 1986 land use classification and remained vacant or changed to the following land use categories: dense residential; medium density residential; high density residential; or non-residential in the 1995 land use classification for the same cells. To ensure that the model is predicting urbanization, the total number of cases in the entire Sierra Vista sub-watershed was reduced to only those that were within a place designation in 1998. It may be noted that 1986 cells may have been outside the place boundaries as delineated if we assume that boundaries expanded in the last 15 years. Even within this subset of data, the overwhelming majority of the land uses as classified remained vacant between 1986 and 1995.

The fit of the model provided in Table 18.2 to the data seems to be excellent given that the difference in the log likelihood statistics between the intercept only model and the final model is 1168 and has a high significance. The pseudo R^2 values are also respectable suggesting that about 90% of the variation in the categorical outcomes is predicted by the independent variables. However, the downside of this model is that it predicts no

Table 18.2 *Parameter Estimates*

		Estimate	Std. Error	Wald	df	Sig.	95% Confidence interval	
							Lower bound	Upper bound
Threshold	[Remained vacant]	21.383	1.379	240.409	1	0.000	18.680	24.086
	[High-density residential]	22.223	2.366	88.257	1	0.000	17.587	26.859
	[Medium-density resid.]	22.814	3.954	33.297	1	0.000	15.065	30.564
	[Low-density residential]	21.463	1.009	452.153	1	0.000	19.485	23.442
Explanatory variables	POPULATION	4.728E-05	0.000	1.231	1	0.267	−3.624E-05	1.308E-04
	DISTANCE from highway	−1.081E-04	0.000	4.600	1	0.032	−2.069E-04	−9.319E-06
	AGE over 60 years	6.410E-07	0.000	0.000	1	0.982	−5.607E-05	5.735E-05
	HISPANIC	2.297E-04	0.000	8.658	1	0.003	7.671E-05	3.828E-04
	BLACK	−1.536E-04	0.000	4.408	1	0.036	−2.969E-04	−1.021E-05
	American Indian	−4.929E-04	0.000	5.501	1	0.019	−9.048E-04	−8.100E-05
	[ADJACENT = Vacant]	20.103	0.290	4796.348	1	0.000	19.534	20.672
	[ADJACENT =Similar use]	17.036	0.000		1		17.036	17.036
	[ADJACENT = dissimilar]	0			0			
Ownership	[PRIVATE = 0]	−1.052	0.285	13.608	1	0.000	−1.611	−0.493
	[PRIVATE = 1]	0			0			
	[STATE = 0]	0.112	0.463	0.059	1	0.808	−0.795	1.020
	[STATE = 1]	0			0			
	[BLM = 0]	0.128	0.568	0.051	1	0.821	−0.985	1.242
	[BLM = 1]	0			0			
	[FEDERAL = 0]	0.272	0.471	0.333	1	0.564	−0.652	1.195
	[FEDERAL = 1]	0			0			

Link function: Logit; Highlighted variables are significant at 95% confidence level.

change much better than change. This is due to the fact that over 90% of the cells did not undergo any change in use between 1986 and 1995.

The parameter estimates presented in Table 18.2 show that the vacant status of adjacent cells is the most important predictor of development followed by ownership of the land. As expected, land in private ownership has the highest probability of being developed. Another spatial variable that has important predictive properties is distance from a highway. Holding other variables constant, cells closer to the highway are more likely to be developed than those that are further away.

Population is a significant predictor of land use change but a further breakdown of population characteristics shows that increase in Hispanic population has resulted in higher probabilities of land use change, all else being equal. In contrast, change in Black and American Indian populations seem to be inversely related to the probability of land use change. Contrary to popular perceptions, the increase in retiree populations (age over 60) does not affect significantly the probabilities of land use change in this region.

The Water Use Component. Once the population, employment and land unit change model was providing expected results, the other critical element of this modelling exercise – use of water – was added. This part closely followed recent hydrological studies that measured the annual recharge, aquifer levels, total pumpage, and baseflow amounts based on 1940 steady-state water budgets (Corell *et al.*, 1996). A schematic of this model is presented in Figure 18.4 (Colour Plate 9). The flows into the aquifer consist of mountain front recharge as a result of annual precipitation and also recharge from the constructed wetlands (surface water infiltration). The outflows are comprised of pumpage for domestic and agricultural uses, evapotranspiration, baseflows to the river, and subsurface outflow from the basin.

The agricultural and domestic pumpage impose the largest demand on groundwater resources (about 11 000 acre-feet annually) followed by riparian vegetation and flows to the stream, respectively. The domestic pumpage in the model is derived from population parameters estimated in the population–economy land use submodel and from current intensity of water use per capita. Agricultural pumpage bears a weak inverse relation to domestic pumpage given that increasing urbanization tends to convert some agricultural land to other urban uses. The model also incorporates the possibility for adding to the aquifer recharge through technological means, such as constructed wetlands. The simulation was run under various conditions of per capita water use, population growth and recharge amounts. Regardless of the speculated parameters in the water use model, almost all simulations resulted in drawing down the aquifer. Only the speed of this decline in ground water levels varied with the supplied parameters. This result is corroborated by scenarios generated by ADWR (Corell *et al.*, 1996). Therefore, most meaningful speculations about the plot of the story of San Pedro seem to converge towards a familiar ending.

18.3.4 Land Use Projection

The baseline projection for the amount of residential land units, provided in Table 18.3, shows the business-as-usual scenario in which past parameters are maintained within the system-dynamic component of the model. Under this scenario, the population growth rate

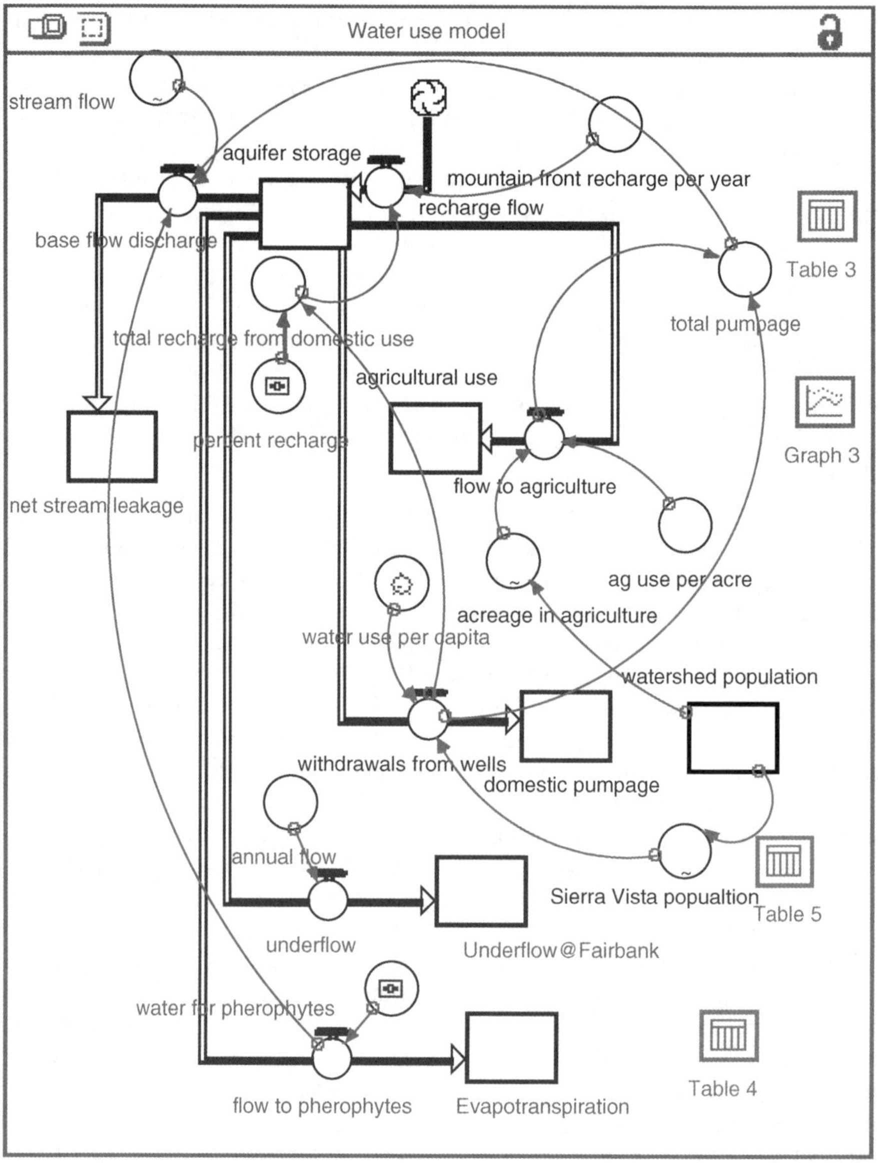

Figure 18.4 (Plate 9) *A system-dynamic model of hydrology in the Sierra Vista sub-watershed*

starts out at about 3.3%, falls to about 2.5% by 2010 and 1.5% by 2030. The residential land unit demand projections are based on density parameters calculated from existing information.

The projected residential land unit demand is mapped according to the probability table generated in the discrete choice model. Figure 18.5 (Colour Plate 10) shows the possible areas that would accommodate future residential units. This projection prohibits

Table 18.3 Projected demand for residential land units

Years	Residential cells (approx. 20-acre parcels)	New additions	Cumulative totals since 1985
1985	1881		
1990	1994	113	113
1995	2122	128	241
2000	2263	141	382
2005	2415	151	534
2010	2572	157	691
2015	2730	158	849
2020	2884	154	1003
2025	3030	145	1149
2030	3163	134	1282
2035	3284	121	1403

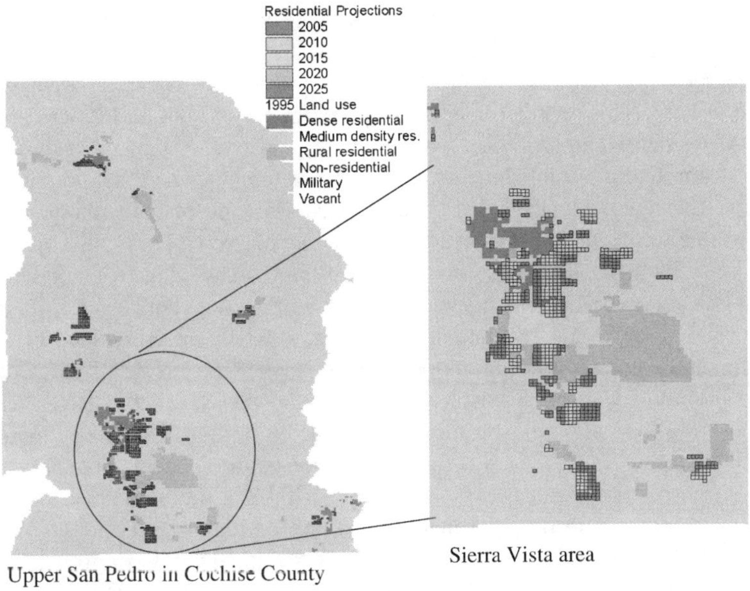

Figure 18.5 (Plate 10) Projected residential development in the San Pedro Watershed region

state land from being converted to private use. Two caveats need to be noted in interpreting the map of projected residential land units. First, residential demand is met entirely within the existing (year 2000) boundaries of census-designated places. Second, although the parameters of the discrete choice model have been derived from a set of detailed demographic and spatial characteristics, the final values are held constant throughout the model projections. This second caveat is not a limitation of the modelling approach, but a limitation of projected data at a disaggregated level. It is possible to rerun the model for different demographic and spatial data and arrive at another probability table that fits the provided data at various points in time.

Figure 18.5 shows the spatial configuration of land units that have the highest probability of being converted to residential use in the next 25 years. Most of the future residential development is concentrated south of Sierra Vista. The trends seem to suggest that the urban conglomeration in and around Sierra Vista will continue towards the US–Mexico border. The growth as projected in this study seems to be less dramatic in its scope and impact on the riparian area along the San Pedro River than expected by many individuals with environmental concerns. However, pumping groundwater continues to be a serious concern regardless of where the projected growth happens within the watershed.

18.4 Constructing a Plot from the San Pedro Model

The story of the Upper San Pedro Watershed as constructed in this study is narrated 50 years or more in the future and begins in 1980. By the mid-1980s, the communities in the upper San Pedro Watershed had already transformed their economic base from agriculture and mining to government, military, services, and tourism. The dramatic landscape comprised of ragged mountains and the Sonoran desert attracted winter visitors, retirees, and tourists. About 46 000 people lived in the urban areas within the watershed in 1980. This population grew to approximately 63 000 in 1990 and 106 000 in 2010. In the residential areas, housing was built with an average density of three units per acre while commercial establishments averaged five units per acre within commercially zoned land. Land was plentiful and cheap, which was one of the principal attractions drawing people to this area. Even with a healthy rate of growth, only 48% of urban land was developed by 2010. Given this abundance of land, many began to build housing at a density of one unit per acre. The planners noted that if average housing density decreases to one per acre, they would still be left with about 20% vacant urban land in 2050. The controls imposed on new housing and the carefully planned infrastructure allowed new houses to be built away from sensitive areas around the San Pedro River. The urban expansion moved southward towards the Mexican border following Interstate 10. Therefore, physical limits of land resources did not feature in their planning decisions. Of course, transportation and energy issues necessitated a more compact development, but there were other more pressing matters that needed their attention.

Although land was plentiful, there were real and critical limits to growth in the San Pedro watershed communities where water was concerned. The aquifer was slowly but surely being depleted as, even in 1990, the recharge was only 78% of total outflows. The spectre of dwindling water supply in the very near future sparked a coordinated programme to recycle water and recharge the aquifer by technological means. The aggressive programme of replenishing the aquifer resulted in recycling 30% of the domestic water, which was recharged back to the aquifer. Sadly, this effort was not nearly enough to arrest the depletion of groundwater. By 2050, there was about 0.5 million acre-feet of water less in the aquifer than there was in 1980 and the deficit continued to grow. Leakage to the river had declined to one-third of the amount in 1980, making San Pedro an intermittent stream. The diminishing flow of water also impacted the riparian corridor adversely. The majestic Cottonwood and Willow trees became a rarity. Many of the animal species that thrived in 1980 are now either extinct or endangered. The river, which once gave life to a thriving ecosystem and provided tourist dollars to the economy, is almost dead in 2050. As jobs disappeared from the tourist economy, people began to move out of the area. Those left behind now asked the state and federal governments to step in and stem the tide of decline. Some of them were the same persons, or the sons and daughters of the same people, who had fought hard initially to keep the state and federal governments out of their communities.

A pertinent question that will haunt the reader is: Can the story be any different? The answer is yes, but it is unlikely. Under almost all scenarios, the outflow from the aquifer is larger than the recharge. However, there are a number of things that can happen to extend the life of the aquifer and flows to the river. For example, if the amount of recycled water recharging the aquifer increases from 30 to 50%, the water deficit in the aquifer is halved. Conversely, if per capita water use is halved, the deficit in the aquifer in 2050 is less than the scenario discussed in the previous paragraph by 40%. If, both the measures are adopted, the aquifer will still run deficits but the leakage to the river will probably continue at a rate of around 6000 acre-feet annually, not much below 2000 levels. This will save the ecosystem and the economy for at least another generation. The targets discussed are achievable but unrealistic given the current attitudes toward water use and growth. It will require a substantial change in the lifestyle of the inhabitants of San Pedro watershed communities. However, it is their opportunity to write a story about saving a river that can perhaps capture the minds and hearts of other communities facing a similar fate.

18.5 Research Conclusions and Significance

This study is among the first to bring system-dynamic and analytical approaches together in a land use model. Although the results as reported provide a projected picture of the likely scenario of continued growth at similar rates moderated only by land prices, several other scenarios can also be tested within this model. For example, the model will be able to project growth at various densities of development. Also, it can incorporate new infrastructure investments such as new highway constructions. The model as developed will be further tested over the next few years to study how well it performs under different parameters and scenarios.

The events and processes that characterize the San Pedro Watershed teach us about the fragility of many natural processes. The human impact on the aquifer has continued for many decades, irreversibly changing the hydrology and ecology of the region. Current efforts to maintain river flow and water supply for human use will perhaps extend the life of the aquifer but it may take decades to revive it back to its steady state. Telling this story serves two purposes. First, it may heighten the awareness of this crisis created by human demand on scarce resources in a manner that changes perceptions and lifestyles of the people in the San Pedro Watershed. Second, it will serve as a constant reminder to all other communities in similar fragile ecosystems that environmental damage is often irreversible and therefore it is best to live within the limits of this environment.

The significance of this research is two-fold. First, it attempts to build a narrative base for integrating different approaches to modelling urban and environmental phenomena. Second, it demonstrates a strategy for building an integrated framework for the effective use of multiple modelling strategies. A number of future research trajectories may be identified to advance the modelling of urban and environmental processes both in the San Pedro region and elsewhere. First, the land use model needs to incorporate the price of water in moderating development and population growth. This is difficult, given that water does not have a market-based price. Methods of contingency valuation can be used to ascertain shadow prices for water and these shadow prices can then inform the land use model. Second, the current study has provided a well-tested strategy that can be modified and finessed for building a similar model for the Mexican San Pedro Watershed. This study has also generated better tools for interpreting satellite imagery, which can now be used to evaluate the land use patterns south of the border. The land use change model for Mexican San Pedro, together with our current results, will provide a more complete picture of the environmental and socioeconomic future of the San Pedro Watershed.

References

Bernstein, R.J., 1983, *Beyond Objectivism and Relativism: Science, Hermeneutics, and Praxis*, Philadelphia, PA: University of Pennsylvania Press.

Bernstein, R.J., 1991, *The New Constellation*, Cambridge, MA: MIT Press.

Bruner, J., 1986, *Actual Minds, Possible Worlds*, Cambridge, MA: Harvard University Press.

Corell, S.W., Corkhill, F., Lovvik, D. and Putnam, F., 1996, *A groundwater flow model of the Sierra Vista Subwatershed of the Upper San Pedro Basin – Southeastern Arizona*, Modeling Report No. 10. Phoenix, AZ: Arizona Department of Water Resources, Hydrology Division.

Costanza, R., Wainger L. and Bockstael, N., 1995, Integrating economic and ecological systems modeling: theoretical issues and practical applications. In: *Economics and Ecological Indicators*, edited by Milon, J.W. and Shogren, J.F., Westport, CT: Prager, pp. 45–66.

Forester, J., 1993, Learning from practice stories: the priority of practical judgment. In: *The Argumentative Turn in Policy Analysis and Planning*, edited by Fischer, F. and Forester, J., Durham, NC: Duke University Press, pp. 186–209.

Krieger, M., 1981, *Advice and Planning*, Philadelphia: Temple University Press.

Landis, J., 1994, The California urban futures model: A new generation of metropolitan simulation models, *Environment and Planning B*, **21**, 399–420.

Landis, J., 1995, Imagining land use futures: applying the California urban futures model, *Journal of the American Planning Association*, **61**, 438–457.

Landis, J. and Zhang, M., 1998a, The second generation of the California urban futures model. Part 1: Model logic and theory, *Environment and Planning B*, **25**, 657–666.

Landis, J. and Zhang, M., 1998b, The second generation of the California urban futures model. Part 2: Specification and calibration results of the land-use change submodel, *Environment and Planning B*, **25**, 795–824.

Majone, G., 1989, *Evidence, Argument, and Persuasion in the Policy Process*, New Haven, CI: Yale University Press.

Neustadt, R. and May, E., 1986, *Thinking in Time: The Uses of History for Decision Makers*, New York: Free Press.

Parry, A. and Doan, R.E., 1994, *Story Re-visions*, New York, NY: Guilford Publications.

Rein, M., 1976, *Social Science and Public Policy*, Harmondsworth, UK: Penguin Educational.

Simon, H., 1976, *Administrative Behavior: A Study of Decision Making Processes in Administrative Organization*, New York, NY: Free Press.

Voinov, A.R., Costanza, L., Wainger, R., Boumans, F., Villa, T., Maxwell and Voinov, H., 1999, The patuxent landscape model: integrated ecological economic modeling of a watershed. *Environmental Modelling and Software*, **14**, 473-491.

Waddel, P., 2000, A behavioral simulation model for Metropolitan policy analysis and planning: residential and housing components of UrbanSim. *Environment and Planning B*, **27**, 247–263.

Wittgenstein, L., 1921, *Tractatus Logico-Philosophicus*, London: Routledge and Kegan Paul.

Part IV

Not 'There' Yet

19

Conclusion: Towards a Research Agenda

David J. Unwin and Peter Fisher

19.1 Introduction

> *... it is GIS's supreme conceit that one can structure a useful representation of geographical knowledge in the absurdly primitive domain of the digital computer, just as it is geography's conceit that one can accomplish the same with pen and paper.*
>
> (Goodchild, 1995, p. 36)

In our introduction to this collection of essays, we highlighted a series of concerns about the nature of any attempt to represent the full richness of geographical knowledge in a digital, computer-based environment such as that offered by contemporary geographical information systems. Reflecting on the co-authored introductory sections and the essays themselves, perhaps two basic lessons emerge. First, although as might be expected there is a concern for operational issues of how the ideas can be put into practice, these concerns are not very different from those of contemporary theory in a much wider context. Second, and again as might be expected, the notions of space, place, scale, and time that the essays highlight have a very long history of debate.

At the workshop that underpins this volume, participants undertook a so-called ‘pyramid exercise’, designed to develop a reasoned research agenda. The technique is well-known in education circles (Gold *et al.*, 1991, pp. 171–3). Initially, all participants are asked to think about and write down, say, four critical items in their research agenda. Next, they meet up in pairs and argue their way from the eight items towards an agreed set of four. The pairs then meet in fours and repeat the process. The exercise can be stopped at any suitable point and the resulting agendas discussed. Experience shows this

Re-presenting GIS Edited by P. Fisher and D. Unwin

to be a good way of reaching a consensus that is not dominated by particular viewpoints and in which all have had their input. In this conclusion, we summarise the issues and report the research agenda items that emerged from this exercise.

19.2 Space and Place

Another, and perhaps more serious, impediment to the use of GIS in social science lies in the current emphasis in its data models on the absolute position of objects, and the inability to represent information about interaction.

(Goodchild, 1995, p. 40)

It is significant that the most common items in the agreed list of research topics were:

- exploring alternative models of space and corresponding visualisation tools;
- distance as a relation;
- adjacency; and
- network geographies.

All represent a call for more work on the representation of 'place'. In the 'absurdly primitive domain of the digital computer' locations in space are usually represented using (x, y) tuples in which the values are expressed in fixed, finite word length. Typically, the space is assumed to be not only metric, but also Euclidean. Yet, quite apart from any social theory, the formal theory of such spaces and their properties can easily be shown to be inadequate as a foundation for measures of distance in geographic spaces (see for example Worboys, 1996; Worboys *et al.*, 1998). For instance, geographic distance is only seldom symmetric and may relate more to time, cost or perception than to 'straight' lines drawn across the globe. It also changes according to the context. In the words of one newspaper columnist:

How far is it to Bethlehem? Not very far, we used to pipe as children. Depends on your point of view, if you ask me. How many shopping days to Christmas, how long is a piece of time and whether Bethlehem is £90, five hours flying time or just a prayer away is entirely a matter of opinion

(Katharine Whitehorn, *The Observer* 21st Dec., 1980 cited in Gatrell, 1983, p. 45)

As we have seen, several strategies have been adopted to address these issues. One is to adopt a 'naïve' stance, similar in spirit to the Naïve Physics Manifesto of the 1970s (Hayes, 1978), and attempt to produce models that reflect such ordinary spatial understanding. In practice, this means that topological relationships are treated as more privileged than distance ones. The paper by Galton (Chapter 10 in this volume) summarises much work in this field. A second approach, not represented in this volume, records the presence or absence of a defined relation between objects in space, or some measure of its strength, but then attempts to recover a projection that enables the 'space' to be mapped using conventional methods. Gatrell (1983) develops this idea, and for recent pedagogic examples see O'Sullivan and Unwin (2003, pp. 336–353). As is shown by Batty (Chapter 11, this volume) the same data on relations between objects can also be analysed as a network (such as the Internet itself, see Dodge and Kitchen, 2001). This is a very general approach, and taking it can have enormous analytic benefits. For example, the spatial stochastic simulation models used to track the 2001 UK foot and mouth disease outbreak, whose results were used to advise on policy, were implemented using a network defined by farm adjacencies (Keeling *et al.*, 2001). A third approach, is that taken by Llobera (Chapter 12 in this volume) and was highlighted as part of any future research agenda:

- exploring the possibility of realising body-centred geographies.

At one level, it is relatively easy to use digital data to create simple views of the landscape that adopt a mobile observer's frame of reference and the approach is implicit in many geographical applications of virtual reality (Fisher and Unwin, 2001). In a sense, what is presented depends on the context provided by the observer's location. The greater research agenda is to extend this work to other contexts, adopting a frame of reference that is both situated and mobile.

19.3 Entitation and Description

> *It is much easier ... when the information being modeled consists of geographical facts (bridges, streets, buildings) than when it consists of geographical interpretations of complex phenomena, like soil, terrain, or urban landscape, or of geographical knowledge and understanding.*
>
> (Goodchild, 1995, p. 36)

If a truly relational view of space is to be adopted, a second representational problem is to describe the objects that create it, and this is the central issue tackled by the authors of the papers in Part I of this book. As they point out, most GIS assume that objects of interest are uncontroversial and have definite, fixed boundaries. This certain world of the GIS is, of course, mostly a fiction and this has been recognised by almost two decades of work on error and uncertainty (Fisher, 1999; Duckham and Sharp, Chapter 8, this volume). As we have seen, there are a number of sources of imprecision that should be of concern to us, to do with what we define as objects, how we delineate them, and how we describe or label them. Some geographic objects are uncontroversial and capable of exact definition, but many are not. Chapter 6 by Fisher, Comber and Wadsworth shows that even the apparently unproblematic idea of land use cannot readily be captured in a digital representation. Ahlqvist (Chapter 7, this volume) examines aspects of the spatial extent of phenomena and alternative representations. Similarly, as pointed out by both Schuurman (Chapter 3) and Harvey (Chapter 4), defining objects has a social dimension and so is not a value or culture free operation. Reading the papers on this theme, it seems clear that, in a technical sense, intrinsically object orientated GIS software is essential. Yet, although 15 years have passed since the classic paper calling for full object orientation in geographic databases (Worboys *et al.*, 1990), it has yet to appear in any successful commercial GIS.

Perhaps as a result of our 'sampling' in the choice of authors, the research issues identified by the pyramid exercise were dominated by concerns about objects such as:

- systems that will support the construction/deduction of objects 'on the fly' in accord with the needs of analysis;
- a theory of spatio-temporal object formation and identity;
- data structures for the emergence of entities from attributes; and
- development of a specification language for spatial ontologies.

Concern over uncertainty in objects was expressed by the perceived need for:

- intrinsically uncertain representations;
- building strategies for formalising and implementing methods for the treatment of uncertainty in a 4D framework; and
- exploration of the relation between natural language of space-time representation in the production of meaning, taking account of the importance of vague concepts.

19.4 Temporality, Change and the Creation of Space and Time

'. . . in general GIS today remains a technology for static data, a major impediment to its use in modeling social and economic systems'.

(Goodchild, 1995, p. 40)

The technology of digital geographies has found the representation of change in time extremely hard to handle. Some progress has been made in the development for academic research of systems such as PCRaster for environmental simulation modelling (Wesseling *et al.*, 1996). Although this allows the creation of models of extraordinary explanatory power from essentially simple, but 'local' physics, its authors would be the first to agree that it is a relatively small step towards what might be required. Moreover, the continued, and tightly coupled, *creation* of space and time conflates all the issues we have discussed in this conclusion into a single representational problem. It is thus hardly surprising that our pyramid exercisers were agreed that the two key issues relating to time remain:

- languages for space-time; and
- integration of model dynamics with GIS.

19.5 Conclusion

All of this presents a rich research agenda, and in the years since the meeting where these papers were presented, new research addressing some of these topics has been presented (Fisher *et al.*, 2004; Worboys, 2005 to name just two). There can be no doubt, however, that much more needs to be done to improve the representation of the many 'geographies' that people see in their world as Geographical Information. Ultimately, whether we call this work Geography, Geographical Information Science, GIS or even Information Science is an irrelevance. What is certain is that the coming decade will see many, many more exciting representations of 'geography'.

References

Dodge, M. and Kitchin, R., 2001, *Mapping Cyberspace*, London: Routledge.

Fisher, P., 1999, Models of uncertainty in spatial data. In: *Geographical Information Systems, Volume 1 Principles and Technical Issues*, edited by Longley, P.A., Goodchild, M.F., Maguire, D.J. and Rhind, D.W., Chichester: John Wiley & Sons, Ltd, pp. 191–205.

Fisher, P. and Unwin, D. (eds), 2001, *Virtual Reality in Geography*, London: Taylor and Francis.

Fisher, P.F., Cheng, T. and Wood, J., 2004, Where is Helvellyn? Multiscale morphology and the mountains of the English Lake District, *Transactions of the Institute of British Geographers*, **29**, 106–128.

Gatrell, A., 1983, *Distance and Space: A Geographical Perspective*, Oxford: Clarendon.

Gold, J.R., Jenkins, A., Lee, R., Monk, J., Riley, J, Shepherd, I. and Unwin, D., 1991, *Teaching Geography in Higher Education: A Manual of Good Practice*, Oxford, UK: Blackwell.

Goodchild, M.F., 1995, Geographic information systems and geographic research. In: *Ground Truth: The Social Implications of Geographic Information Systems*, edited by Pickles, J., New York: Guilford, pp. 31–50.

Hayes, P., 1978, The naïve physics manifesto. In: *Expert Systems in the Microelectronic Age*, Michie, D. (ed.), Edinburgh: Edinburgh University Press, pp. 242–270.

Keeling, M.J., Woolhouse, M.E.J., Shaw, D.J., Matthews, L., Chase-Topping, M., Haydon, D.T., Cornell, S.J., Kappey, J., Wilesmith, J. and Grenfell, B.T., 2001, Dynamics of the 2001 UK foot and mouth epidemic: stochastic dispersal in a heterogeneous landscape, *Science*, **294**, 817–831.

O'Sullivan, D. and Unwin, D., 2003, *Geographic Information Analysis*, Hoboken, NJ: John Wiley & Sons, Ltd.

Wesseling, C.G., Karssenberg, D., Van Deursen, W.P.A. and Burrough, P.A., 1996, Integrating dynamic environmental models in GIS: the development of a dynamic modelling language, *Transactions in GIS*, **1**, 40–48.

Worboys, M.F., 1996, Metrics and topologies for geographic space, In: *Advances in Geographic Information Systems Research II: Proceedings of the International Symposium on Spatial Data Handling, Delft*, edited by Kraak, M.J. and Molenaar, M., London: Taylor and Francis. pp. 365–376.

Worboys, M.F., 2005, Event-oriented approaches to geographic phenomena, *International Journal of Geographical Information Science*, **19**, 1–28.

Worboys, M.F., Hearnshaw, H.M. and Maguire, D.J., 1990, Object-oriented data modelling for spatial databases, *International Journal of Geographical Information Systems*, **4**, 369–383.

Worboys, M.F., Mason, K. and Lingham, J., 1998, Computational techniques for non-Euclidean planar spatial data applied to migrant flows. In: *Innovations in Geographical Information Systems: 5*, edited by Carver, S., London: Taylor and Francis, pp. 35–45.

Index

Re-presenting GIS Edited by P. Fisher and D. Unwin